Spherical Geometry and Its Applications

Textbooks in Mathematics
Series editors:
Al Boggess and Ken Rosen

CRYPTOGRAPHY: THEORY AND PRACTICE, FOURTH EDITION

Douglas R. Stinson and Maura B. Paterson

GRAPH THEORY AND ITS APPLICATIONS, THIRD EDITION

Jonathan L. Gross, Jay Yellen and Mark Anderson

A TRANSITION TO PROOF: AN INTRODUCTION TO ADVANCED MATHEMATICS

Neil R. Nicholson

COMPLEX VARIABLES: A PHYSICAL APPROACH WITH APPLICATIONS, SECOND EDITION

Steven G. Krantz

GAME THEORY: A MODELING APPROACH

Richard Alan Gillman and David Housman

FORMAL METHODS IN COMPUTER SCIENCE

Jiacun Wang and William Tepfenhart

AN ELEMENTARY TRANSITION TO ABSTRACT MATHEMATICS

Gove Effinger and Gary L. Mullen

ORDINARY DIFFERENTIAL EQUATIONS: AN INTRODUCTION TO THE FUNDAMENTALS, SECOND EDITION

Kenneth B. Howell

SPHERICAL GEOMETRY AND ITS APPLICATIONS

Marshall A. Whittlesey

COMPUTATIONAL PARTIAL DIFFERENTIAL PARTIAL EQUATIONS USING MATLAB®, SECOND EDITION

Jichun Li and Yi-Tung Chen

AN INTRODUCTION TO MATHEMATICAL PROOFS

Nicholas A. Loehr

DIFFERENTIAL GEOMETRY WITH MANIFOLDS, SECOND EDITION

Stephen T. Lovett

MATHEMATICAL MODELING WITH EXCEL

Brian Albright and William P. Fox

https://www.crcpress.com/Textbooks-in-Mathematics/book-series/CANDHTEXBOOMTH

Spherical Geometry and Its Applications

Marshall A. Whittlesey

CRC Press is an imprint of the
Taylor & Francis Group, an **informa** business
A CHAPMAN & HALL BOOK

CRC Press
Taylor & Francis Group
6000 Broken Sound Parkway NW, Suite 300
Boca Raton, FL 33487-2742

First issued in paperback 2022

ISBN 13: 978-1-03-247537-0 (pbk)
ISBN 13: 978-0-367-19690-5 (hbk)

DOI: 10.1201/9780429328800

Library of Congress Cataloging-in-Publication Data

Names: Whittlesey, Marshall A., author.
Title: Spherical geometry and its applications / Marshall A. Whittlesey.
Description: Boca Raton : CRC Press, Taylor & Francis Group, 2020.
Identifiers: LCCN 2019020861 | ISBN 9780367196905
Subjects: LCSH: Geometry, Solid--Textbooks. | Sphere--Textbooks.
Classification: LCC QA457 .W5275 2020 | DDC 516.24/4--dc23
LC record available at https://lccn.loc.gov/2019020861

Visit the Taylor & Francis Web site at
http://www.taylorandfrancis.com

and the CRC Press Web site at
http://www.crcpress.com

Contents

Preface

It has been at least fifty years since spherical geometry and spherical trigonometry have been a regular part of the high school or undergraduate curriculum. It is an unusual mathematics program that has a course in them today, except as a topic in a survey of geometry. This work is an attempt to bring a comprehensive coverage of spherical geometry and its applications to a modern audience that is unfamiliar with it.

The study of geometry on the sphere dates back at least two thousand years. Educated people have long understood that the earth is round, and use of spherical geometry has been valuable in navigation, surveying, and other work involving understanding the nature of the earth. It is also apparent to the nighttime observer that the stars and planets in the sky appear to lie on a sphere about the earth. Their positions can be used to keep track of time and seasons. A traveler can also use them to keep track of his/her position on the surface of the earth. A basic fact in the field of navigation is that the angular elevation of the North Star (Polaris) above the horizon is the same as the observer's latitude. But if that star is not visible, other stars can be used to determine position with methods that are more complex. More recently, spherical geometry has found use in the field of plate tectonics. In the twentieth century it was shown that the surface of the earth is covered by a series of fairly rigid plates which move relative to each other. These motions cause earthquakes and volcanoes. Spherical geometry helps us understand how these plate movements work. Lastly, spherical geometry has found application in the field of crystallography. Crystals are objects in which atoms or molecules are carefully arranged in a regular pattern. Typically, these patterns include both lines and planes of atoms and molecules. Measurement of the angles among these lines and planes on a crystal yields clues as to the internal molecular structure of the crystal. Internal structure can also be studied by studying the pattern of scattering of an x-ray fired into the crystal. Spherical geometry can be helpful in analyzing these patterns.

Hopefully, these applications will persuade the reader that spherical geometry is worth mastering. My reasons for writing a new book on the subject are several. It is intended to be a comprehensive coverage of spherical geometry and its applications in a mathematically rigorous manner. I think that spherical geometry is perhaps a better route by which to introduce a student to an axiomatic system of non-Euclidean ideas, since a number of facts about geometry on the two-dimensional sphere differ notably from those with which the student is familiar in plane geometry. The applications of spherical geometry are also quite accessible, making it easier to persuade the student that geometry on a curved surface has some value. All mathematicians should know something about spherical geometry, but for the high school teacher, broad knowledge of the applications of mathematics in the natural sciences is particularly important.

Spherical geometry was a significant part of the mathematics curriculum until the 1950s. Many standard books in trigonometry in this period include topics in both plane and spherical trigonometry. Its use in navigation was probably important enough for it to be regarded as worth teaching to a general audience. But high school mathematics became increasingly geared toward calculus after the 1950s, and spherical geometry is difficult to fit in if one wishes to learn calculus in high school. When I began working on this book, there appeared to be no recent book in English covering spherical geometry exclusively. In the meantime, Glen Van Brummelen published his book [VB2012]. His book covers much more of the history of the subject than mine, but avoids a number of technical points in order to focus on the overall beauty of the subject. I attempt to fill in all technical details but spend less time on the history of the subject. For classroom use, the instructor might find either of these approaches better, depending on the emphasis of the course.

This book is intended as a course in spherical geometry for mathematics majors. Prerequisites include knowledge of plane and solid Euclidean geometry, trigonometry, and coordinate geometry. The properties of all the trigonometric functions and their inverses are essential, including identities such as the double-angle and half-angle formulas, sum-to-product formulas, and the laws of sines and cosines. The student should also be familiar with methods of logic and formal proof similar to that obtained either in a proof-based high school geometry class or a course in transition to proof-based mathematics. It helps if the student is familiar with modular arithmetic, as it is necessary to add angles modulo 2π. These prerequisites are enough to understand most of the book. But I do use other more advanced methods when convenient. I make brief use of Taylor series to compare the spherical Pythagorean theorem to the planar Pythagorean theorem. I make some use of calculus and the logarithm function in discussion of the Mercator projection. In Chapter 7, it helps to be familiar with the notions of mappings and related ideas such as domain, range, and what it means for a mapping to be injective or surjective. I assume some familiarity with basic linear algebra in the sections on crystallography and the stereographic projection. In Chapter 8, it helps if the student is familiar with the complex numbers and vectors in three dimensions, including the dot and cross product and their properties.

My approach here is to motivate the basic properties on the sphere as a three-dimensional object informally with pictures in Chapter 2. Here the idea of the great circle is introduced as the intersection of a sphere with a plane through its center. Distance on the sphere is measured via the central angle of an arc. Angles between great circles are measured via the dihedral angles of the planes containing the circles. I also discuss how to calculate the surface area of a sphere with elementary methods. In Chapter 3, I create an axiomatic system for spherical geometry similar to that sometimes used in plane geometry: first we have axioms involving incidence, then involving distance, and then involving measure of angles. These axioms avoid reference to the sphere as a subject of three-dimensional space. I feel this exercise is good for a class where the instructor wishes to teach proof in a geometric setting.

I have tried to set up an axiomatic system for spherical geometry with as few axioms as possible. The usual problem arises: one can spend a lot of time proving propositions that are not very interesting! The instructor should feel free to skim over sections 9 and 10 in a first course and simply take for granted propositions there which seem clear to the student. For example, I have avoided studying the proofs of Proposition 10.8 and Proposition 10.9 in an introductory class. In section 11, I prove the basic results about spherical triangles. In section 12, I prove the theorems about congruence of triangles. Here the differences between plane and spherical geometry begin to emerge: some of the theorems are the same, but some are dramatically different. In section 13, I delve into inequalities in spherical triangles. Again, some theorems are like those for plane triangles and some are not. A highlight of this section is what I call the spherical exterior angle theorem: that the measure of an exterior angle of a spherical triangle is less than the sum of the measures of the opposite interior angles but greater than their difference. While these inequalities are not really new I am not aware that any other work states this useful theorem in this form. Lastly in section 14, I discuss the areas of spherical triangles and polygons.

Chapter 4 considers the main formulas of spherical triangle trigonometry and introduces their basic applications. Tools from three-dimensional geometry are avoided.

Chapter 5 shows how to use spherical trigonometry in the field of spherical astronomy. The student learns about coordinate systems for the sky and how to change coordinates. I then discuss the problem of determining the time when the sun and stars rise and set. This problem is more complex than most people would guess. It is not too difficult to obtain answers that are within ten minutes or so of the actual time, but answers that are within a minute of the actual time require considerably more effort and can be omitted from a course.

Chapter 6 applies spherical trigonometry to polyhedra in three dimensions. I feature the formula for the angle between faces in a Platonic solid. I also briefly show how to use that formula to determine the number of four-dimensional regular polytopes.

In Chapter 7, I discuss the mappings of a sphere to itself and projections to a plane. Here is where I discuss the application in plate tectonics.

In Chapter 8, I show how to prove the main formulas for spherical triangles with quaternion and vector methods. I think this chapter helps feature the quaternions as a tool in three-dimensional geometry that most math majors are not exposed to.

I have tried to write a book by which one can understand spherical geometry from various levels of background knowledge. But I also make clear in the text that more sophisticated tools are sometimes appropriate, especially where I show how to use the quaternions to streamline proofs and broaden our perspective on the subject. In the hope of appealing to the largest audience possible, I have taken the step of presenting several proofs of some theorems. Most modern approaches to spherical geometry include considerable usage of vectors, and while this has some advantages, intuitive motivation for certain theorems is often lost. As an example, vector proofs of the spherical law of sines often start with a quadruple vector product identity that must be checked. For a theorem as intuitively simple as the law of sines, this seems like overkill. As with many synthetic proofs in plane and spatial geometry, some synthetic proofs have logical gaps, or extra cases, which the reader is invited to fill in. Different approaches to a subject often offer different insights.

The text features many exercises inspired by propositions from an ancient text on spherical geometry called the *Sphaerica* of Menelaus of Alexandria. This text dates to about 100 AD and originally appeared in Greek. That text has been lost but it was translated and reworked into Arabic in various versions. These are the oldest versions of the text that we now have. My student Rani Hermiz translated a manuscript of *Sphaerica* into English. (See [He2015].) That translation appears online at the web site of the library of California State University San Marcos. All my references to propositions in *Sphaerica* use the numbering system used in [Kr1936] and [He2015]. A critical translation is now available in English: see [RP2017]. I hope readers will find the content of the ancient propositions intriguing. It should be clear that these ancient mathematicians were extremely capable people who could do much with limited mathematical technology. Many of the propositions seem to appear in no modern work.

There are a number of online resources in spherical geometry. John Sullivan of the University of Illinois has a Java applet for demonstrating certain features in spherical geometry such as parallel transport. James King of the University of Washington has a web page with many links to spherical geometry pages. Tevian Dray of Oregon State University has a spherical drawing program Spherical Easel. The web site for the U.S. Naval Observatory and www.timeanddate.com provide resources for determining rise and set for various celestial objects including the sun and the moon. The Department of Mathematics and Statistics at Saint Louis University has an online feature of mathematics and the art of M.C. Escher which includes a page on spherical geometry. The reader should also be aware of the Lenart Sphere as a useful teaching and learning tool. There are surely other resources I am not aware of.

A number of people have been helpful in the creation of this book. I am grateful for valuable commentary on the manuscript from Kenneth Rosen (AT&T Labs) and Glen Van Brummelen (Quest University). The people at CRC Press, including Robert Ross, Michele Dimont, Suzanne Lassandro, and Shashi Kumar, have been terrific to work with. I am most grateful to Rani Hermiz for his translation of *Sphaerica.* I owe a debt of gratitude to Thomas Banchoff (Brown University) for mentioning *Sphaerica* to me in the first place and for many conversations about spherical geometry over the years. I would also like to thank Professors Stephen Nelson (Tulane University), V. Frederick Rickey (US Military Academy), Francis Ford (Providence College), and David Robbins (Trinity College). I am also grateful to students from my classes who made corrections or suggestions: Megan Amely, Edgar Ayala, Marilena Beckstrand, Kenny Courser, Abigail Dunlea, Rani Hermiz, Jesus Hernandez, Robert Kendrick, Yuan Lin (Annie) Lee, Sean Malter, Jonathan Singh, Gregory Guayante, Jill Richard, and Xiaodan Xu.

I would like to thank my department and my deans Victor Rocha and Katherine Kantardjieff for approving professional leave time to work on this book.

Chapter 1

Review of three-dimensional geometry

The geometric objects being discussed in this book all sit in three-dimensional real Euclidean space, which we refer to throughout as *space*. We can study space with various tools. The "synthetic" tools include notions of points, lines, planes, rays, segments, angles, triangles, congruence, interior/exterior, and betweenness. "Metric" geometry adds notions of distance and angle measurement. (See [Mo1963] for a study in the differences in the synthetic and metric approaches to plane geometry.) If desired, one may also impose a Cartesian system of coordinates on space and make use of algebraic formulas. Sometimes usage of the notion of vector is helpful.

In each section of this chapter we review key tools of three-dimensional Euclidean geometry.

We will understand a *set* as being a collection of objects. In geometry, the objects in the set are typically "points." The notion of set is perhaps the most basic structure in geometry.

If X and Y are sets of points, then the set of points which are in both X and Y is the *intersection* of X and Y, denoted $X \cap Y$. The set consisting of all points which lie in either X or Y is the *union* of X and Y, denoted $X \cup Y$.

1 Geometry in a plane

The notions of point, line, and plane are central to the geometry of space. We here summarize key properties of them that we will need. We should understand that a line and a plane are sets of points.

Given two distinct points A and B in space, there exists a unique line passing through them; the line through A and B is denoted by $\overleftrightarrow{AB}$. Points which lie on a single line are said to be *collinear*. The *distance* between A and B is denoted by AB. There exists a coordinate labeling of the line with real numbers such that the coordinate of A is zero, the coordinate of B is positive, and the absolute value of the difference between the coordinates of two points on the line is the distance between the points. The *ray* $\overrightarrow{AB}$ is the set of all points on $\overleftrightarrow{AB}$ whose coordinate is greater than or equal to zero. The point A is called the *endpoint* or *vertex* of the ray. If A, B, and C are distinct points on a line, we say that B and C are on the *same side* of A if $\overrightarrow{AB}$ and $\overrightarrow{AC}$ are the same ray. If A, B, and C are distinct collinear points, then we say that A is between B and C if $\overrightarrow{AB}$ and $\overrightarrow{AC}$ are not the same ray (and then we say that $\overrightarrow{AB}$ and $\overrightarrow{AC}$ are *opposite* rays).

The set of all points between A and B taken together with the points A and B is called the *segment* $\overline{AB}$ (also referred to as the segment between A and B). The *length* of $\overline{AB}$ is the distance between A and B. Two segments are said to be *congruent* if they have the same length.

A set of points is said to be *convex* if for every pair of points in the set, the points on the segment between them also belong to the set.

An *angle* is the union of two rays which have the same endpoint, but which are not part of the same line. The endpoint is called the *vertex* of the angle and the two rays without the vertex are the *sides* of the angle. To every angle is associated a *measure* between 0 and π radians (or between 0 and 180 degrees). In this book we shall generally use radian measure as a default, but will also make use of degree measure when appropriate with certain applications. An angle measure given without units should be assumed to be a radian measure. If an angle is the union of two rays $\overrightarrow{BA}$ and $\overrightarrow{BC}$ then we use the notation $\angle ABC$ to denote the angle formed by these two rays, and $m\angle ABC$ to denote its measure. Two angles are said to be *congruent* if they have the same measure. A *right* angle is an angle with measure $\frac{\pi}{2}$ radians (90 degrees). Two lines that intersect are said to be *perpendicular* if the angles formed at their point of intersection are right angles. An angle with measure less than that of a right angle is said to be *acute*. An angle with measure greater than that of a right angle is said to be *obtuse*. If the sum of the measures of two angles equals $\frac{\pi}{2}$ radians (90 degrees), the two angles are said to be *complementary*, and are *complements* of each other. If the sum of the measures of two angles equals π radians (180 degrees), the angles are said to be *supplementary*, and are *supplements* of each other. Two angles are said to be in a *linear pair* if they have the same vertex, have one side in common, and the other two sides are opposite rays. The angles in a linear pair are supplementary.

Suppose that one angle is formed by the union of two rays $\overrightarrow{r_1}$ and $\overrightarrow{r_2}$, and another angle is formed by the union of rays $\overrightarrow{r_3}$ and $\overrightarrow{r_4}$. If $\overrightarrow{r_1}$ is opposite to $\overrightarrow{r_3}$ and $\overrightarrow{r_2}$ is opposite to $\overrightarrow{r_4}$, then we say that the angles are *vertical* angles.

A pair of vertical angles must be congruent.

Given a line lying in a plane, every point in the plane belongs either to the line or to one of two convex sets called a *half-plane.* If a point in one half-plane and another point in the other half-plane are connected with a line segment, this segment intersects the line. The line is said to be the *edge* of each half-plane. If two points in the plane are in the same half-plane associated with the line, they are said to be on the *same side* of the line. If two points in the plane are in different half-planes associated with the line, they are said to be on *opposite sides* of the line.

Suppose that line $\overleftrightarrow{AB}$ is the edge of a half-plane h. Then given a real number m between 0 and π radians (0 and 180 degrees) there exists a unique ray r in h with vertex at A such that the angle formed by r and $\overrightarrow{AB}$ has measure equal to m.

Suppose that B is between A and C in a given plane, and that D and E lie in the plane on the same side of $\overleftrightarrow{AB}$. Then $\angle ABD$ and $\angle ACE$ are said to be *corresponding* angles.

Given a point P and a line ℓ lying in a single plane, there exists a unique line m passing through P perpendicular to the given line. If m meets ℓ at the point P^ℓ, then we say that P^ℓ is the *projection* of P to ℓ.

Two distinct lines in a plane intersect in either a single point, or in no point. If two lines in a plane do not meet, we say that the lines are *parallel* lines. If line ℓ_1 is parallel to line ℓ_2, we write $\ell_1 \| \ell_2$. If two distinct lines lying in a plane are both perpendicular to a third line, the first two lines are parallel. With the corresponding angles of the previous paragraph, if $\prec ABD \cong \angle ACE$ then we have that $\overleftrightarrow{BD} \parallel \overleftrightarrow{CE}$.

The (Euclidean) *parallel postulate* states that given a line and a point not on the line, there exists a unique line through the given point parallel to the given line.

The parallel postulate has a long and important history in geometry. Euclid took a similar statement as a basic proposition not to be proven, but many mathematicians in the millennia after him thought it important (and non-obvious) enough that it ought to be proven from other more basic propositions in plane geometry. In the nineteenth century it turned out that the uniqueness assertion in the parallel postulate above cannot be proven from other more generally accepted postulates in plane geometry. The basic reason is that there are other two-dimensional surfaces with natural notions of "point" and "line" satisfying all the accepted postulates about points and lines in the plane, except for the parallel postulate. The sphere will turn out to be a surface without parallelism.

Suppose that A, B, and C are three points which do not lie on a single line. Then the union of the three segments $\overline{AB}$, $\overline{BC}$, and $\overline{AC}$ constitutes the *triangle* $\triangle ABC$. Each of the three points is said to be a *vertex* (plural: *vertices*) of the triangle, the three segments are the *sides* of the triangle, and the angles $\angle ABC$, $\angle ACB$, and $\angle BAC$ are the *(interior) angles* of the triangle. If one of the angles of a triangle is in a linear pair with a second angle, the second angle is said to be an *exterior angle* of the triangle.

If one of the angles of the triangle is a right angle, then the triangle is said to be a *right triangle.* The side opposite the right angle is called the *hypotenuse* and the other sides are called *legs.* If two sides of a triangle are congruent then the triangle is *isosceles.* In an isosceles triangle, the angles opposite the congruent sides are congruent. Conversely, if two angles of a triangle are congruent, the opposite sides are congruent.

Two triangles are said to be *congruent* if their vertices can be put into a one-to-one correspondence such that corresponding sides and angles are congruent. There are several important propositions concerning how to show triangles are congruent. First, *SAS congruence* states that if two pairs of corresponding sides and the angle between them are congruent, the triangles are congruent. Next, *SSS congruence* states that if all corresponding sides are congruent, then the triangles are congruent. Lastly, if two pairs of corresponding angles and a side between are congruent, the triangles are congruent (*ASA congruence*). In fact, the triangles are congruent even if the side is not between the angles (*SAA congruence* or *AAS congruence*). In general, an "SSA correspondence" between does not guarantee congruence of the triangles, unless the angle happens to be a right angle. (The so-called "hypotenuse-leg theorem" states that if two right triangles have congruent hypotenuses and one pair of corresponding legs is congruent, then the triangles are congruent.) The "hypotenuse-angle theorem" states that if two right triangles have congruent hypotenuses and one pair of corresponding acute angles is congruent, then the triangles are congruent. (The hypotenuse-angle theorem is simply the SAA congruence theorem applied to a right triangle.)

The SAA congruence theorem can be viewed as a consequence of the theorem that the sum of the measures of the angles of a triangle is π radians (180 degrees). Suppose that there is an SAA correspondence between $\triangle ABC$ and $\triangle DEF$: $\angle ABC \cong \angle DEF$ and $\angle BAC \cong \angle EDF$. Then $\angle BCA \cong \angle EFD$ also because the sum of the measures of the angles in each triangle is the same. Thus we must also have an ASA correspondence between the triangles; hence they are congruent.

However, this proof of SAA congruence suffers somewhat from the fact that the angle sum of a triangle in the plane is $180°$. This latter theorem depends on the parallel postulate. It turns out that SAA congruence can be proven without the assumption of the parallel postulate. In the exercises the reader will see how to do this.

There are also a number of theorems concerning inequalities in triangles. In any triangle, if two sides (respectively, angles) are unequal in measure, then the opposite angles (respectively, sides) are also unequal in measure, and in the same order. The *triangle inequality* states that the sum of the lengths of two sides of a triangle must be larger than the length of the third side. An exterior angle of a triangle is larger in measure than either of the two opposite interior angles. A consequence of this is the fact that in a right triangle, the non-right angles are acute.

Proposition 1.1 (Hinge Theorem) *Suppose that* $\triangle ABC$ *and* $\triangle DEF$ *are planar triangles with* $\overline{AB} \cong \overline{DE}$ *and* $\overline{BC} \cong \overline{EF}$. *Then* $m\angle ABC < m\angle DEF$ *if and only if* $\overline{AC}$ *is shorter than* $\overline{DF}$.

Proposition 1.2 *If* P *is a point,* ℓ *is a line, and* Q *is the foot of the perpendicular from* P *to* ℓ, *then the distance* PQ *is less than* PR *for any point* R *in* ℓ *different from* P.

Suppose that the points of two geometric objects are in a one-to-one correspondence. Then the objects are said to be *similar* if the lengths of corresponding segments are all in the same proportion, and the measures of all corresponding angles are the same. For triangles, we have the AAA similarity property: if corresponding angles are congruent, the triangles are similar. Furthermore, we have the SAS similarity property: if two triangles $\triangle ABC$ and $\triangle DEF$ satisfy $\angle B \cong \angle E$ and $AB/BC = DE/EF$, then the triangles are similar. The SAS similarity theorem depends on the parallel postulate.

The most important consequence of the similarity properties of planar triangles is:

Theorem 1.3 (Pythagorean theorem) *If a right triangle has sides of length* a, b, *and* c, *where* c *is the length of the hypotenuse, then* $a^2 + b^2 = c^2$.

With the Pythagorean theorem in the plane, the hypotenuse-leg congruence theorem for right triangles is not difficult to deduce. (See Exercise 9.) However, the Pythagorean theorem depends on the parallel postulate, and it turns out that the hypotenuse-leg theorem can be proven without the assumption of the parallel postulate. This is also discussed in the exercises.

In any triangle, the sum of the measures of the angles is π radians ($180°$). An immediate consequence is that the measure of an exterior angle is equal to the sum of the measures of the opposite interior angles.

Suppose that $A_1, A_2, \ldots, A_n$ is a set of n points in a plane such that no three consecutive points on the list are collinear. Consider the segments $\overline{A_1A_2}$, $\overline{A_2A_3}$, $\overline{A_3A_4}$, $\ldots$, $\overline{A_{n-1}A_n}$, and $\overline{A_nA_1}$. If no two of them intersect anywhere but at their endpoints, their union is a *polygon* of n *sides.* A *quadrilateral* is a polygon of four sides and a *pentagon* is a polygon of five sides. A *parallelogram* is a quadrilateral where the non-intersecting (opposite) sides are parallel. If the opposite sides in a parallelogram are congruent, the parallelogram is called a *rhombus.* A rectangle is a quadrilateral such that the angle between any two sides with a common endpoint is a right angle. A *square* is a rectangle whose sides all have the same length. A *trapezoid* is a quadrilateral where one pair of two opposite sides is parallel (the *bases*) and the other pair of opposite sides is not parallel. The distance between the parallel sides is the *height* of the trapezoid. A trapezoid is said to be *isosceles* if the non-parallel sides are congruent. Given a trapezoid, there exists a line parallel to both bases halfway between the bases. This line intersects the non-parallel sides in one point each. The line segment between these two points is called the *midline* or *median* of the trapezoid. The length of the midline is the arithmetic mean (half of the sum) of the length of the bases.

Let O be a point in a plane and r a positive real number. The *circle* with *center* O and *radius* r is the set of all points in the plane at distance r from O.

We shall need the following lemma when we discuss areas in §7.

Lemma 1.4 *Suppose ℓ is a line in a plane, s is a line segment, and M is a point on s. Suppose that the line perpendicular to s at M intersects ℓ in a point N. Let a_1 be the length of s, b_1 the distance from M to ℓ, a_2 the length of the projection of s to ℓ, and b_2 the length of the segment from M to N. Then $a_1 b_1 = a_2 b_2$.*

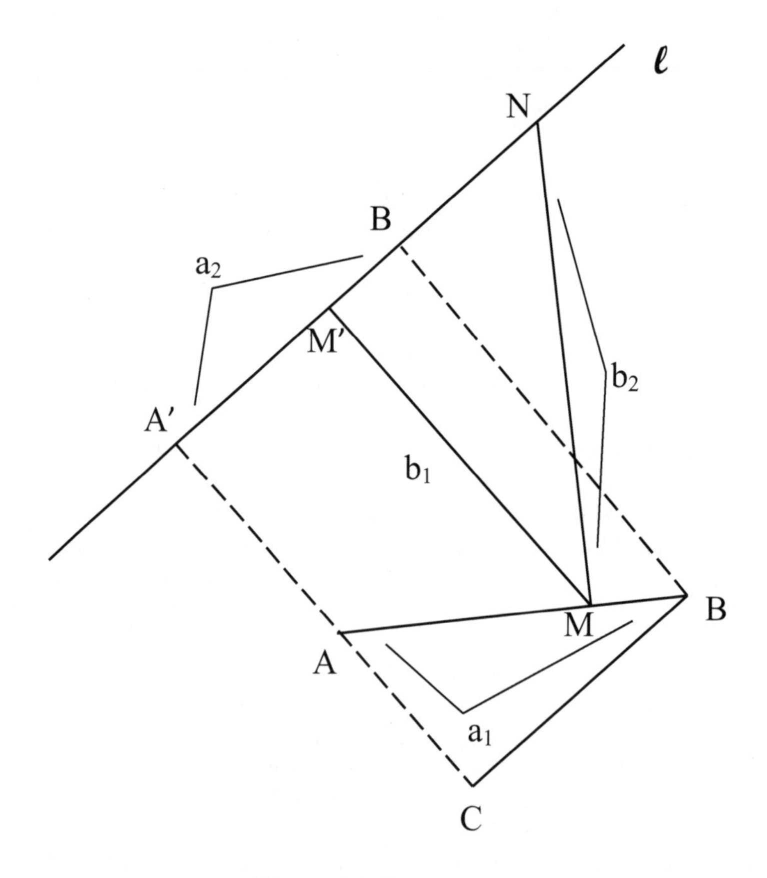

Figure 1.1: Lemma 1.4.

Proof. Because the point N of intersection exists, the line containing s cannot be perpendicular to ℓ. If s is parallel to ℓ, then $a_1 = a_2$ and $b_1 = b_2$ and so $a_1b_1 = a_2b_2$ by multiplying these equations. If s is not parallel to ℓ, then let A be the endpoint of s which is closer to ℓ and let B be the other endpoint. Let A', B', and M' be the projections of A, B, and M to ℓ and let C be the projection of B to $\overleftrightarrow{AA'}$. Since s is not parallel to ℓ, the points A, B, and C are not collinear; hence, they form a triangle. Similarly, M, M', and N form a triangle. We claim that $\triangle ABC \sim \triangle NMM'$. To see this, note that the triangles have a right angle at C and M'. We have that $\angle M'MN$ is complementary to $\angle M'MA$ since $\overleftrightarrow{MN}$ is perpendicular to $\overleftrightarrow{AB}$ at M. Also, $\angle M'MA$ is congruent to $\angle B'BA$ as corresponding angles of the parallel lines $\overleftrightarrow{BB'}$ and $\overleftrightarrow{MM'}$ (these lines being parallel since both are perpendicular to ℓ). Since C is the projection of B to $\overleftrightarrow{AA'}$, and A' and B' are the projections of A and B to ℓ, respectively, the quadrilateral $CA'B'B$ is a rectangle. Thus $\angle CBB'$ is right, and so $\angle B'BA$ and $\angle CBA$ are complementary. Summing up what we have found, $m\angle M'MN = \frac{\pi}{2} - m\angle M'MA = \frac{\pi}{2} - m\angle B'BA = m\angle CBA$. Thus in $\triangle ABC$ and $\triangle NMM'$, two pairs of corresponding angles are congruent. We conclude that these triangles are similar. Since the ratios of the lengths of corresponding sides are the same, $AB/MN = BC/MM'$, i.e., $a_1/b_2 = a_2/b_1$, or $a_1b_1 = a_2b_2$, as desired. $\diamondsuit$

The reader will find another approach to Lemma 1.4 in §4, Exercise 1.

The area enclosed by a rectangle is equal to the base times the height. The area enclosed by a triangle is half of the base times the height. For a circle of radius r, the area enclosed is πr^2.

Exercises §1

1. Prove: If $\angle ABC$ is acute (respectively, obtuse) then the projection of A to $\overleftrightarrow{BC}$ is on the same (respectively, opposite) side of B as C.

2. Let plane angle $\angle RST$ be an angle with vertex S. Let R' be chosen so that the ray $\overrightarrow{SR'}$ is perpendicular to the ray $\overrightarrow{ST}$ and R and R' are on the same side of the line $\overleftrightarrow{ST}$. Similarly suppose that T' is chosen so that $\overrightarrow{ST'}$ is perpendicular to $\overrightarrow{SR}$ and T and T' lie on the same side of line $\overleftrightarrow{RS}$. Prove that $\angle R'ST'$ and $\angle RST$ are supplementary angles (i.e., the sum of their measures is π).

3. In a parallelogram, show that the opposite sides are congruent and the diagonals bisect each other. In a rhombus, show that the diagonals also are perpendicular and bisect the angles.

4. In a trapezoid, let the height be h and let the length of the bases be b_1 and b_2. By using a formula for the area of a triangle, argue that the area of the trapezoid is $\frac{h}{2}(b_1 + b_2)$.

5. A *circular sector* is a region bounded by an arc of a circle and two radii of that circle. If the arc has radian measure θ and the circle has radius r, argue that the area is $\frac{1}{2}r^2\theta$.

6. An *annulus* is a region bounded by two circles with the same center but different radii. An *annular sector* is the portion of an annulus between two radii of the outer circle of the annulus. If the inner and outer arcs on the annulus have length s_1 and s_2, respectively, and the difference of the radii is ℓ, show that the area of the annular sector is $\frac{\ell}{2}(s_1 + s_2)$.

7. Prove Proposition 1.2.

8. In this exercise we indicate another proof of the SAA congruence theorem. Suppose that $\triangle ABC$ and $\triangle DEF$ satisfy $\overline{AB} \cong \overline{DE}$, $\angle C \cong \angle F$, and $\angle B \cong \angle E$. Without using the fact that the sum of the measures of the angles in the triangle is π (i.e., without using the parallel postulate), prove that $\triangle ABC \cong \triangle DEF$. To do this, note that if $BC < EF$ then there exists a point G in $\overline{EF}$ such that $\overline{BC} \cong \overline{EG}$. Proceed to a contradiction via the exterior angle theorem.

9. Explain how the hypotenuse-leg theorem follows quickly from two facts: (1) SSS congruence and (2) the Pythagorean theorem.

10. Explain how to prove the hypotenuse-leg theorem by using the exterior angle theorem.

2 Geometry in space

The following propositions are basic to our understanding of the way points, lines, and planes interact in space.

Proposition 2.1 *If a line contains two distinct points of a plane, the line lies entirely in the plane.*

Proposition 2.2 *Three noncollinear points determine a unique plane.*

Proposition 2.3 *A line and a point not on the line determine a unique plane which contains both the line and the point.*

Proposition 2.4 *If two distinct planes intersect, their intersection must be a line.*

If two lines in space do not intersect, then either they lie in the same plane (in which case they are parallel) or they do not lie in the same plane (in which case we say they are *skew*). A set of points which lies in some plane is said to be *planar*.

Proposition 2.5 *Given a plane in space, every point in space not on the plane belongs to one of two disjoint convex sets. Each of these two sets is called a* half-space. *If a point in one half-space and another point in the other half-space are connected with a line segment, this segment intersects the plane.*

If two points are chosen from the same half-space, then we say that the two points are on the *same side* of the plane. If two points are chosen such that one lies in one half-space and the other lies in the other half-space, we say that the two points lie on *opposite sides* of the plane.

Definition 2.6 *We say that a line ℓ is perpendicular to a plane p if there exists a point Q such that ℓ intersects p at Q and ℓ is perpendicular to all lines in p which pass through Q.*

Proposition 2.7 *Given a point P and a plane p in space, there exists a unique line ℓ through P perpendicular to p.*

In Proposition 2.7, the (unique) point P^p where the line meets the plane is called the *foot of the perpendicular* from P to p (or the *projection* of P to p) and PP^p is called the *distance* from the point P to the plane p. If $\overline{PQ}$ is a segment then the *projection* of $\overline{PQ}$ to p is the segment $\overline{P^pQ^p}$.

The following proposition is valuable in proving that a line is perpendicular to a plane.

Proposition 2.8 *If a line is perpendicular to each of two distinct intersecting lines where the latter two lines intersect, then the first line is perpendicular to the plane containing the other two lines.*

Proposition 2.9 *If line ℓ is perpendicular to plane p at Q, then any line containing the point Q which is perpendicular to ℓ must lie in p.*

Theorem 2.10 *If P is a point, p a plane, and P^p is the foot of the perpendicular from P to p, then the distance PP^p is less than PR for any point R in p different from P.*

Proof. Since PP^p is perpendicular to p, it is perpendicular to any line in plane p which passes through P^p. In particular, $\overleftrightarrow{PP^p}$ is perpendicular to $\overleftrightarrow{P^pR}$. Thus $\triangle PP^pR$ is a right triangle with a right angle at P^p. The angle opposite $\overline{PP^p}$ is acute. The angle opposite $\overline{PR}$ is a right angle. Since the longer side is opposite the larger angle, $PP^p < PR$. $\diamondsuit$

We have similar theorems and definitions concerning a point P and a line ℓ in space.

Proposition 2.11 *Given a point P and a line ℓ in space, there exists a unique plane p through P perpendicular to ℓ.*

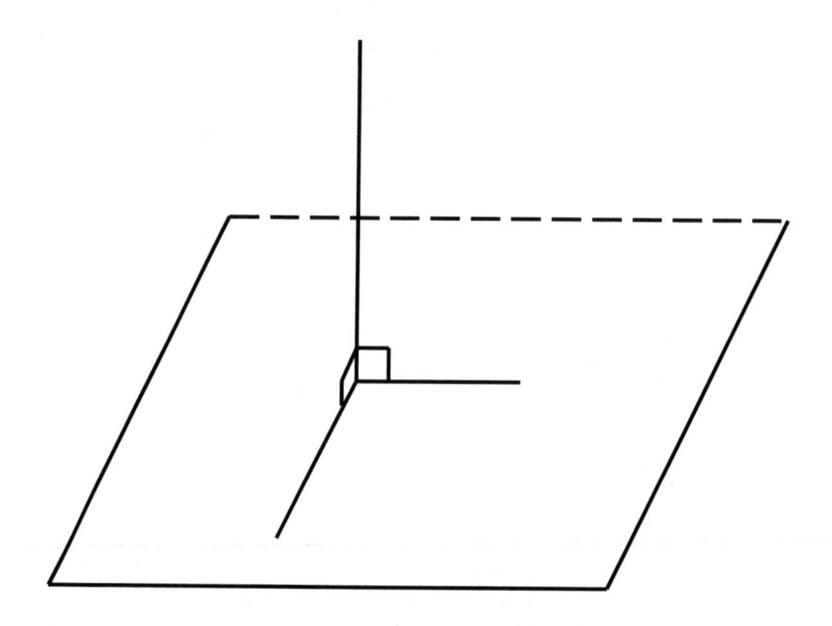

Figure 1.2: Propositions 2.8 and 2.9.

In Proposition 2.11, the point P^ℓ where the line meets the plane is called the *foot of the perpendicular* from P to ℓ (or the *projection* of P to ℓ) and PP^ℓ is called the *distance* from the point P to the line ℓ. If $\overline{PQ}$ is a segment, then its *projection* to ℓ is the segment $\overline{P^\ell Q^\ell}$.

We review the notion of dihedral angle in space. Recall that if a line lies in a plane, then the points of the plane not on the line consist of two pieces called *half-planes.*

Definition 2.12 *Suppose two half-planes in space have the same edge but are not in the same plane. Then the union of two such half-planes along with their common edge is called a* dihedral angle*; the two half-planes are called the* sides *of the dihedral angle and the line of intersection is the* edge *of the dihedral angle. If A is a point in one side of the dihedral angle, D lies in the other side and the edge is the line $\overleftrightarrow{BC}$, then the dihedral angle is denoted by $\angle A-BC-D$. The* interior *of $\angle A-BC-D$ consists of the set of all X such that X and A are on the same side of plane BCD and X and D are on the same side of plane ABC.*

Definition 2.13 *A* plane angle *of a dihedral angle is formed as follows: a point V in the edge of the dihedral angle is the vertex of the plane angle. Then in each side of the dihedral angle there exists a ray with vertex V perpendicular to the edge of the dihedral angle. The union of these two rays is the plane angle with vertex V.*

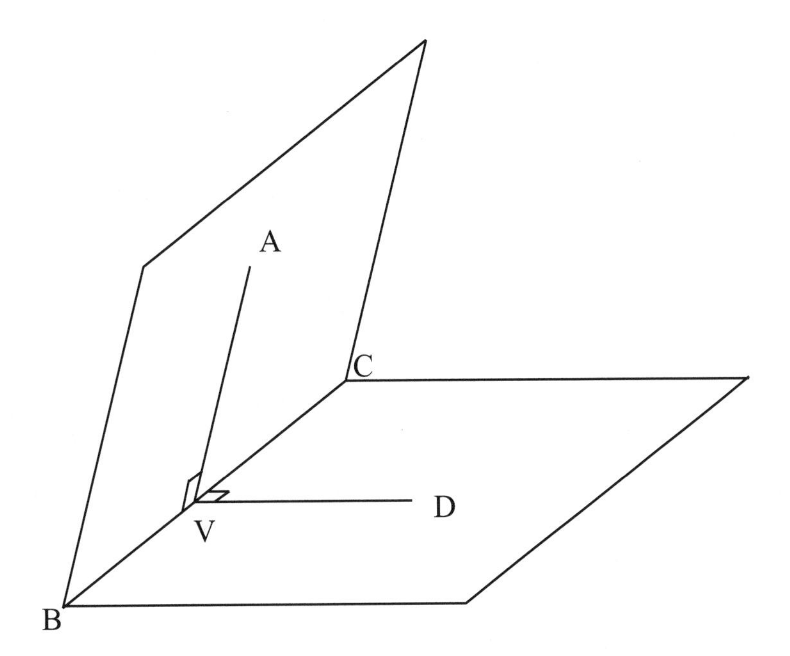

Figure 1.3: A dihedral angle $\angle A - BC - D$ and its plane angle $\angle AVD$.

Proposition 2.14 *An angle is a plane angle of a dihedral angle if and only if it is the intersection of the dihedral angle with a plane perpendicular to the edge of the dihedral angle.*

Proposition 2.15 *The plane angles of a dihedral angle are all congruent.*

Proposition 2.15 can be proven by observing that any two plane angles have corresponding sides which are parallel.

Definition 2.16 *The* measure *of a dihedral angle is given by the measure of any of its plane angles. We say that two planes are* perpendicular *if a dihedral angle formed between them is a right angle. Two dihedral angles are* congruent *if they have the same measure.*

Theorem 2.17 *If planes p and q are perpendicular, then the projection of any point of p to q must also lie in p.*

Theorem 2.18 *Suppose that a line is perpendicular to one plane and is contained in a second plane. Then the planes are perpendicular.*

Theorem 2.19 *Suppose that two intersecting planes are each perpendicular to a third plane. Then the third plane is perpendicular to the intersection of the first two planes.*

Exercises §2

1. If two distinct lines are perpendicular to the same plane, then the lines must be parallel.

2. Suppose that h is a half-plane whose edge lies in a plane p. If two dihedral angles are thus formed which are congruent, prove that h is perpendicular to p.

3. Suppose that planes p_1 and p_2 are orthogonal. Let P be a point in the intersection of p_1 and p_2 and let r_1 and r_2 be perpendicular rays based at P such that r_1 and r_2 lie in p_1 and p_2, respectively. Assume neither r_1 nor r_2 lies in the line of intersection of p_1 and p_2. Let p_3 be the plane containing r_1 and r_2. Prove that p_3 is perpendicular to either p_1 or p_2.

4. Suppose that P is a point in the interior of a dihedral angle and let O be a point in the edge of the angle. Prove that every point of $\overrightarrow{OP}$ is in the interior of the dihedral angle.

3 Plane trigonometry

Plane right triangle trigonometry is heavily based on the following fact: if $\triangle ABC$ and $\triangle DEF$ are plane right triangles with right angles at B and E, respectively, and $m\angle CAB = m\angle FDE$, then $AB/AC = DE/DF$, $BC/AC = EF/DF$, and $BC/AB = EF/DE$. (Verification of this via similar triangles is left to Exercise 1.) Then, given an angle measure θ we may define the cosine, sine, and tangent of angle θ as $\cos(\theta) = XY/XZ$, $\sin(\theta) = YZ/XZ$, and $\tan(\theta) = YZ/XY$, where $\triangle XYZ$ is some triangle with a right angle at Y and $m\angle ZXY = \theta$. Then Exercise 1 shows that these side ratios do not depend on the right triangle $\triangle XYZ$ chosen. Hence a mathematician would say that the trigonometric functions sin, cos, and tan are "well-defined."

A slightly different approach is needed when trigonometric functions of non-acute angles are to be defined. We assume the plane has an xy coordinate system with origin O. Let θ be an angle measure. Let r be any positive real number. We rotate the x-axis counterclockwise about the origin through an angle measure of θ and suppose that the point $(r, 0)$ rotates to a point with coordinates (x, y). Then $\cos(\theta)$ and $\sin(\theta)$ are defined to be the values x/r and y/r, respectively. It must be checked that this definition does not depend on the value of r. If θ is negative, we would rotate the x-axis clockwise through an angle measure equal to the absolute value of θ. Note that because the legs of a right triangle are always shorter than the hypotenuse, we must have for any angle θ, $-1 \le \sin(\theta) \le 1$ and $-1 \le \cos(\theta) \le 1$.

We then may define $\tan(\theta) = \sin(\theta)/\cos(\theta)$, $\cot(\theta) = \cos(\theta)/\sin(\theta)$, $\sec(\theta) = 1/\cos(\theta)$, and $\csc(\theta) = 1/\sin(\theta)$, except for a value of θ where a denominator is zero.

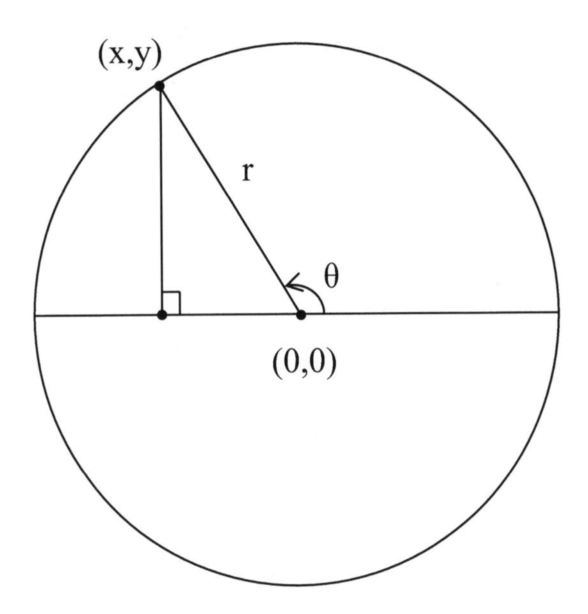

Figure 1.4: The sine and cosine: $\cos(\theta) = x/r, \sin(\theta) = y/r$.

Suppose that through a rotation of angle θ, the point $(r, 0)$ rotates to (x, y). Then a rotation through the supplementary angle $\pi - \theta$ moves $(r, 0)$ to $(-x, y)$. This leads to the identities:

$$\sin(\pi - \theta) = \sin(\theta) \tag{1.1}$$
$$\cos(\pi - \theta) = -\cos(\theta) \tag{1.2}$$

Similarly, if through a rotation of angle θ, the point $(r, 0)$ moves to (x, y), then a rotation through the complementary angle $\frac{\pi}{2} - \theta$ moves $(r, 0)$ to (y, x). This leads to the identities:

$$\sin(\frac{\pi}{2} - \theta) = \cos(\theta) \tag{1.3}$$
$$\cos(\frac{\pi}{2} - \theta) = \sin(\theta) \tag{1.4}$$

Similar arguments lead to

$$\sin(-\theta) = -\sin(\theta) \tag{1.5}$$
$$\cos(-\theta) = \cos(\theta) \tag{1.6}$$

Division of (1.5) and (1.6) leads to

$$\tan(-\theta) = -\tan(\theta) \tag{1.7}$$
$$\cot(-\theta) = -\cot(\theta). \tag{1.8}$$

The definitions of the sine and cosine together with the Pythagorean theorem lead immediately to the identity:

$$\cos^2(\theta) + \sin^2(\theta) = 1 \tag{1.9}$$

We recall the sum and difference formulas for the sine and cosine

$$\sin(x \pm y) = \sin(x)\cos(y) \pm \cos(x)\sin(y) \tag{1.10}$$

$$\cos(x \pm y) = \cos(x)\cos(y) \mp \sin(x)\sin(y) \tag{1.11}$$

from which follow the double angle formulas

$$\cos(2x) = \cos^2(x) - \sin^2(x) = 2\cos^2(x) - 1 = 1 - 2\sin^2(x) \tag{1.12}$$

$$\sin(2x) = 2\sin(x)\cos(x) \tag{1.13}$$

and the half-angle formulas

$$\begin{aligned} \sin^2(\frac{x}{2}) &= \frac{1-\cos(x)}{2} \\ \cos^2(\frac{x}{2}) &= \frac{1+\cos(x)}{2} \\ \tan^2(\frac{x}{2}) &= \frac{1-\cos(x)}{1+\cos(x)}. \end{aligned} \tag{1.14}$$

If we add together the sum and difference formulas for the cosine we obtain $\cos(x+y)+\cos(x-y) = 2\cos(x)\cos(y)$. The variable substitutions $A = x+y$ and $B = x - y$ lead to the "sum-to-product" formula $\cos(A) + \cos(B) = 2\cos(\frac{A+B}{2})\cos(\frac{A-B}{2})$. Similar methods allow us to conclude a whole set of identities called the sum-to-product formulas:

$$\cos(A) + \cos(B) = 2\cos(\frac{A+B}{2})\cos(\frac{A-B}{2}) \tag{1.15}$$

$$\cos(A) - \cos(B) = 2\sin(\frac{A+B}{2})\sin(\frac{B-A}{2}) \tag{1.16}$$

$$\sin(A) + \sin(B) = 2\sin(\frac{A+B}{2})\cos(\frac{A-B}{2}) \tag{1.17}$$

$$\sin(A) - \sin(B) = 2\cos(\frac{A+B}{2})\sin(\frac{A-B}{2}) \tag{1.18}$$

Let $\triangle ABC$ be a planar triangle with vertices at A, B, and C. We also use the letters A, B, and C to denote the measures of the angles at the vertices A, B, and C, respectively. We let a, b, and c denote the lengths of the sides $\overline{BC}$, $\overline{AC}$, and $\overline{AB}$. Then we recall the planar law of sines

$$\frac{a}{\sin(A)} = \frac{b}{\sin(B)} = \frac{c}{\sin(C)} \tag{1.19}$$

and the planar law of cosines

$$c^2 = a^2 + b^2 - 2ab\cos(C). \tag{1.20}$$

Given a real number L, we are occasionally confronted with the need to solve an equation $\cos(x) = L$ for x. Such an equation will have solutions if and only if $-1 \leq L \leq 1$. However, even then the solution is not unique unless we make an artificial restriction on what values are allowed for x. It is customary to demand that $0 \leq x \leq \pi$, in which there is a unique solution, and we write $x = \cos^{-1}(L)$ (or $x = \arccos(L)$). In the context of triangle trigonometry, this is pleasant if the x desired is the measure of an angle of a triangle because the measure of such an angle must be between 0 and π. Similarly, given any real number L, the equation $\cot(x) = L$ has a solution which is unique if we demand that $0 < x < \pi$. (We write $x = \cot^{-1}(L) = arccot(L)$.)

The situation is not quite so pleasant for the sine and tangent functions. If we need to solve $\sin(x) = L$ for x then there is a unique solution if we demand $-\frac{\pi}{2} \leq x \leq \frac{\pi}{2}$ and we write $y = \sin^{-1}(L) = \arcsin(L)$.) Unfortunately there is usually not a unique solution for $0 \leq x \leq \pi$ because of (1.1): any acute angle whose measure x satisfies $\sin(x) = L$ has a corresponding supplement which is also a solution.

If $\tan(x) = L$, then this can always be solved for x (regardless of the value of L) and we write $x = \tan^{-1}(L) = \arctan(L)$ for $-\frac{\pi}{2} < x < -\frac{\pi}{2}$. It is not generally customary to seek x so that $0 < x < \pi$ since the tangent is not defined at $x = \frac{\pi}{2}$ but in particular cases this may be appropriate.

We recall from plane geometry that in a right triangle $\triangle ABC$ where $A = \frac{\pi}{6}$, $B = \frac{\pi}{2}$, $C = \frac{\pi}{3}$ and the hypotenuse has length 1, then $AB = \frac{\sqrt{3}}{2}$ and $BC = \frac{1}{2}$. This allows us to calculate $\cos(\frac{\pi}{6}) = \sin(\frac{\pi}{3}) = \frac{\sqrt{3}}{2}$ and $\cos(\frac{\pi}{3}) = \sin(\frac{\pi}{6}) = \frac{1}{2}$.

In this book it will be useful to have the values of trigonometric functions for angles which are multiples of $\frac{\pi}{10} = 18°$. We derive some of these as follows. Note that if $x = \frac{\pi}{10}$, $2x + 3x = 5x = \frac{\pi}{2}$. Using (1.3), we get $\sin(2x) = \cos(3x)$, so using (1.13),(1.11), and (1.12) we obtain

$$\begin{aligned} 2\sin(x)\cos(x) &= \cos(2x)\cos(x) - \sin(2x)\sin(x) \\ 2\sin(x)\cos(x) &= (1 - 2\sin^2(x))\cos(x) - 2\sin^2(x)\cos(x), \end{aligned}$$

so gathering on the left and factoring we have $(4\sin^2(x) + 2\sin(x) - 1)\cos(x) = 0$. Now $\cos(x) \neq 0$ because x is strictly between 0 and $\frac{\pi}{2}$. Thus $4\sin^2(x) + 2\sin(x) - 1 = 0$. Solving for $\sin(x)$ by using the quadratic formula, $\sin(x) = (-2 \pm \sqrt{4 - 4(4)(-1)})/8 = (-2 \pm 2\sqrt{5})/8 = (-1 \pm \sqrt{5})/4$. Since $0 < x < \frac{\pi}{2}$, $\sin(x) > 0$, so $\sin(x) = \sin(\frac{\pi}{10}) = \sin(18°) = (-1 + \sqrt{5})/4$. It is left to the exercises to show that $\cos(\frac{\pi}{5}) = \cos(36°) = (1 + \sqrt{5})/4$. We also there present a different geometric approach to calculating these values.

Let us recall the customary approaches to "solving triangles" in the plane: that is, given some of the measures of sides and angles in a triangle, determine the measures of the remaining sides and angles. We suppose that the three angles have measures A, B, and C, and the opposite sides have measures a, b, and c, respectively.

Given the measures of all three sides of the triangle, the use of the planar law of cosines (1.20) delivers the value of $\cos(C)$, and hence that of C. Permutations of the values of a, b, c, A, B, C in (1.20) allows for the determination of A and B. A unique solution exists provided we have been given positive values of a, b, and c such that $a+b > c$, $a+c > b$, and $b+c > a$, as is required for any plane triangle.

Given the measures of two sides and the included angle (say a, b, and C) the law of cosines (1.20) delivers a unique value of c. Once the third side is found, the other two angles are found by the process in the previous paragraph. The only restrictions needed for a unique solution are that a, b be positive and $0 < C < \pi$.

Given the measures of two angles and the side between them, the value of the third angle is found from the fact that $A + B + C = \pi$. The planar law of sines (1.19) then delivers the values of the remaining two sides. The only restriction required for a unique solution is that the sum of the measures of the two given angles must be less than π.

The last scenario, where we are given the measures of two sides and an angle opposite one of them (say a, b, and A), is the most complex. This is the only case where we may have more than one solution (at most two) and the only case where, to most observers, the existence of a solution is not determined by a quick glance at the numbers given. Hence this scenario is dubbed the "ambiguous case." We here lay out a procedure by which one may find all solutions to a triangle in the ambiguous case.

Algorithm for solving the ambiguous case SSA.

Suppose that the known sides of a plane triangle are a and b and that the known angle is A.

(a) Determine all possible values for B: $\sin(B) = b\sin(A)/a$.

(b) If $\sin(B)$ is found to be larger than 1, there are no solutions.

(c) If $\sin(B)$ is found to be less than or equal to 1, then there are one (if $\sin(B) = 1$) or two (if $\sin(B) \leq 1$) possible values for B. We discard any value for B found where $A + B \geq \pi$.

(d) For each value of B emerging from part (c), we obtain one solution for the triangle, and find $C = \pi - A - B$ and $c = a\sin(C)/\sin(A)$.

Theorem 3.1 *Given real numbers $a, b > 0$ and A such that $0 < A < \pi$, the set of all possible solution triangles is the set of all triangles emerging from the above algorithm.*

Proof. First we prove that a known solution triangle with elements a, b, c, A, B, and C must emerge from this algorithm. Since the triangle satisfies the planar law of sines, B must satisfy the equation in part (a), where $\sin(B) \leq 1$ (in part (c)). Since the sum of the measures of the angles $A + B + C < \pi$, we must have $A + B < \pi$ in part (c). Furthermore, in part (d), C must equal

$\pi - A - B$ and by the law of sines, $c = a\sin(C)/\sin(A)$. So a known solution triangle must emerge from the algorithm.

Conversely, suppose that we are given some a, b, c, A, B, and C which emerge from the algorithm. We claim there exists a triangle having these values for the measures of its sides and angles. Since A, B, and c are all positive with $A + B < \pi$, there must be a unique triangle $\triangle ABC$ with elements A, B, and c, but where the other elements are possibly different from a, b, and C: we call them $\tilde{a}$, $\tilde{b}$, and $\tilde{C}$. From the algorithm part (d), we know that $A + B + C = \pi$; we also know that in $\triangle ABC$, $A + B + \tilde{C} = \pi$. Thus $C = \tilde{C}$. From the algorithm part (d), $c = a\sin(C)/\sin(A)$. Applying the law of sines to $\triangle ABC$, $c = \tilde{a}\sin(\tilde{C})/\sin(A)$, which we now know equals $\tilde{a}\sin(C)/\sin(A)$ since $C = \tilde{C}$. Thus $a = \tilde{a}$. From the algorithm part (a), $\sin(B) = b\sin(A)/a$. Applying the law of sines to $\triangle ABC$, $\sin(B) = \tilde{b}\sin(A)/\tilde{a}$, which we now know is $\tilde{b}\sin(A)/a$, since $a = \tilde{a}$. But then $b\sin(A)/a = \tilde{b}\sin(A)/a$, so $b = \tilde{b}$, as desired. So the elements a, b, c, A, B, and C which emerge from the algorithm form a triangle with these side and angle measures. ◇

Key facts from the study of areas: given two sides of a triangle and an included angle C, the area is $\frac{1}{2}ab\sin(C)$. A parallelogram with two sides of length a and b, and angle C between has area $ab\sin(C)$.

Exercises §3

1. Show (using similar triangles) that if $\triangle ABC$ and $\triangle DEF$ are plane right triangles with right angles at B and E, respectively, and $m\angle CAB = m\angle FDE$, then $AB/AC = DE/DF$, $BC/AC = EF/DF$, and $BC/AB = EF/DE$. Conclude that the notions of $\sin(\theta)$ and $\cos(\theta)$ for general angle θ are well-defined. (That is, the value of r used in the definition does not matter.)

2. Following the method of the text, prove the formulas (1.16),(1.17),(1.18).

3. Explain why formulas (1.5) and (1.6) are true.

4. Calculate the following in terms of radicals.

   (a) $\cos(36°)$ (b) $\cos(18°)$ (c) $\sin(36°)$ (d) $\sin(54°)$ (e) $\sin(72°)$ (f) $\cos(162°)$ (g) $\cos(108°)$ (h) $\sin(126°)$ (i) $\sin(144°)$ (j) $\cos(216°)$ (k) $\cos(288°)$ (l) $\sin(234°)$ (m) $\sin(342°)$ (n) $\csc(18°)$ (o) $\sec(18°)$ (p) $\tan(18°)$ (q) $\csc(18°)$ (r) $\sec(36°)$ (s) $\csc(36°)$ (t) $\tan(36°)$ (u) $\sin(66°)$ (v) $\cos(63°)$ (w) $\sin(99°)$ (x) $\cos(9°)$ (y) $\sin(3°)$ (z) $\cos(3°)$

5. Let $\triangle ABC$ be an isosceles triangle with vertex A: i.e., length AB is the same as length AC. Suppose the measure of the angle at A is $\frac{\pi}{5}$ (i.e., $36°$). Suppose $AB = 1$ and let $x = BC$. Let D be a point in side $\overline{AC}$ such that $\overrightarrow{BD}$ bisects $\angle ABC$.

   (a) Prove that $AD = BD$.

   (b) Prove that $BD = BC$.

(c) Explain why $\frac{1}{x} = \frac{x}{1-x}$.

(d) Solve the equation in (c) for x.

(e) Use part (d) to conclude that $\sin(18°) = (-1+\sqrt{5})/4$.

6. Solve the following triangles using the algorithm for solving the ambiguous case. Draw a picture to illustrate the situation.

   (a) $a = 1, A = 40°, b = 5$

   (b) $a = 4, A = 40°, b = 5$

   (c) $a = 7, A = 40°, b = 5$

   (d) $a = 1, A = 130°, b = 6$

   (e) $a = 5, A = 130°, b = 6$

   (f) $a = 8, A = 130°, b = 6$

7. Consider the following step in the algorithm for solving the ambiguous case: "We discard any value for B found where $A + B \geq \pi$." Prove that if $\angle A$ is acute or $a \neq b$ this is the same as discarding any value of B found where $a - b$ and $A - B$ are not both positive, both zero, or both negative. What happens in the other cases?

8. Let A, B, and C be the measures of the angles in a triangle and let a, b, and c be the lengths of the opposite sides. Use the law of sines to transform the quantity $\frac{a-b}{c}$ into an expression involving angles. Then conclude that

$$\frac{a-b}{c} = \frac{\sin\frac{1}{2}(A-B)}{\cos\frac{1}{2}C}. \tag{1.21}$$

9. Use the technique of Exercise 8 to prove that

$$\frac{a+b}{c} = \frac{\cos\frac{1}{2}(A-B)}{\sin\frac{1}{2}C}. \tag{1.22}$$

(Equations (1.21) and (1.22) are called Mollweide's equations.)

10. Prove that

$$\frac{\tan\frac{1}{2}(A-B)}{\tan\frac{1}{2}(A+B)} = \frac{a-b}{a+b}. \tag{1.23}$$

when both fractions are defined. (This is the planar *law of tangents.*)

11. Suppose we write $s = (a+b+c)/2$. Use (1.20) to show that

$$\sin(\frac{C}{2}) = \sqrt{\frac{(s-a)(s-b)}{ab}} \quad (1.24)$$

$$\cos(\frac{C}{2}) = \sqrt{\frac{s(s-c)}{ab}}. \quad (1.25)$$

12. Prove that

$$\tan(\frac{C}{2}) = \sqrt{\frac{(s-b)(s-a)}{s(s-c)}}. \quad (1.26)$$

where $s = (a+b+c)/2$.

13. Using Exercise 12, prove that given three positive real numbers a, b, and c such that $a+b > c$, $a+c > b$, and $b+c > a$, there exists a plane triangle whose side lengths are a, b, and c.

14. Using Exercise 12, show that

$$r = \sqrt{\frac{(s-a)(s-b)(s-c)}{s}}, \quad (1.27)$$

where r is the radius of the inscribed circle of the triangle.

15. Using Exercise 14, prove that the area of a triangle is

$$\sqrt{s(s-a)(s-b)(s-c)}. \quad (1.28)$$

(This is Hero's formula.)

16. Suppose that x and y are positive numbers such that $x+y < \pi$. Prove that

$$\frac{\cot(x)+\cot(y)}{2} \geq \cot(\frac{x+y}{2}), \quad (1.29)$$

where equality occurs if and only if $x = y$.

4 Coordinates and vectors

In this section we briefly review ideas from rectangular coordinates and vectors in space which we will need. Every point in space can be written as an ordered triple (x, y, z), where x, y, and z are real numbers. These are the so-called *Cartesian* coordinates of the point.

Let $P_1 = (x_1, y_1, z_1)$ and $P_2 = (x_2, y_2, z_2)$ be two points in space. Then the *distance* between them is calculated via the *distance formula*

$$d(P_1, P_2) = \sqrt{(x_2-x_1)^2 + (y_2-y_1)^2 + (z_2-z_1)^2}.$$

Since a sphere of radius r and center (x_1, y_1, z_1) consists of all points (x, y, z) such that $\sqrt{(x-x_1)^2+(y-y_1)^2+(z-z_1)^2} = r$, i.e.,

$$(x-x_1)^2+(y-y_1)^2+(z-z_1)^2 = r^2, \tag{1.30}$$

we shall say that (1.30) is the equation of the sphere with center (x_1, y_1, z_1) and radius r. For convenience, most of the time we will assume the sphere has center at the origin $(0, 0, 0)$ so the equation would be

$$x^2+y^2+z^2 = r^2.$$

We will also need the notation of vectors. A *vector* in space may be thought of as an arrow in space pointing from one point to another. The arrow has an *initial* point P_1 and a *terminal* point P_2 and is denoted by $\vec{P_1P_2}$. Vectors may also be expressed with coordinates $\langle x, y, z\rangle$ which may be determined as follows: if P_1 and P_2 have coordinates as given above, then $\vec{P_1P_2}$ has coordinates $\langle x_2-x_1, y_2-y_1, z_2-z_1\rangle$ found by subtracting the Cartesian coordinates of P_1 from the Cartesian coordinates of P_2 (terminal coordinates minus initial coordinates). Another vector pointing in the same direction for the same distance as $\vec{P_1P_2}$ will be identified as the same vector as $\vec{P_1P_2}$ and hence has the same coordinates. (See Figure 1.5.)

The *length* of a vector $\mathbf{r}$ with coordinates $\langle x, y, z\rangle$ is defined to be

$$|\mathbf{r}| = \sqrt{x^2+y^2+z^2}. \tag{1.31}$$

The motivation behind this definition is as follows. Suppose a vector $\mathbf{r}$ has initial point $P_1 = (x_1, y_1, z_1)$ and terminal point $P_2 = (x_2, y_2, z_2)$. Then $\mathbf{r}$ has coordinates $\langle x_2-x_1, y_2-y_1, z_2-z_1\rangle$ and hence by (1.31) has length $|\mathbf{r}| = \sqrt{(x-x_1)^2+(y-y_1)^2+(z-z_1)^2}$. But by the distance formula, this is the distance between P_1 and P_2 in space. So the length of a vector is the length of its associated arrow. The length is also sometimes called the *norm*, the *absolute value*, or the *modulus*.

Two vectors $\mathbf{r}_1 = \vec{P_1Q_1} = \langle x_1, y_1, z_1\rangle$ and $\mathbf{r}_2 = \vec{P_2Q_2} = \langle x_2, y_2, z_2\rangle$ can be *added* to obtain the vector $\mathbf{r}_1+\mathbf{r}_2 = \langle x_1+x_2, y_1+y_2, z_1+z_2\rangle$, the *sum* of $\mathbf{r}_1$ and $\mathbf{r}_2$. The arrow associated with the sum is found as follows: move the arrow for $\mathbf{r}_2$ without changing its length or direction so that its initial point coincides with the terminal point of $\mathbf{r}_1$. Then the vector $\mathbf{r}_1+\mathbf{r}_2$ has an arrow which points from the initial point of $\mathbf{r}_1$ to the terminal point of $\mathbf{r}_2$. (See Figure 1.5.)

A vector $\mathbf{r} = \langle x, y, z\rangle$ may be multiplied by a real number c called a *scalar* to obtain another vector we denote by $c\mathbf{r}$ with coordinates $\langle cx, cy, cz\rangle$. This *scalar multiplication* of a real number with a vector has a geometric interpretation: if $c > 0$, the vector $c\mathbf{r}$ has an arrow with the same initial point and direction as $\mathbf{r}$ but whose length is $c|\mathbf{r}|$, i.e., $c\mathbf{r}$ is c times as long as $\mathbf{r}$. If $c < 0$, $c\mathbf{r}$ is a vector $|c|$ times as long as $\mathbf{r}$ pointing in the opposite direction as $\mathbf{r}$. If we write $\mathbf{i} = \langle 1, 0, 0\rangle$, $\mathbf{j} = \langle 0, 1, 0\rangle$ and $\mathbf{k} = \langle 0, 0, 1\rangle$ then $\langle x, y, z\rangle$ can be written as $x\mathbf{i}+y\mathbf{j}+z\mathbf{k}$.

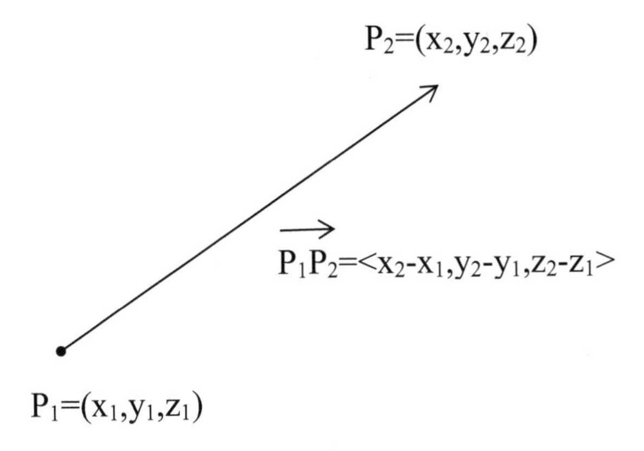

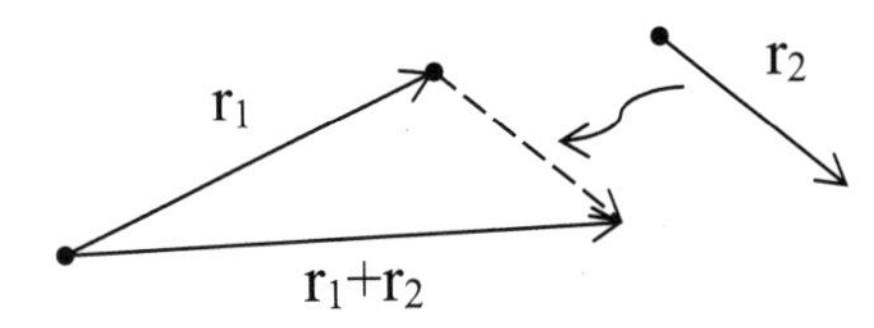

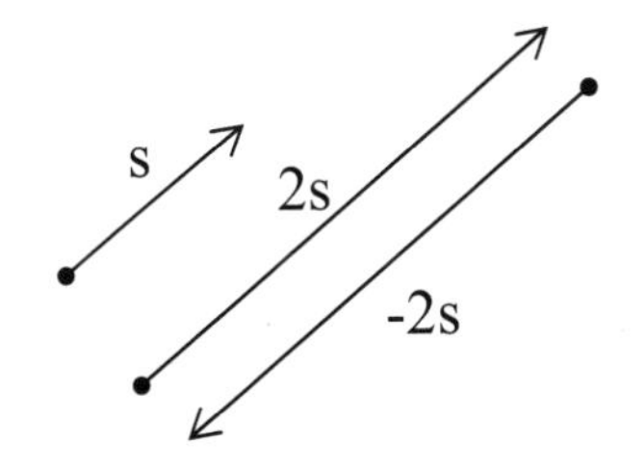

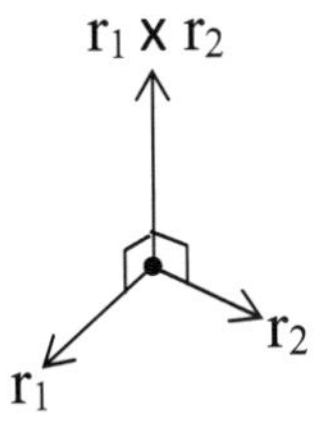

Figure 1.5: Vectors: coordinates, sum, scalar multiplication, and cross product.

The *angle* between two vectors $\vec{OP}$ and $\vec{OQ}$ with the same initial point is the angle $\angle POQ$. If two vectors do not have the same initial point, then we may form an angle between them by moving one without changing its length or direction so that they do have the same initial point.

Another operation with vectors we shall need is called the *dot product* or *scalar product*. Given vector $\mathbf{r_1} = \langle x_1, y_1, z_1\rangle$ and $\mathbf{r}_2 = \langle x_2, y_2, z_2\rangle$ the dot product $\mathbf{r}_1 \cdot \mathbf{r}_2$ is the real number

$$\mathbf{r}_1 \cdot \mathbf{r}_2 = x_1x_2 + y_1y_2 + z_1z_2. \tag{1.32}$$

For the dot product, we recall the following key theorem.

Theorem 4.1 *If θ is the angle between the vectors $\mathbf{r}$ and $\mathbf{s}$ then*

$$\mathbf{r} \cdot \mathbf{s} = |\mathbf{r}||\mathbf{s}| \cos(\theta). \tag{1.33}$$

Note that the angle between the vectors is a right angle when their dot product is zero; in this case we say that the vectors are perpendicular.

We also have an operation called the *cross product* or *vector product*. If $\mathbf{r_1}$ and $\mathbf{r_2}$ are as above then

$$\begin{aligned} \mathbf{r_1} \times \mathbf{r_2} &= \langle y_1z_2 - y_2z_1, x_2z_1 - x_1z_2, x_1y_2 - x_2y_1\rangle & (1.34)\\ &= (y_1z_2 - y_2z_1)\mathbf{i} + (x_2z_1 - x_1z_2)\mathbf{j} + (x_1y_2 - x_2y_1)\mathbf{k}. & (1.35)\end{aligned}$$

Note that the cross product can be expressed with the determinant

$$\begin{vmatrix} \mathbf{i} & \mathbf{j} & \mathbf{k} \\ x_1 & y_1 & z_1 \\ x_2 & y_2 & z_2 \end{vmatrix}.$$

(See Figure 1.5.) One key property of the cross product $\mathbf{r} \equiv \mathbf{r_1} \times \mathbf{r_2}$ is that it is perpendicular to both of $\mathbf{r_1}$ and $\mathbf{r_2}$, and the direction of $\mathbf{r}$ is found from $\mathbf{r_1}$ and $\mathbf{r_2}$ via the right hand rule. That is, if one turns $\mathbf{r_1}$ into $\mathbf{r_2}$ with the right hand, the thumb will point in the direction of $\mathbf{r_1} \times \mathbf{r_2}$. We also have

$$|\mathbf{r_1} \times \mathbf{r_2}| = |\mathbf{r_1}||\mathbf{r_2}| \sin(\theta), \tag{1.36}$$

where θ is the angle between $\mathbf{r_1}$ and $\mathbf{r_2}$. A consequence of this is that $|\mathbf{r_1} \times \mathbf{r_2}|$ is the area of the parallelogram determined by the vectors $\mathbf{r_1}$ and $\mathbf{r_2}$.

If $\mathbf{r_3} = (x_3, y_3, z_3)$ then properties of determinants (or algebraic manipulation) show that $\mathbf{r_1} \cdot \mathbf{r_2} \times \mathbf{r_3} = \mathbf{r_2} \cdot \mathbf{r_3} \times \mathbf{r_1} = \mathbf{r_3} \cdot \mathbf{r_1} \times \mathbf{r_2}$. The absolute value of each of these gives the volume of the parallelepiped with three edges $\mathbf{r_1}, \mathbf{r_2}, \mathbf{r_3}$. (See Figure 1.6.) This is because $|\mathbf{r_1} \cdot \mathbf{r_2} \times \mathbf{r_3}| = |\mathbf{r_1}||\mathbf{r_2} \times \mathbf{r_3}||\cos(\theta)|$, where θ is the angle between $\mathbf{r_1}$ and the normal to the plane containing $\mathbf{r_2}$ and $\mathbf{r_3}$. But $|\mathbf{r_1}||\cos(\theta)|$ is the length of the perpendicular from the terminal point of $\mathbf{r_1}$ to the plane containing $\mathbf{r_2}$ and $\mathbf{r_3}$. This is the height of the parallelepiped. Since $|\mathbf{r_2} \times \mathbf{r_3}|$ is the area of a base parallelogram, we obtain the volume of the appropriate parallelepiped.

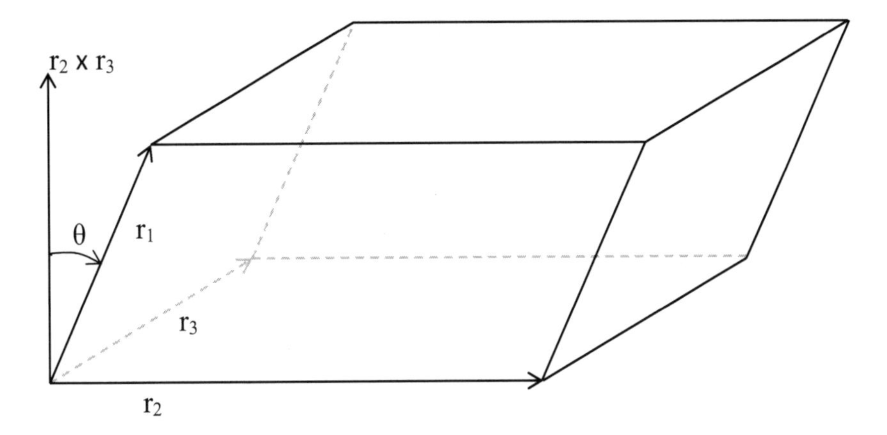

Figure 1.6: The solid parallelepiped formed by $\mathbf{r_1}, \mathbf{r_2}$, and $\mathbf{r_3}$.

Exercises §4

1. Use vectors to prove Lemma 1.4 with vectors as follows. Choose vectors pointing along the four line segments and prove that appropriately chosen pairs of cross products are equal.

Chapter 2

The sphere in space

We begin our study of spherical geometry by noting two different ways of presenting the subject, both of which we consider in this book.

We may first consider a sphere the way most people understand it: as a round object that lives in three-dimensional space. By doing so, we may use the tools of three-dimensional geometry, trigonometry, and vectors to learn about the sphere. This we will call an *extrinsic* approach to spherical geometry. An understanding of it is particularly important in applications. On the other hand, we may study spherical geometry via the *intrinsic* properties of the sphere — that is, properties of the sphere that can be thought of without reference to the larger three-dimensional space in which a sphere sits. It will be our main goal to do this in Chapter 3.

We begin by looking at the extrinsic properties of the sphere in this chapter. As we do so, it will be important to understand how to see these properties as intrinsic, and hence motivate the axiom system that we set up in Chapter 3.

5 Great circles

Definition 5.1 *Let O be a point in space and r a positive real number. The* sphere *with* center O *and* radius r *is the set of all points in space which lie at distance r from O.*

The following proposition summarizes the properties of intersections between spheres and planes in space.

Proposition 5.2 *The intersection between a sphere and a plane in space is a circle, a point, or the empty set.*

In Figure 2.1 we may see these possibilities geometrically by considering the line in space through the center O of the sphere perpendicular to the plane

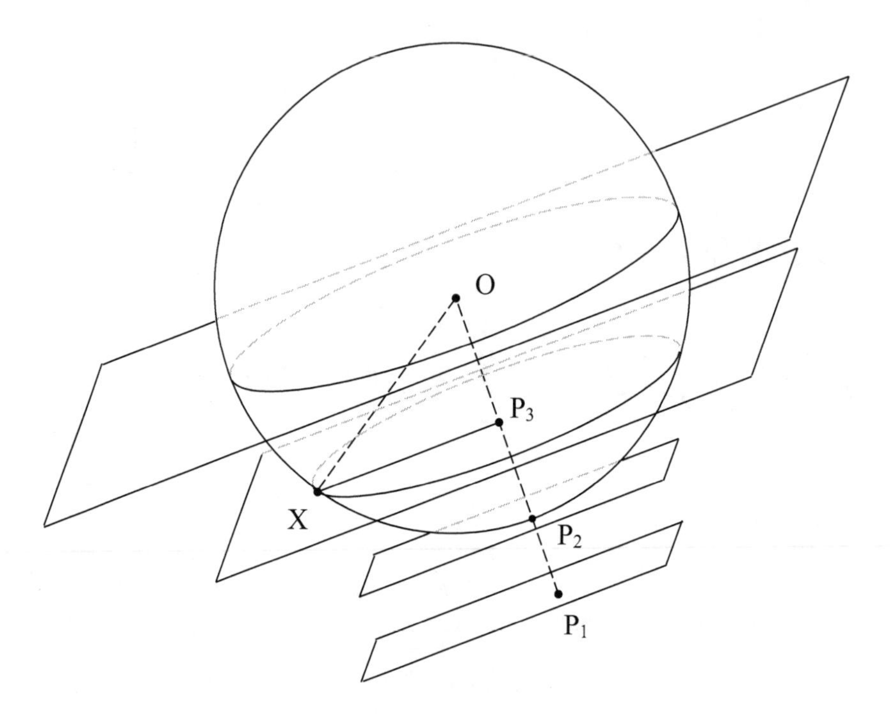

Figure 2.1: The intersection of a plane with a sphere.

at point P. We distinguish four cases: P is outside the sphere, on the sphere, interior to the sphere but different from O, or equal to O. In the first case, every point in the plane is outside the sphere because $P = P_1$ is the closest point on the plane to O — hence every point on the plane is at a greater distance from O than the radius. If $P = P_2$ is on the sphere, then the plane is tangent to the sphere, hence meets it in a single point. In the case of $P = O$, the intersection of the sphere with the plane is the set of all points in the plane at a fixed distance from O. This is a circle. Lastly, suppose $P = P_3$ is inside the sphere, but different from O. Then let X be any point on the intersection of the plane and the sphere, let r be the radius of the sphere, and let d be the distance between P_3 and O. Then $\triangle OP_3X$ has a right angle at P_3, so the length of $\overline{P_3X}$ must be $\sqrt{r^2 - d^2}$. Thus the points on both the plane and sphere all lie at the same distance from P_3, so lie on a circle. To be complete, we must show that all points in the plane at distance $\sqrt{r^2 - d^2}$ from P_3 are also on the sphere; this is left as an exercise. So the intersection of the plane and the sphere is a circle of radius $\sqrt{r^2 - d^2}$.

A similar analysis works for the intersection of a line with a sphere: it consists of zero, one, or two points.

Definition 5.3 *(See Figure 2.2.) Let s be a sphere with center O, p a plane in space and ℓ the line passing through O perpendicular to p.*

If $p = p_1$ passes through O then the intersection of p_1 and s is called a great circle *of the sphere. The two points where ℓ meets the sphere are called the* poles *of the great circle. We will say that p_1 is the plane of the great circle.*

If $p = p_2$ lies at distance $0 < d < r$ from O, then the intersection of p_2 and s is called a small circle*. The two points where ℓ meets the sphere are called the* poles *of the small circle. We will say that p_2 is the plane of the small circle.*

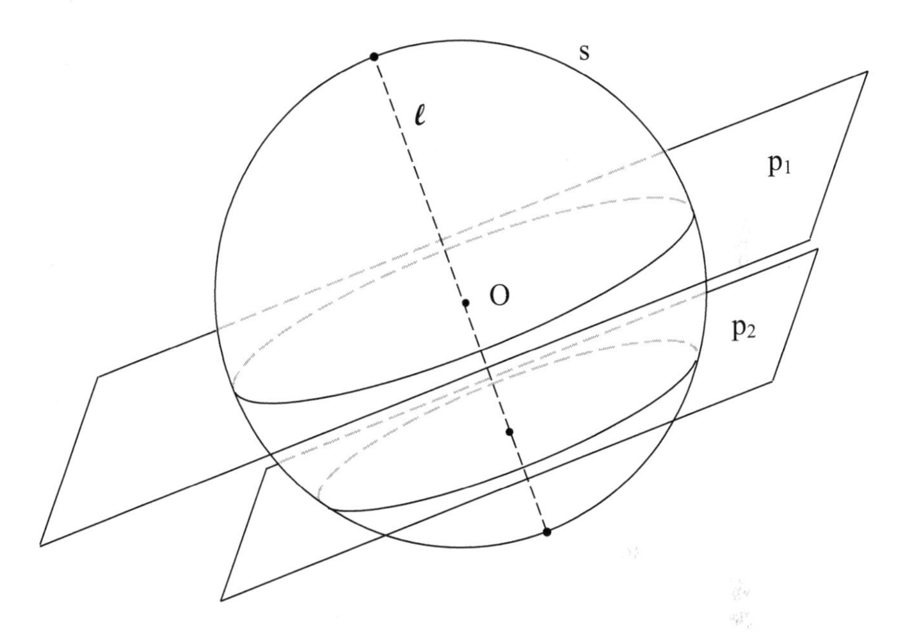

Figure 2.2: Great circles, small circles, and poles.

Definition 5.4 *Two distinct points on a sphere are said to be* antipodal *if the line passing through them passes through the center of the sphere. Each point is said to be the* antipode *of the other.*

If points A and B are antipodal then we write $B = A^a$ (and $A = B^a$). Note that the poles of a great circle are antipodal (as are the poles of a small circle).

Two points on a great circle which are antipodal divide a great circle into two semicircles which are called *great semicircles.*

As an example, the surface of the earth is (approximately) a sphere. The equator of the earth is a great circle, and the north and south poles of the earth are its poles. The circles of constant latitude strictly between zero and ninety degrees are small circles; again, the poles are the north and south poles of the earth. Each meridian forms a semicircle which is a great semicircle.

Great circles are significant because they are the "lines" of spherical geometry. That is, they serve many of the same functions in spherical geometry that lines do in plane geometry. We will see that the shortest path between two points on a sphere follows the route of a great circle, and that most of the time, two points determine a unique great circle passing through them.

Two great circles arise by intersecting the sphere with two planes through the center of the sphere. These two planes intersect in a line through the center of the sphere, which meets the sphere in two points. Thus we have the fundamental result:

Proposition 5.5 *Two distinct great circles meet in exactly two points, and these two points are antipodal. These two points divide each great circle into two great semicircles.*

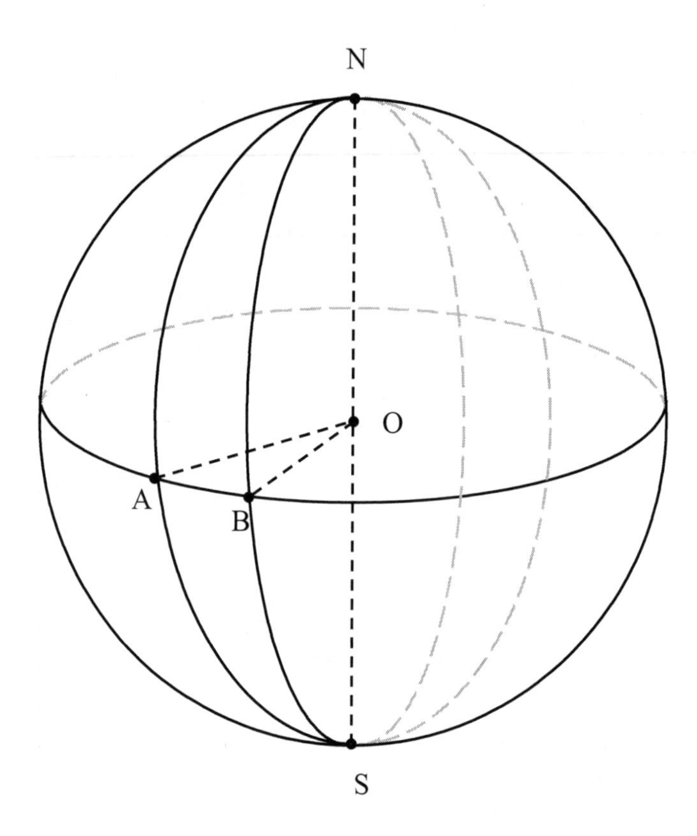

Figure 2.3: Pairs of great circles meeting at antipodal points N and S. Measure of arc. Great semicircle NAS, arc $\overset{\frown}{AB}$, and lune $NASBN$.

Suppose we are given two great semicircles with common antipodal end points N and S. If the semicircles are not on the same great circle, they form a two-sided object called a *lune*. The semicircles are said to be the *sides* of the lune, and N and S are the *vertices* of the lune.

The reader will recall that two points in a plane determine a unique line containing them. What about great circles?

Considering Figure 2.3, it is clear that this is not always the case on the sphere. Two antipodal points N and S lie on a line through the center of the sphere. There are many planes containing that line, each of which determines a great circle containing the antipodal points. But it turns out that this is the only exception:

Proposition 5.6 *If two distinct points on a sphere are not antipodal then there exists a unique great circle passing through them.*

We may see why by thinking about what properties the plane containing the great circle would have to satisfy. Such a plane would have to contain the two distinct non-antipodal points (A and B) on the sphere along with the center O of the sphere. Three points determine a unique plane in space if they do not all lie on the same line. If these three points lay on a line, then because one of the points is the center of the sphere, the two points on the sphere would have to be antipodal by definition. Thus the plane containing the three points is unique, and hence the great circle containing them is also unique.

In plane geometry, two points determine a unique "line segment" consisting of points "between" the given points, together with the two original points. On the sphere, the situation is again somewhat different. Given two points on a great circle, there are two circular arcs between them on that great circle. Which one shall be "the" segment between the two points?

Again, there are two cases to consider. The points are either antipodal or not. If they are antipodal (e.g., N and S in Figure 2.3), there are many great circles to choose from which contain the two points. In this case, there is no natural choice to make of a great circle arc between the points. We will have to specify a third point to determine a semicircle. That is, there is only one great semicircle from N to S passing through A (denoted NAS), and only one great semicircle from N to S passing through B, etc. If two given points are not antipodal, then there is a unique great circle through them and two arcs which together make up that great circle, one shorter and one longer. Typically we shall choose the shorter of the two to denote "the" arc between the given points. The points on that arc are then said to be "between" the given points (except that the original two points, the "endpoints" of the arc, are not said to be "between" the endpoints). If A and B are the endpoints then we denote the arc as $\overset{\frown}{AB}$.

In a plane a line segment between two points has a length which is the distance between the two points. On the sphere we have a similar phenomenon. We define spherical distance between two points to be the measure of the shortest arc of a great circle between those points. Typically by "measure" we mean the measure of the central angle of that arc (e.g., in Figure 2.3, $\angle AOB$ is the central angle of the great circle arc from A to B). The "measure" of the arc is closely related to the length of the arc in space. If an arc of a circle of radius r has central angle of radian measure θ then its length is $r\theta$. Thus on a sphere of radius 1, the length of the arc is the same as its measure.

A great semicircle will have a measure of π radians (180°). For units of arc measure we use either degrees or radians depending on the situation. In order to avoid speaking of radians and degrees at the same time, we often shall speak of a pair of points as being a "semicircle" apart (π radians, 180°) or a "quarter circle" apart ($\frac{\pi}{2}$ radians, 90°).

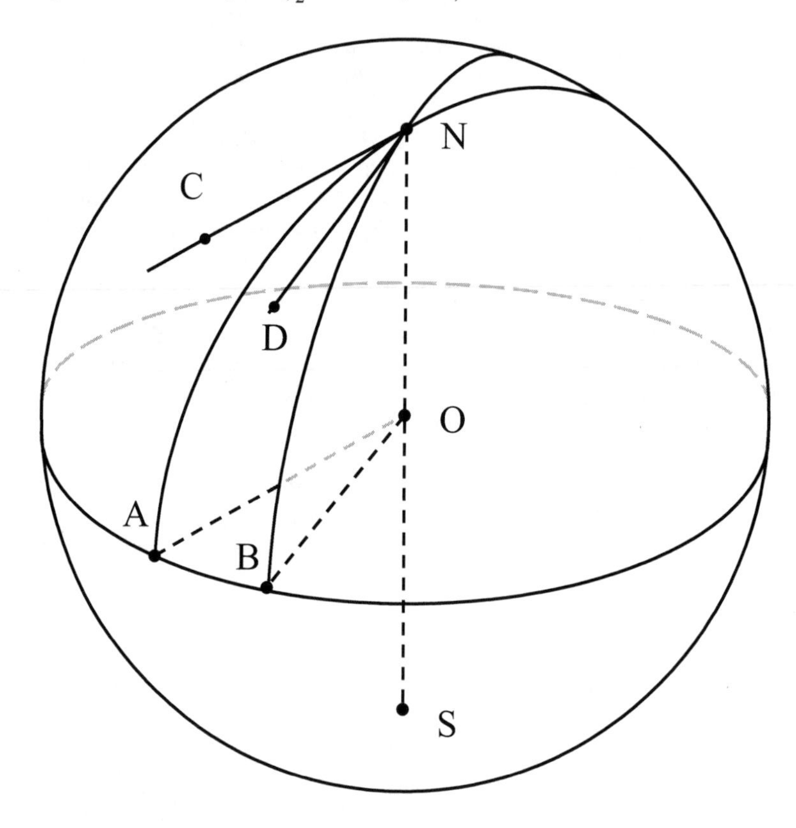

Figure 2.4: Measure of a spherical angle: $m \prec ANB = m \overset{\frown}{AB} = m\angle AOB = m\angle CND = m\angle A - ON - B$.

Figure 2.4 illustrates how we will understand the measure of angles between great circles on the sphere. The great circle arcs $\overset{\frown}{NA}$ and $\overset{\frown}{NB}$ lie on the sides of what we will understand to be a spherical angle. The rays $\overrightarrow{NC}$ and $\overrightarrow{ND}$ are tangent to these two arcs at N. The measure of the angle between these two rays ($\angle CND$) will be the measure of the angle between the arcs. But since ray $\overrightarrow{OA}$ is parallel to the ray $\overrightarrow{NC}$ and $\overrightarrow{OB}$ is parallel to $\overrightarrow{ND}$, the measures of $\angle CND$ and $\angle AOB$ are the same. We above noted that the measure of arc $\overset{\frown}{AB}$ equals the measure of $\angle AOB$. Then $m \overset{\frown}{AB} = m\angle AOB = m\angle CND$ is the measure of what we will call the "spherical angle" $\prec ANB$. Note also that since $\angle AOB$ is a plane angle of the dihedral angle $\angle A-ON-B$, the measure of spherical angle $\prec ANB$ is the measure of the dihedral angle $\angle A-ON-B$. Thus the measure of the angle between two great circles is the measure of the angle between the planes containing them in space. (Figure 2.5 illustrates this situation with the view from above N.)

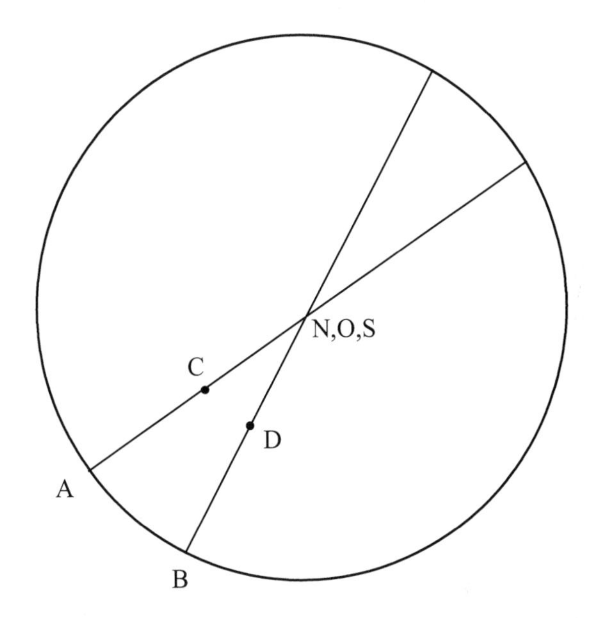

Figure 2.5: Measure of a spherical angle, top view.

A lune has two angles — one based at each of its vertices. These two angles have the same measure. This can be seen in Figure 2.3: lune $NASBN$ has two angles $\prec ANB$ and $\prec ASB$, both of which have measure equal to the measure of arc $\overset{\frown}{AB}$. The *measure* of a lune is defined to be the same as the measure of its angles.

Our definition of the notion of poles of a great circle arose from the fact that a line perpendicular to the plane of a great circle at its center meets the sphere in two points. We shall want to have an understanding of properties of the poles which depends solely on intrinsic properties of the sphere:

Proposition 5.7 *A great circle of a sphere s in space is the set of all points of s which lie a spherical distance of a quarter circle from a pole of the great circle.*

Proof. The great circle is the intersection of a plane p with sphere s and the poles of the great circle are the intersection of a line ℓ with s, where ℓ is perpendicular to p at O. First we show that any point of the great circle is at distance $\frac{\pi}{2}$ from a pole. Any point A of the great circle is a point of p and a pole N of the great circle lies on ℓ. Since ℓ is perpendicular to p, $\overrightarrow{ON}$ is perpendicular to $\overrightarrow{OA}$. By definition the spherical distance between A and N is the measure of $\angle AON$, which is $\frac{\pi}{2}$. Conversely, suppose point A is at spherical distance $\frac{\pi}{2}$ from pole N. By definition, $\angle AON$ has measure $\frac{\pi}{2}$, so $\overrightarrow{OA}$ is perpendicular to $\overrightarrow{ON}$. By Proposition 2.9, we conclude that A is a point of p, so belongs to the great circle. $\diamondsuit$

Proposition 5.7 is useful in that it provides a description of the relationship between a great circle and its poles that depends only on an intrinsic property (distance) on the sphere, and not on lines and planes in space. A similar characterization is desired for small circles.

Proposition 5.8 *A small circle on a sphere is the set of all points on the sphere at a fixed spherical distance $\rho < \frac{\pi}{2}$ from one of its poles P.*

See Exercise 4 to justify this via three-dimensional geometry.

Definition 5.9 *The point P found in Proposition 5.8 is called the* (spherical) center *of the small circle, and the quantity ρ is called the* (spherical) radius *of the small circle.*

It is not hard to see that the points on the small circle are also equidistant from the antipode of P; see Exercise 5.

In plane geometry we learn that a line in a plane splits the plane into two sets of points off the line called half-planes. These sets are convex in the sense that given two points in a half-plane, the line segment between them also lies in the half-plane.

To see that this idea carries over to the sphere, one simply notes that a plane in space splits space into two convex sets called half-spaces. Given a great circle, we consider the two half-spaces associated with its plane. If we intersect these half-spaces with the sphere, we obtain two so-called hemispheres into which the sphere is divided by the great circle. These hemispheres are the “sides” of the great circle.

Are these hemispheres “convex”? What does “convex” mean on the sphere, if anything? We take as our understanding of convexity on the sphere that the hemisphere is “spherically convex” if the shortest great circle arc between any two (non-antipodal) points also lies in that hemisphere. That this is the case may be seen as suggested in Figure 2.6.

Given two points in a hemisphere, we take a great circle containing them, which meets the edge of the hemisphere in two antipodal points. These antipodal points split the original great circle into two great semicircles, one of which is in our hemisphere, and the other of which is in the opposite hemisphere. Our two points must lie in the semicircle in our hemisphere, but not at the endpoints. Thus the shorter arc between them lies in that semicircle, as desired.

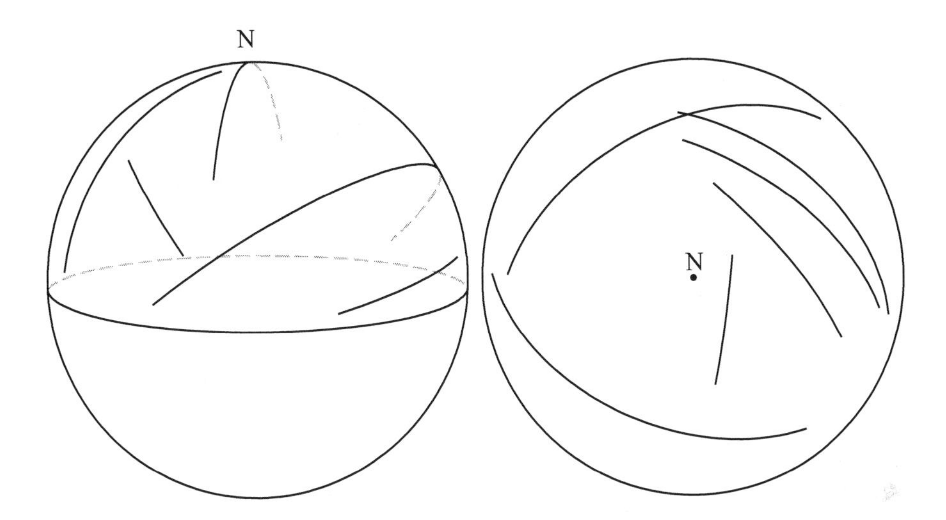

Figure 2.6: A hemisphere is spherically convex — side and top views. The shorter great circle arc between two points of the hemisphere lies in the hemisphere.

A great circle is spherically convex since a spherical arc between two of its non-antipodal points lies on the given great circle. Furthermore, a spherical arc is also spherically convex.

We conclude with one more observation about hemispheres as suggested by Figure 2.7. There we have a hemisphere, a great circle which is the edge of that hemisphere, and a pole N of the great circle contained in the hemisphere. Then the hemisphere consists of all points which are within a quarter circle of the pole N. Any great circle passing through N meets the edge of the hemisphere in two points which cut the great circle into two great semicircles, one of which contains N in the middle. Thus the points of the semicircle lie within a quarter circle of N (except for the endpoints).

Exercises §5

For each of these exercises, use whatever knowledge and technique from three-dimensional geometry is necessary.

1. For each of the following values of r and θ, determine the length of an arc of a great circle whose measure is θ on a sphere of radius r.

   (a) $r = 50$, $\theta = \frac{\pi}{4}$ radians

   (b) $r = 2000$, $\theta = 50°$

   (c) $r = 100$, $\theta = 140°$

2. Suppose that point A is on one side of a great circle. Show that its antipode is on the other side of the great circle.

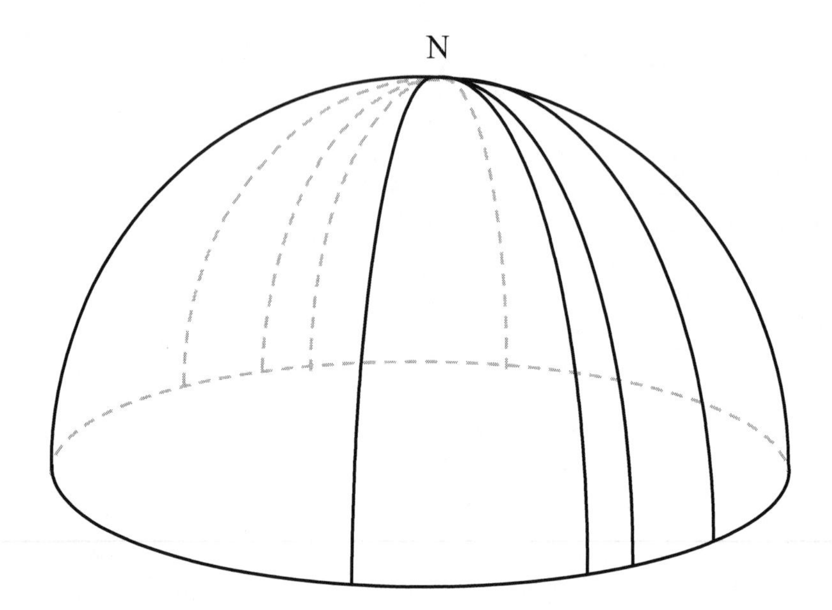

Figure 2.7: A hemisphere is the set of points less than a quarter circle from a pole N of its edge.

3. Suppose that A is a point of a sphere with antipode A^a and B is any other point of the sphere. Explain why $m\,\widehat{BA} + m\,\widehat{BA^a} = \pi$.

4. Use three-dimensional geometry to justify Proposition 5.8.

5. Let c be a small circle with center P and spherical radius ρ. Show that c is the set of all points on the sphere which are at spherical distance $\pi - \rho$ from P^a, the antipode of P.

6. Given three points on a sphere which do not lie on a single great circle, argue that they lie on a unique small circle.

7. Let A and B be two points on a sphere. Using a similar theorem in space, explain why the set of all points Γ on the sphere which are at the same spherical distance from A and B must be a great circle perpendicular to any great circle passing through A and B. If A and B are not antipodal, then Γ is called the *perpendicular bisector* of $\widehat{AB}$). If A and B are antipodal, the great circle is called the *polar circle* or *polar* of A and B. (Compare this extrinsic approach to the problem to an intrinsic approach in §12, Exercise 10.)

8. Suppose that a small circle has two points on opposite sides of a great circle. Argue that the small circle and great circle intersect in two distinct points.

9. Suppose that a sphere has radius r and a small circle on it has spherical radius ρ. Show that the perimeter of the small circle in space is $2\pi r \sin(\rho)$. Conclude that if a small circle arc from point A to point B has arc measure ϕ radians, then the length (in space) of that arc is $r\phi \sin(\rho)$.

10. Referring to problem 9, we define the *spherical length* of the small circle arc given to B to be $\phi \sin(\rho)$. Show that the spherical distance from A to B is less than $\phi \sin(\rho)$.

6 Distance and angles

Plane geometry has the good fortune that the most important theorem regarding distances (the Pythagorean theorem) can be stated without the use of trigonometric functions. Spherical geometry is not so fortunate. Even the most basic propositions about relationships in right triangles require the use of trigonometric functions. We illustrate with an example.

Proposition 6.1 *Suppose two spherical arcs form an angle with measure θ. Let x be the spherical distance between two points (one on each arc) at spherical distance ϕ from the vertex. Then* $\sin(\frac{x}{2}) = \sin(\phi)\sin(\frac{\theta}{2})$.

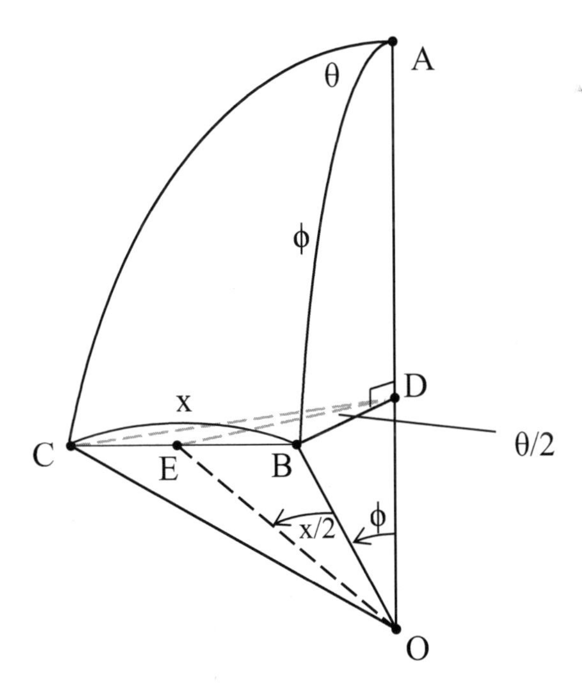

Figure 2.8: Proposition 6.1.

Proof. Suppose the sphere has center O and radius r. Since $\overset{\frown}{AB}$ and $\overset{\frown}{AC}$ have the same measure ϕ, there is a plane perpendicular to $\overleftrightarrow{OA}$ (at D) passing through both B and C. By symmetry, we also have $DB = DC$ and $OB = OC = r$. Let E be the midpoint of the line segment $\overline{BC}$. Then $\triangle DEB$ and $\triangle OEB$ both have a right angle at E. We get $m\angle EOB = \frac{1}{2}m\angle COB = \frac{x}{2}$ by the definition of the measure of $\overset{\frown}{BC}$. Also, $m\angle EDB = \frac{1}{2}m\angle BDC$. But $\angle BDC$ is a plane angle of $\angle B - OA - C$, so has measure θ by definition of the measure of spherical angle $\prec BAC$. So $m\angle EDB = \theta/2$. By definition of arc measure, $m\angle BOD = m\angle BOA = \phi$. By plane right triangle trigonometry applied to $\triangle OBD$, $BD = OB\sin(\phi) = r\sin(\phi)$. Applying plane trignometry to $\triangle BED$, $EB = BD\sin(\frac{\theta}{2}) = r\sin(\phi)\sin(\frac{\theta}{2})$. Applying plane trigonometry to $\triangle OEB$, we find $EB = r\sin(\frac{x}{2})$. Thus $r\sin(\frac{x}{2}) = EB = r\sin(\phi)\sin(\frac{\theta}{2})$ and the conclusion follows. (This argument assumes ϕ is acute; a similar argument holds in the cases where it is right or obtuse.) $\diamondsuit$

Proposition 6.1 can be thought of as our first result about what we will call spherical triangles. We have three points A, B, and C with three spherical arcs $\overset{\frown}{AB}$, $\overset{\frown}{AC}$, and $\overset{\frown}{BC}$. The union of three such arcs is a spherical triangle. The proposition assumes the measures of $\overset{\frown}{AB}$ and $\overset{\frown}{AC}$ are known, as is the measure of the angle between them, and relates them to the measure of the third arc. In the exercises, the interested reader can see how to use techniques of three-dimensional geometry to prove a number of other properties of spherical triangles in space.

Exercises §6.

1. Let O be the center of a sphere and let A, B, and C be distinct points of the sphere, no pair of which is antipodal. Thus we have three arcs $\overset{\frown}{AB}$, $\overset{\frown}{AC}$, and $\overset{\frown}{BC}$ whose measures we denote by c, b, and a, respectively. We let A, B, and C denote the measures of the spherical angles $\prec BAC$, $\prec ABC$, and $\prec ACB$, respectively. Suppose that the spherical angle $\prec ACB$ is right. (That is, the plane containing O, A, and C is perpendicular to the plane containing O, B, and C.) Then we will think of A, B, and C as forming a "spherical triangle" $\triangle^s ABC$ with a right angle at C. Suppose that D is between O and A, E is between O and B, and F is between O and C. (See Figure 2.9.) Also assume that $\overline{DE}$ and $\overline{DF}$ are perpendicular to $\overline{OD}$.

   (a) Explain why $\angle DFE$ and $\angle OFE$ are right angles.

   (b) Explain why the measure of $\angle FDE$ is the same as the measure of the spherical angle $\prec BAC$.

   (c) Using the setting with the structure $ODEF$ introduced at the beginning of this section, prove that $\cos(c) = \cos(a)\cos(b)$, $\sin(A) = \sin(a)/\sin(c)$, $\cos(A) = \tan(b)/\tan(c)$, and $\tan(A) = \tan(a)/\sin(b)$.

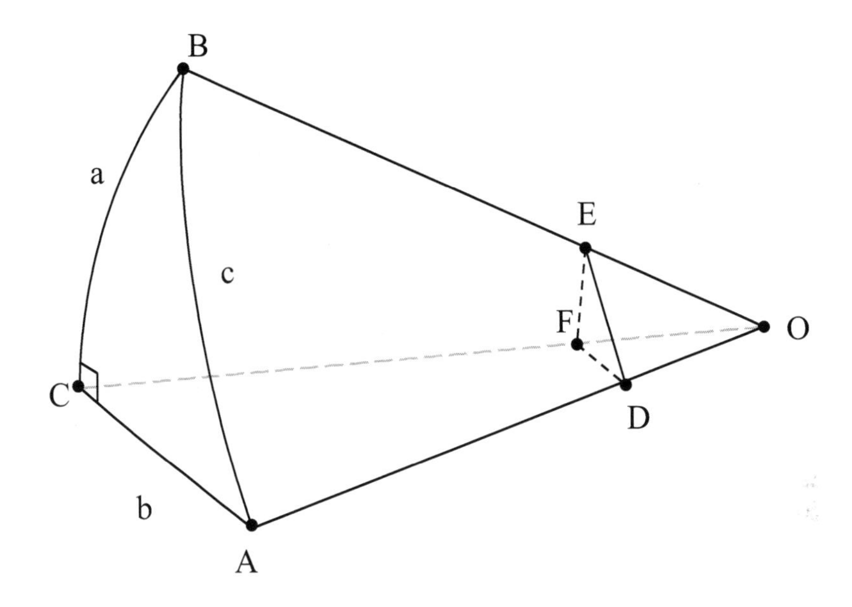

Figure 2.9: Exercise 1.

(d) Prove that $\sin(A) = \cos(B)/\cos(b)$ and $\cos(c) = \cot(A)\cot(B)$.

7 Area

We shall determine areas of a curved surface by approximating it with pieces of surfaces for which the area is known. The idea is that if a sequence of surfaces approaches the surface with unknown area, then the area of the unknown surface is the limiting value of the surfaces in the sequence. An entirely rigorous discussion of this usually involves integral calculus (and real analysis). However, for readers who are interested in an understanding of area that does not require these nonelementary methods, we here provide a discussion which will hopefully be persuasive. We begin by providing an approach to determining the area of the whole sphere.

In space, suppose we are given a planar set of points b and a point P not in the plane of the circle. (See Figure 2.10.) Then we recall that the *cone* with *base* b and *vertex* P is the union of all line segments with one endpoint at P and one endpoint in b.

Suppose that the base is a circle. Then we say that the cone is *circular*. We say that the cone is a *right circular cone* if the segment between P and the center of the circle is perpendicular to the plane of the circle. For a right circular cone, the length of the segment between P and a point in the circle is always the same length called the *slant height*. The radius of the circle is called the *radius* of the cone.

Suppose that the base of the cone is a polygon. Then we say that the cone is *polygonal.* A polygonal cone is *regular* if the base is a regular polygon. We say that a regular polygonal cone is *right* if the segment between the vertex and the center of the polygon is perpendicular to the plane of the polygon.

A *frustum* of a cone is that portion of the cone obtained as follows: Let p be a plane parallel to the base between P and the base. Then the frustum obtained is the set of points of the cone on p, or on the side of p opposite from the vertex. The *bases* of the frustum are (1) the base of the original cone, and (2) the intersection of p with the cone.

A frustum of a right regular polygonal cone is the union of n isosceles trapezoids and their interiors, where n is the number of sides in the polygon. The *slant height* ℓ of the frustum is the height of the trapezoid.

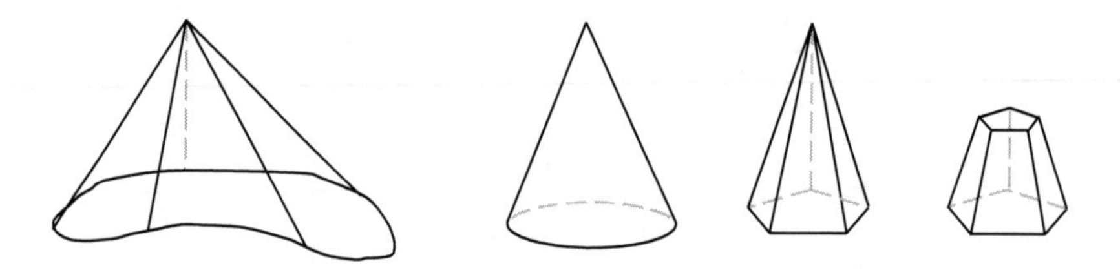

Figure 2.10: Cone, right circular cone, polygonal cone, and polygonal frustum.

Proposition 7.1 *Suppose that a frustum of a right polygonal cone has bases of n sides each. Suppose that the perimeter of the bases are p_1 and p_2, and the frustum has slant height ℓ. Then the lateral area (i.e., the area without the bases) of the frustum is $\frac{\ell}{2}(p_1 + p_2)$.*

Proof. The frustum is the union of n trapezoids. Each has height ℓ and bases of length p_1/n and p_2/n, since the bases have n sides each of equal length. By §1, Exercise 4, the area of each trapezoid is $\frac{\ell}{2}(\frac{p_1}{n} + \frac{p_2}{n})$. Since there are n of them the total area is $n\frac{\ell}{2}(\frac{p_1}{n} + \frac{p_2}{n})$, or $\frac{\ell}{2}(p_1 + p_2)$. $\diamondsuit$

Proposition 7.2 *Suppose that a frustum of a right circular cone has two bases with perimeter s_1 and s_2 and slant height ℓ. Then the (lateral) surface area of the cone is $\frac{\ell}{2}(s_1 + s_2)$.*

Proof. We approximate the circular base of the one with a regular polygon of n sides formed by spacing n points equally around the perimeter of the base, and obtain the formula for the area of the polygonal frustum from Proposition 7.1. For large values of n, the slant height ℓ_n of the polygonal frustum is close to the slant height ℓ of the circular frustum, and the perimeters p_1^n, p_2^n of the bases of the polygonal frustum are close to the perimeters s_1, s_2 of the bases of the circular frustum, so the formula $\frac{\ell_n}{2}(p_1^n + p_2^n)$ obtained from Proposition 7.1 approaches the value $\frac{\ell}{2}(s_1 + s_2)$, as desired. $\diamondsuit$

We may see this another way as follows. We unroll the frustum on a plane; the lateral surface lays out onto the sector of an annulus. The lengths of the circular arcs are s_1 and s_2. Because the slant height of the frustum is ℓ, the difference of the two radii of the annulus is ℓ. By §1, Exercise 6, the area rolled out is $\frac{\ell}{2}(s_1 + s_2)$.

Note that if the radii of the bases are r_1 and r_2, respectively, then $s_1 = 2\pi r_1$ and $s_2 = 2\pi r_2$, so the formula from Proposition 7.2 gives a surface area of

$$\pi\ell(r_1 + r_2) \tag{2.1}$$

We now proceed to derive the formula for the surface area of a sphere of radius r. We view the sphere as being obtained by revolving a semicircle of radius r about its diameter. If we view the semicircle as being approximated by a union of small chords of the semicircle, then the whole sphere is approximated by revolving the chords around the axis of the diameter of the semicircle.

Suppose that an arc of a semicircle is revolved around a diameter of that semicircle. The result is a surface called a *zone* of the sphere.

Theorem 7.3 *Suppose that an arc of a semicircle of radius r is revolved around the diameter of the semicircle. Suppose that the projection of the arc on the diameter has length d. Then the zone generated has surface area $2\pi rd$.*

Proof. (See Figure 2.11.) Let us suppose that $n-1$ distinct points are chosen

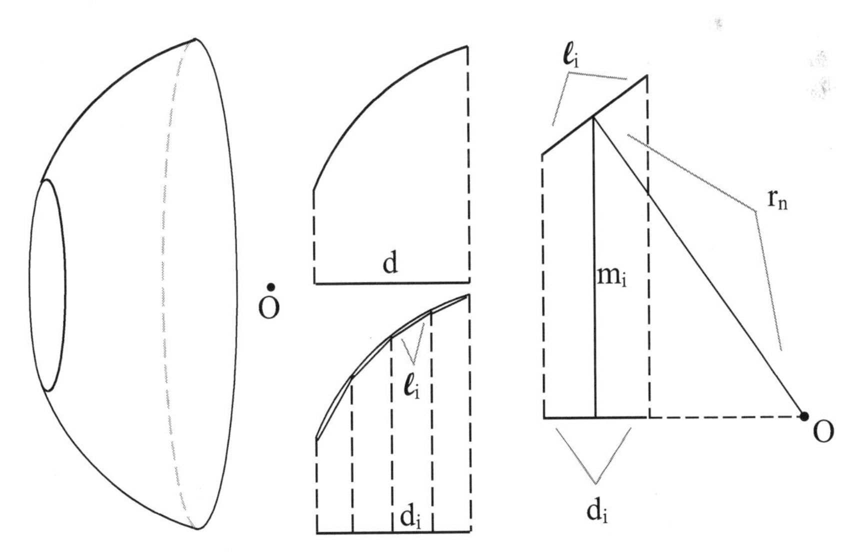

Figure 2.11: A zone, the arc being revolved, its approximation by segments, and the trapezoid of Theorem 7.3.

equally spaced on the arc being revolved. We connect the points in order — along with the endpoints of the arc — to form n chords of the arc of equal length. If we view the arc as being approximated by the n chords, then the zone is approximated by revolving the n chords around the axis of the diameter of the semicircle. Each chord along with its projection to the diameter form a trapezoid. For $i = 1, 2, \ldots, n$, let ℓ_i denote the length of the ith chord, let m_i be the length of the midline of the ith trapezoid, and let d_i be the length of the projection of the ith chord on the diameter. Since every chord has the same length, the midpoint of each is at the same distance r_n from the center of the semicircle. Then the ith chord generates a surface area of $2\pi\ell_i m_i$, according to (2.1) and the formula for the length of the midline of a trapezoid (the length of the midline is the average of the lengths of the bases). By Lemma 1.4, $2\pi\ell_i m_i = 2\pi d_i r_n$. Then the union of all the chords generates a surface of area

$$\sum_{i=1}^{n} 2\pi\ell_i m_i = \sum_{i=1}^{n} 2\pi d_i r_n = 2\pi d r_n, \tag{2.2}$$

where $\sum_{i=1}^{n} d_i = d$ because the union of the projections of the chords is the same as the projection of the arc, which has length d. Now as the number of arcs n is allowed to become arbitrarily large, r_n approaches r, so the surface area generated by the union of the arcs approaches $2\pi rd$; this is the surface area of the zone. ◊

Corollary 7.4 *The surface area of a sphere of radius r is $4\pi r^2$.*

Proof. Using Theorem 7.3, we let the arc be the whole semicircle. Since the projection of the whole semicircle has length $2r$, the surface area of the whole sphere is $2\pi rd = 2\pi 2r^2 = 4\pi r^2$. ◊

We conclude by discussing the area enclosed by a lune. Since a lune itself is merely the union of two semicircles, it technically has area zero. But it is common to make a small abuse of language and write "area of a lune" when we mean "area enclosed by a lune." We will also do so as long as there is no reason for confusion.

Proposition 7.5 *In a sphere of radius r, the area of a lune whose angle has radian measure θ is $2\theta r^2$.*

The simplest justification for this involves use of the rotational symmetry of the sphere. If we rotate any lune on its axis, we obtain a lune of the same size and shape and hence the same area. Then the proportion of the area of the sphere that lies in the lune is the same as the proportion of θ to 2π. Thus the area of the lune is $\frac{\theta}{2\pi}4\pi r^2 = 2\theta r^2$, as desired.

Exercises §7.

1. Suppose that an arc of a semicircle is revolved about the diameter of the semicircle through an angle of θ (measured in radians). If the semicircle

has radius r and the projection of the arc on the diameter has length d, prove that the surface area of the region generated (called a *zone of a lune*) is $rd\theta$.

2. Prove that on a sphere of radius r, the area of the interior of a small circle of spherical radius θ is $4\pi r^2 \sin^2(\frac{\theta}{2})$.

3. Let r be the radius of the earth. What proportion of the earth's surface is visible at an altitude of a above the surface of the earth?

4. Two spheres of radius 3 and 4 have centers 5 units apart. Determine the area of each sphere which is contained inside the other.

5. What proportion of the earth's surface lies between the latitudes of $45°N$ and $45°S$?

8 Spherical coordinates

We now introduce two closely related kinds of coordinates on a sphere: polar coordinates and latitude-longitude coordinates. Suppose that s is a sphere of radius r. We shall specify coordinates (ϕ, θ) for points on s. The coordinates shall be determined as follows, referring to Figure 2.12. Choose a point P of s, which we will call the (north) pole of our coordinate system, and a great semicircle PQP^a (called the *zero meridian* or *prime meridian,*) where P^a is the antipode of P and Q is any point of s other than P and P^a.

Here is how we specify polar coordinates (ϕ, θ) for a point X on the sphere. We let ϕ be the measure of the great circle arc from X to P. (If $X = P$ or $X = P^a$ then $\phi = 0$ or $\phi = \pi$, respectively.) The angle θ is the measure through which we rotate greate semicircle PQP^a to obtain great semicircle PXP^a. (We must specify whether a positive θ is a rotation in the clockwise or counterclockwise direction when viewed from above P. If $\theta > 0$ is a clockwise rotation by θ, then a negative value of θ gives a counterclockwise rotation by angle $-\theta$.) A full rotation by measure 2π brings PQP^a back to itself. Thus θ is determined only up to integer multiples of 2π, so as in the plane θ is often restricted to intervals of length 2π, e.g., $-\pi < \theta \leq \pi$ or $0 \leq \theta < 2\pi$. We could also use degree measure for ϕ and θ.

If $X = P$ or $X = P^a$ then θ could have any value.

A great semicircle where θ is constant is called a *meridian*. The great circle where ϕ equals $\frac{\pi}{2}$ is called the *equator*. From this we also obtain *latitude-longitude coordinates*, closely related to the polar coordinates above. Given polar coordinates (ϕ, θ), the associated latitude-longitude coordinates for s are $(\ell_1, \ell_2) \equiv (\frac{\pi}{2} - \phi, \theta)$. Here the *latitude* variable ℓ_1 is zero on the equator E. Then ℓ_1 is positive in the hemisphere of s containing P and negative in the hemisphere of s containing P^a, so $-\frac{\pi}{2} \leq \ell_1 \leq \frac{\pi}{2}$. Furthermore, absolute value of ℓ_1 turns out to be the measure of an arc from X perpendicular to the equator. The *longitude* variable ℓ_2 is zero on the *zero meridian*. As with

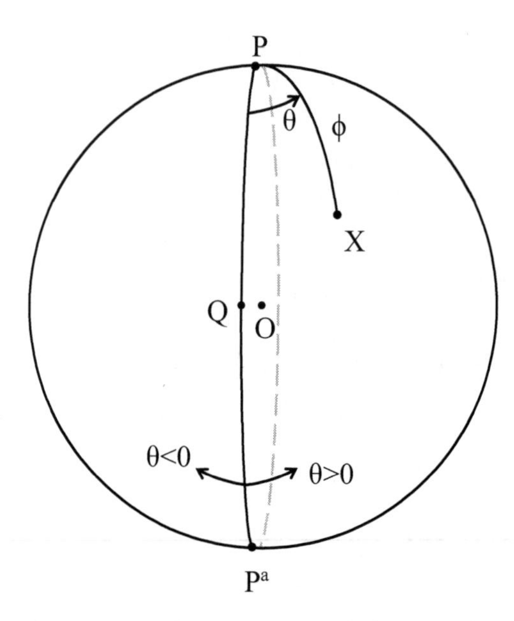

Figure 2.12: Spherical polar coordinates.

polar coordinates, we continue to say that a *meridian* is any great semicircle where longitude is constant.

The natural example of latitude-longitude coordinates are the latitude-longitude coordinates for the surface of the earth. Here P is the geographic north pole of the earth. The zero meridian (usually called the *prime meridian*) is the great semicircle between the geographic north and south poles which passes through Greenwich, England. Longitude is measured positively (usually in degrees) in the direction east of the zero meridian.

Since any point in space must lie on a sphere of some (possibly zero) radius r about the origin, the ordered triple (r, ϕ, θ) serves as what is known as a triple of spherical coordinates for a point in space.

Note that different ordered triples may determine the same point X. If $r = 0$, the point X must be O regardless of the values of ϕ or θ since only O is at distance zero from O. Also, given $r > 0$, any point where $\phi = 0$ must be P, regardless of the value of θ. Also, if $\phi = \pi$, X must be P^a regardless of the value of θ. Lastly, as noted above, given a triple (r, ϕ, θ), if we add an integer multiple of 2π to θ, we arrive at the same point. Our choice above that $-\pi < \theta \leq \pi$ is a matter of convenience. So a point in space does not have a unique set of spherical coordinates. However, we have the following theorem.

Theorem 8.1 *Suppose that (ϕ, θ) are polar coordinates for a sphere of radius r. Suppose that x, y, and z axes are chosen for space in the following manner. Let the center of the sphere be the origin. Let the positive z axis point in the* *direction of the pole of the coordinate system. Let the positive x axis point in the direction of the point $(\frac{\pi}{2}, 0)$ and let the positive y axis point in the direction of the point $(\frac{\pi}{2}, \frac{\pi}{2})$. Then a point A in space on the sphere of radius r with xyz coordinates (x, y, z) and polar coordinates (ϕ, θ) satisfies*

$$(x, y, z) = (r \sin\phi \cos\theta, r \sin\phi \sin\theta, r \cos\phi). \tag{2.3}$$

Proof.

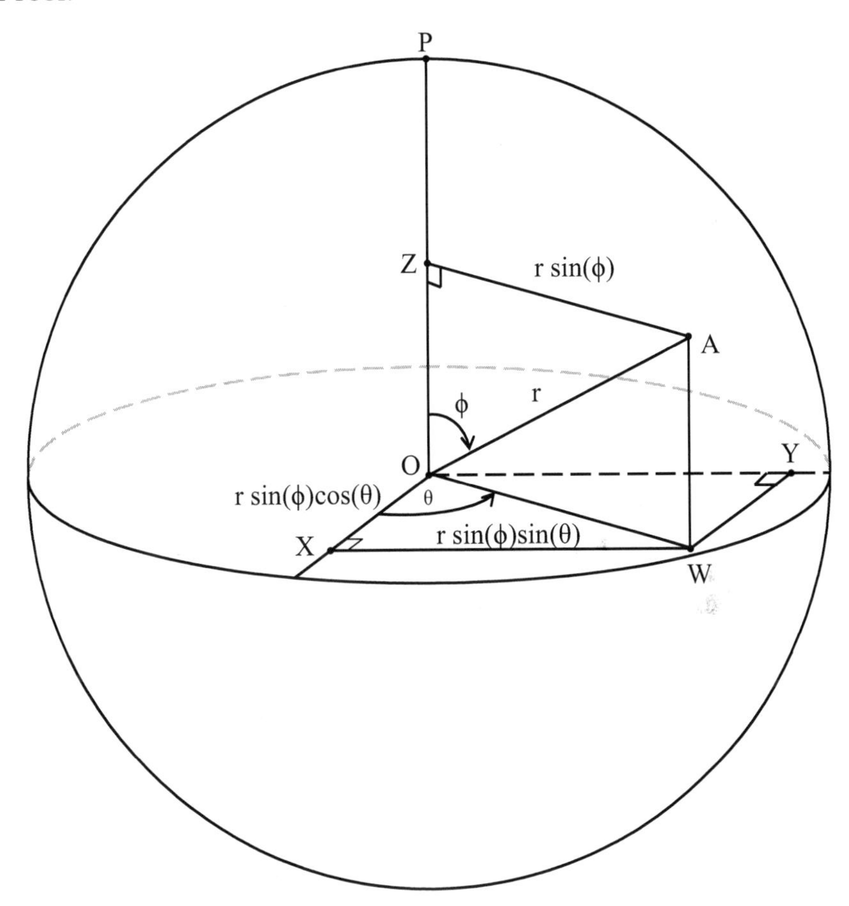

Figure 2.13: Theorem 8.1.

The discussion above shows how every (ϕ, θ) determines a point A on the sphere. Let X, Y, and Z be the projections of the point A to the corresponding axes. Then $\triangle AZO$ is a right triangle with a right angle at Z. By definition of arc measure, $m\angle AOZ = m\ \widehat{AP}$, which is ϕ by definition. Since $\triangle AZO$ is a right triangle and the length of the segment $\overline{OA}$ is r, the coordinate of Z is $r\cos\phi$ and the length of $\overline{AZ}$ is $r\sin\phi$. Thus $z = r\cos\phi$, as desired. Next, let W be the projection of A to the XY plane. Then the length of $\overline{OW}$ is the same as the length of $\overline{AZ}$, which is $r \sin \phi$. The projections of W to the x and y axes are X and Y, respectively. Next, the absolute value of θ is the measure of planar angle $\angle XOW$. Then since W lies at radius $r \sin \phi$ from O and $\overline{OW}$ forms a signed angle of θ with the positive x axis, its xy coordinates must be $x = r \sin \phi \cos \theta$ and $y = r \sin \phi \sin \theta$. ◇

Exercises §8

1. Extend §6, Exercise 1c as follows: if a point X has latitude-longitude coordinates (a, b) with respect to some system of cooordinates on the sphere, then the distance c from X to the point with coordinates $(0, 0)$ satisfies $\cos(c) = \cos(a) \cos(b)$.

Chapter 3

Axiomatic spherical geometry

9 Basic axioms

In this section we set up a system of axioms for spherical geometry. We assume the reader possesses some familiarity with the axiomatic method used for plane geometry. Recall that in the plane we have several undefined terms (e.g., "point," "line," and "plane"). Then we have three kinds of statements: definitions, axioms, and theorems. A definition is a statement that says exactly what a term means. An axiom (also known as a "postulate") is a statement which is assumed about the terms being discussed. That is, an axiom is a statement whose truth is taken for granted without proof. An axiom should set out some fundamental property of a system. In contrast, a theorem is a statement whose truth is established via a proof by making use of the axioms and other theorems which have already been proven. (The terms "lemma" and "corollary" refer to types of theorems; the term "proposition" can refer to either a theorem or an axiom.)

An axiom tends to be a statement that is self-evident, but this is not necessary. Sometimes one must assume statements whose truth is not obvious. Even if some axioms are not evident, exhibition of the axioms makes clear what the assumptions are in a discussion. On the other hand, a theorem tends to be a statement whose truth is not obvious, but this is also not necessary.

In axiomatic spherical geometry, we will also have undefined terms: "point," "great circle," and "sphere." Our understanding of what these terms mean will come solely from the axioms. Theorems shall be proven solely based on the assumptions from the axioms, and from theorems previously proven. In particular, a sphere will not be assumed to consist of points equidistant from a center in space. In fact, it will not be assumed to have any such thing as a center. A great circle will not be assumed to be the intersection of a plane with a sphere.

The reader may wonder why we would do this. Are not the results of the previous chapter an adequate enough beginning to our study of spherical geometry? There are several answers. First, all mathematics makes use of assumptions which are not proven. It is simply good manners to be explicit about what those assumptions are. In the previous sections we did not do this thoroughly. We assumed all sorts of facts about the nature of space without explicitly stating them. Second, the axiomatic method will base assumptions solely on intrinsic properties of the sphere, where the previous section made use of properties of space where the sphere lives.

Once we have axioms for spherical geometry, we should check that the sphere of radius r in space satisfies these axioms, in order to conclude that theorems that arise from the axioms apply to the sphere of radius r in space. For most axioms this verification will be simple (in fact, in most cases it will turn out that we have already checked them in §5) but in other cases it will require some work.

Our system of axioms attempts to minimize the number of axioms. Consequently most of the results in this section and the next are obvious enough that the reader may not find them very interesting. The reader may profit by skimming the next two sections for now and only returning later.

In plane geometry a line is found to be in one-one correspondence with the set of real numbers $\mathbf{R}$. This correspondence allows the creation of a notion of distance between points. The great circle is treated similarly on the sphere. For a circle of radius 1 we see that if we label some starting point with 0 we may proceed around that circle one way or the other: one direction (say counterclockwise) around may be considered a positive direction and the other may be considered negative. We may label each point with a real number which is the signed distance around the circle from the origin point — positive if traced around the circle counterclockwise and negative if traced clockwise. What makes the situation different from the line is that after tracing around the circle a multiple of $\pm 2\pi$ we return to the starting point. (See Figure 3.1.) Thus 0 is the same point as $\pm 2\pi$. Tracing around again, 0 is the same point as $\pm 4\pi$. In fact, with this labeling, a point with label x is the same as the point with label $x \pm 2\pi, x \pm 4\pi, x \pm 6\pi, \ldots$. We call the set $\{\ldots x - 6\pi, x - 4\pi, x - 2\pi, x, x + 2\pi, x + 4\pi, x + 6\pi, \ldots\}$ an *equivalence class* of real numbers modulo 2π. Any element of this set is said to be a *representative* of the equivalence class. We create the set called $\mathbf{R}/2\pi$ (read: "R modulo 2π" or "R mod 2π") which is the set of all equivalence classes of real numbers modulo 2π. If x and y are two real numbers, we will write $x = y(\bmod 2\pi)$ if $x - y$ is an integer multiple of 2π. Then x and y are both a representative of the same equivalence class of real numbers modulo 2π. Furthermore, it will make sense to add and subtract elements "modulo 2π" in $\mathbf{R}/2\pi$. Let X be a class in $\mathbf{R}/2\pi$ with a representative x and Y be a class with representative y. We write $X = Y(\bmod 2\pi)$ if $x = y(\bmod 2\pi)$. We define $X + Y$ to be the class with representative $x + y$ and $X - Y$ to be the class with representative $x - y$. One must check that the operations $'+'$ and $'-'$ do not depend on the choice of x and y (and we say that $'+'$ and $'-'$ are "well defined"). For this one must check that if $x_1 = x_2 (\mathrm{mod}\, 2\pi)$ and $y_1 = y_2 (\mathrm{mod}\, 2\pi)$, then $x_1 + y_1 = x_2 + y_2 (\mathrm{mod}\, 2\pi)$ and $x_1 - y_1 = x_2 - y_2 (\mathrm{mod}\, 2\pi)$.

Given any equivalence class X in $\mathbf{R}/2\pi$ there exists a unique x, $-\pi < x \leq \pi$ such that x is a representative of X. Given a real y which is a representative of X, the process of obtaining such an x is called "reduction" of a number modulo 2π, and we speak of "reducing" $y (\mathrm{mod}\, 2\pi)$ to obtain x.

Axiom 9.1 (A-1) [1] *A sphere and a great circle are sets of points. There are at least two points on the sphere.*

Axiom 9.2 (A-2) *There is a one-one correspondence between the points of a great circle and the points of* $\mathbf{R}/2\pi$.

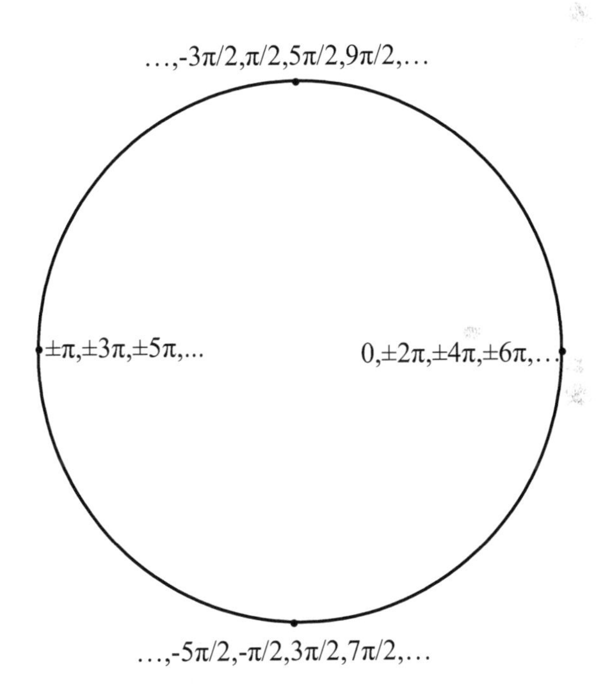

Figure 3.1: The labeling of a circle with $\mathbf{R}/2\pi$.

This axiom sets up a coordinate on the great circle. There will in general be more than one such one-one correspondence, but once we have one, we will not use another one-one correspondence for that great circle unless it is in some sense "consistent" with the first. (See Proposition 9.14.)

We next define what it means for two points on the sphere to be antipodal.

[1] The axioms for spherical geometry will be labeled A-1, A-2, A-3, etc.

Definition 9.3 *Two points on a sphere are said to be* antipodal *if there exists a great circle passing through them whose one-one correspondence takes these two points to elements of* $\mathbf{R}/2\pi$ *which differ by* $\pi \pmod{2\pi}$*. Each point is said to be an antipode of the other.*

In the standard three-dimensional model sphere, this will simply mean what it meant in Definition 5.4: that the line segment between the two points on the sphere passes through the center of the sphere. But Definition 9.3 succeeds in defining the notion of antipodal using only intrinsic properties of the sphere. This will not depend on what great circle contains the two points, via another axiomatic property for spherical geometry:

Axiom 9.4 (A-3) *A point on the sphere has no more than one antipode.*

We may then speak of <u>the</u> antipode of a point. If B is the antipode of A, then we write $B = A^a$ and $A = B^a$. We leave it to Exercise 1 to check the simple fact that the antipode of the antipode of a point is the given point.

Proposition 9.5 *If a great circle contains a point, it also contains the antipode of that point.*

(See Exercise 2.)

Proposition 9.6 *Suppose two points are antipodal. Given <u>any</u> great circle which contains the two points, its one-one correspondence takes the two points to elements in* $\mathbf{R}/2\pi$ *which differ by* π.

That is, if two points are a semicircle apart on some great circle, they are a semicircle apart on any great circle containing them.

Proof. Suppose the two points A and B lie on a great circle on which the one-one correspondences for them differs by a value other than $\pi \pmod{2\pi}$. Then there is a third point C different from A and B whose label under the one-one correspondence differs by π from the label for A. So C is an antipode for A by definition. But then A has two antipodes, a contradiction of Axiom 9.4. ◊

Axiom 9.7 (A-4) *If two distinct points on a sphere are not antipodal then there exists a unique great circle passing through them.*

Definition 9.8 *If two distinct points A and B are not antipodal, the unique great circle passing through A and B is denoted by* $\bigcirc AB$.

If two distinct points are antipodal, then by definition of antipodal there is at least one great circle containing them. But there could be more than one such great circle.

Proposition 9.9 *Through any point there exists at least one great circle.*

For the proof, see Exercise 3.

Proposition 9.10 *Every point has an antipode.*

For the proof, see Exercise 4. Our next axiom concerns what happens when two great circles meet.

Axiom 9.11 (A-5) *Two distinct great circles meet in at least one point.*

We can easily prove more than that:

Proposition 9.12 *Two distinct great circles meet in exactly two points which are antipodal.*

Proof. Since the two circles meet in a point, each must also pass through the antipode of that point by Proposition 9.5. If there were a pair of non-antipodal points in the intersection, there could only be one great circle passing through them, by Axiom 9.7. ◇

Definition 9.13 *Let A and B be two points, Γ a great circle containing them with one-one correspondence ℓ to $\mathbf{R}/2\pi$. Let δ be the representative of the equivalence class of $\ell(A) - \ell(B)$ which lies in $(-\pi, \pi]$. Then the distance $d(A, B)$ between A and B is the absolute value of δ.*

Note that $d(A, B) = d(B, A)$ because $(\ell(A) - \ell(B))$ is the negative of $(\ell(B) - \ell(A))$ (details are left to Exercise 5). The reader may note one problem with Definition 9.13: what happens if the two points have more than one great circle containing them? Could we get two values for the distance between them? The answer is no, because by Axiom 9.7, this could only occur if the two points are antipodal, and by Axiom 9.6, $\ell(A) - \ell(B)$ must be equal to π modulo 2π.

Note that since $0 \leq d(A, B) \leq \pi$ then $d(A, B) \pm 2\pi$ represents the same element in $\mathbf{R}/2\pi$. Note also that $\pm d(A, B) = \ell(A) - \ell(B) (\bmod 2\pi)$ so

$$\ell(A) = \ell(B) \pm d(A, B) (\bmod 2\pi). \tag{3.1}$$

We have assumed by axiom that every great circle has a coordinate ℓ which induces the distance function of Definition 9.13. But this coordinate is not always convenient: sometimes it is important to be able to choose which point is zero, and decide which way around the circle is positive:

Proposition 9.14 *Given one-one correspondence ℓ for a great circle, the functions $-\ell$ and $\ell + k$ (where k is a constant) are also one-one correspondences on the same great circle which induce the same distance $d(\cdot, \cdot)$.*

The proof is in Exercise 7. We speak of these one-one correspondences as being "equivalent" to ℓ because of the fact that they induce the same distance on the circle.

Proposition 9.15 *Given a point A on great circle Γ, there are exactly two points on Γ at distance d for $0 < d < \pi$ and exactly one such point if $d = 0$ (point A) or π (the antipode of A). The two points at distance d are antipodal if $d = \frac{\pi}{2}$.*

Proof. Let ℓ be the one-one correspondence for Γ. The points at distance d from A are those which under ℓ correspond to $\ell(A) \pm d$ modulo 2π. If $d = 0$ or $d = \pi$ there is exactly one such point (A and an antipode, respectively). If $0 < d < \pi$ $(\ell(A) \pm d)(\bmod 2\pi)$ are distinct elements of $\mathbf{R}/2\pi$, so correspond to distinct points on Γ since ℓ is a one-one correspondence. If $d = \frac{\pi}{2}$, $\ell(A) + d - (\ell(A) - d) = \pi$, so the points are antipodal. ◇

Definition 9.16 *Given three distinct points A, B, and C, we say that B is between A and C if A and C are not antipodal, B is on $\bigcirc AC$, and $d(A, C) = d(A, B) + d(B, C)$.*

Proposition 9.17 *Suppose A, B, and C are points. Then B is between A and C if and only if A, B, and C lie on a great circle with one-one correspondence ℓ such that $0 < d(A, B) < d(A, C) < \pi$ and*
(1) $\ell(C) = \ell(A) + d(A, C)$ and $\ell(B) = \ell(A) + d(A, B)(\bmod 2\pi)$
or
(2) $\ell(C) = \ell(A) - d(A, C)$ and $\ell(B) = \ell(A) - d(A, B)(\bmod 2\pi)$.

That is, B is between A and C if the labels $\ell(A), \ell(B), \ell(C)$ are "increasing" or "decreasing" modulo 2π without spanning more than π. Note that (1) (respectively, (2)) holds for ℓ if and only if (1) (respectively, (2)) holds for a one-one correspondence $\ell + k$. Also, (1) holds for ℓ if and only if (2) holds for $-\ell$. So in Proposition 9.17 the one-one correspondence ℓ can be replaced by any of those mentioned in Proposition 9.14.

Proof. Suppose B is between A and C. Then A and C are distinct and not antipodal, so determine the great circle $\bigcirc AC$ with one-one correspondence ℓ and B is on $\bigcirc AC$. As noted in (3.1), we must have $\ell(C) = \ell(A) \pm d(A, C)$ and $\ell(B) = \ell(A) \pm d(A, B)(\bmod 2\pi)$, in one of the four possible combinations of the $\pm$. Since A and C are distinct and not antipodal, $0 < d(A, C) < \pi$. Since $d(A, C) = d(A, B) + d(B, C)$, $d(A, B) < d(A, C) < \pi$. Since A and B are distinct, $0 < d(A, B)$.

Without loss of generality we may assume $\ell(C) = \ell(A) + d(A, C)(\bmod 2\pi)$; the case with "$-$" is similar (or we may replace ℓ by $-\ell$ by Proposition 9.14). If $\ell(B) = \ell(A) - d(A, B)(\bmod 2\pi)$, then $\ell(C) + \ell(B) = 2\ell(A) + d(A, C) - d(A, B)(\bmod 2\pi)$, and since B is between A and C, $d(A, C) - d(A, B) = d(B, C)$. Now $d(B, C) = \pm(\ell(B) - \ell(C))$ reduced modulo 2π. If $d(B, C) = \ell(B) - \ell(C)(\bmod 2\pi)$, then $\ell(C) + \ell(B) = 2\ell(A) + \ell(B) - \ell(C)(\bmod 2\pi)$, so $2\ell(C) = 2\ell(A)(\bmod 2\pi)$, so $2(\ell(C) - \ell(A)) = 0(\bmod 2\pi)$, so $\ell(C) - \ell(A) = 0$ or π modulo 2π. Thus A and C are identical or antipodal, a contradiction.

If $d(B, C) = \ell(C) - \ell(B)(\bmod 2\pi)$, then a similar argument shows that $2(\ell(B) - \ell(A)) = 0(\bmod 2\pi)$, so A and B are identical or antipodal, another contradiction since $0 < d(A, B) < \pi$.

Conversely, suppose A, B, and C lie on a great circle with one-one correspondence ℓ, $0 < d(A, B) < d(A, C) < \pi$ and (1) holds. Then A, B, and C are distinct and A and C are not antipodal. Subtracting these two equations, $\ell(C) - \ell(B) = d(A, C) - d(A, B)(\bmod 2\pi)$. Now $0 < d(A, C) - d(A, B)$

and $d(A,C) - d(A,B) < d(A,C) < \pi$. So $\ell(C) - \ell(B)$ reduced modulo 2π is $d(A,C) - d(A,B) > 0$, so $d(B,C) = d(A,C) - d(A,B)$, or $d(A,C) = d(A,B) + d(B,C)$, as desired. $\diamondsuit$

Proposition 9.17 can be used to prove properties of betweenness such as:

Proposition 9.18 *If C is between A and D and B is between A and C then B is between A and D and C is between B and D.*

Proof. If C is between A and D then we apply Proposition 9.17 to conclude either (1) or (2) there. By Proposition 9.14 we may assume (1) so that $\ell(D) = \ell(A) + d(A,D)$ and $\ell(C) = \ell(A) + d(A,C)$ modulo 2π with $0 < d(A,C) < d(A,D) < \pi$. We again apply Proposition 9.17 to the assumption that B is between A and C; then we are forced to conclude that $\ell(B) = \ell(A) + d(A,B) (\bmod 2\pi)$ with $0 < d(A,B) < d(A,C) < \pi$. But then $\ell(D) = \ell(A) + d(A,D)$ and $\ell(B) = \ell(A) + d(A,B) (\bmod 2\pi)$ for $0 < d(A,B) < d(A,C) < d(A,D) < \pi$, so by Proposition 9.17, B is between A and D. Thus $d(A,D) = d(A,B) + d(B,D)$ and we already had by assumption that B is between A and C so $d(A,C) = d(A,B) + d(B,C)$. Then $\ell(D) = \ell(A) + d(A,D) = \ell(B) - d(A,B) + d(A,D) = \ell(B) + d(B,D) (\bmod 2\pi)$ and $\ell(C) = \ell(A) + d(A,C) = \ell(B) - d(A,B) + d(A,C) = \ell(B) + d(B,C) (\bmod 2\pi)$. Furthermore $\pi > d(B,D) = d(A,D) - d(A,B) > d(A,C) - d(A,B) = d(B,C) > 0$. Then by Proposition 9.17, C is between B and D. $\diamondsuit$.

Definition 9.19 *Given two non-antipodal points A and B, the* (spherical) arc *$\overset{\frown}{AB}$ is the set of points consisting of A, B, and all points on $\bigcirc AB$ between A and B. The* measure *of $\overset{\frown}{AB}$ is $d(A,B)$.*

Definition 9.20 *Two spherical arcs are said to be* congruent *if they have the same measure. If the two arcs are $\overset{\frown}{AB}$ and $\overset{\frown}{CD}$ then we write $\overset{\frown}{AB} \cong \overset{\frown}{CD}$.*

Our understanding of the notion of poles in the previous section of a great circle arose from the fact that a line perpendicular to the plane of a great circle meets the sphere in two points. We shall want to have an understanding of properties of the poles which depend solely on intrinsic properties of the sphere and not its placement relative to space:

Axiom 9.21 (A-6) *For any great circle there exists a point such that the great circle consists of points at spherical distance a quarter circle from the given point.*

Such a point is defined to be a *pole* of the great circle.

Proposition 9.22 *The antipode of a pole in Axiom 9.21 is the only other point which satisfies the same property.*

For a proof, see Exercise 9.

Axiom 9.21 guarantees that a great circle consists of points at a quarter circle from the pole. But could there be any points at a quarter circle from the pole which are not on the great circle? The answer is no:

Theorem 9.23 *A great circle is the set of all points on the sphere at spherical distance of a quarter circle from its pole.*

Proof. By the definition of pole, all points on the great circle (call it Γ) are at distance $\frac{\pi}{2}$ from a pole. Now conversely, suppose that a point Q is a quarter circle from the pole P. Then there exists a great circle passing through P and Q. Then $\bigcirc PQ$ is distinct from Γ since P is not on Γ. So $\bigcirc PQ$ meets Γ at two points A and A^a. Then on Γ, Q is at distance $\frac{\pi}{2}$ from P by definition of Q and A, A^a are both at distance $\frac{\pi}{2}$ by definition of P. By Proposition 9.15, Q must be the same as either A or A^a. This places Q on Γ. $\diamondsuit$

Theorem 9.24 *Every point is a pole of some great circle.*

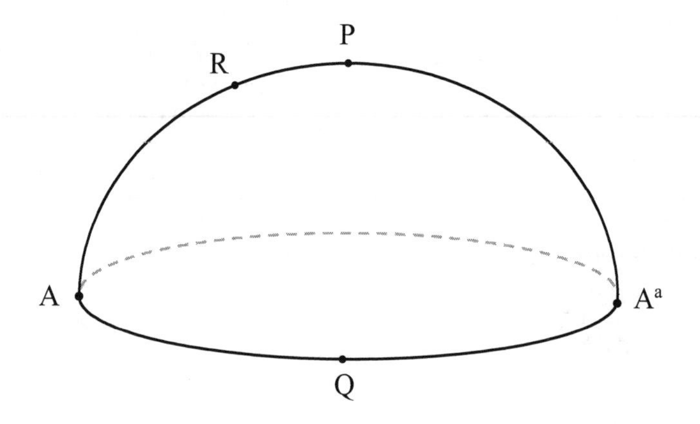

Figure 3.2: Diagram for proof of Theorem 9.24.

Proof. (See Figure 3.2.) Let the given point be P. There exists a great circle passing through P by Proposition 9.9. Let Q be one of its poles. There exist two (antipodal) points A and A^a on this great circle at distance $\frac{\pi}{2}$ from P (using the one-one correspondence between the great circle and $\mathbf{R}/2\pi$). Since Q is at distance $\frac{\pi}{2}$ from A, it is not antipodal, so there exists a unique great circle passing through Q and A. Note that all of A, A^a and Q are at distance $\frac{\pi}{2}$ from P. We claim that $\bigcirc AQ$ is the set of all points at spherical distance $\frac{\pi}{2}$ from P. Let R be a pole of $\bigcirc AQ$. We claim $R = P$ or $R = P^a$. Since R is a pole of $\bigcirc AQ$, it is a quarter circle from Q. Since Q is a pole of $\bigcirc PA$ and by Theorem 9.23, R lies on $\bigcirc PA$. On $\bigcirc PA$, R is also a quarter circle from A since it is a pole of $\bigcirc AQ$, so by Proposition 9.15, R must be P or P^a. $\diamondsuit$

Definition 9.25 *Given a point on the sphere, a great circle of which the point is a pole is called a* polar circle *or* polar *of the point.*

By Theorem 9.23, a point has a unique polar.

Proposition 9.26 *If A, B, C are points on a sphere, $\overset{\frown}{BA}$ and $\overset{\frown}{BC}$ are both quarter circles, and A and C are neither the same nor antipodal then B is a pole of great circle $\bigcirc AC$.*

Proof. Point B is the pole of some great circle (by Theorem 9.24) which contains A and C. But there is only one such circle $\bigcirc AC$ since A and C are neither the same nor antipodal. $\diamondsuit$

Proposition 5.8 gives us a way to define the term "small circle" which depends only on intrinsic properties of the sphere.

Definition 9.27 *A small circle on a sphere is the set of all points on the sphere at a fixed spherical distance $\rho < \frac{\pi}{2}$ from a point P. The point P is called the* (spherical) center *of the small circle, and the quantity ρ is called the* (spherical) radius *of the small circle. The set of all points at distance less (greater, respectively) than ρ from P is called the interior (exterior, respectively) of the small circle.*

It is not hard to see that the points on the small circle are also equidistant from the antipode of P; see Exercise 16.

Definition 9.28 *Let A and B be two points on a sphere which are not antipodal. Then the* spherical ray *$\overset{\hookrightarrow}{AB}$ is the set of all points C on the great circle $\bigcirc AB$ such that C is on $\overset{\frown}{AB}$ or B is between A and C. We say that A is the* vertex *or* endpoint *of the spherical ray. The* open ray *with vertex A through B consists of all points on $\overset{\hookrightarrow}{AB}$ except for A.*

See Figure 3.3. We leave several natural properties of rays to the exercises.

Proposition 9.29 *If A and B are not antipodal and*

$$d(A, B) = \ell(B) - \ell(A) (\text{mod } 2\pi)$$

or

$$d(A, B) = \ell(A) - \ell(B) (\text{mod } 2\pi),$$

respectively, then $\overset{\hookrightarrow}{AB}$ consists of all points C satisfying $\ell(C) = \ell(A) + x$ or $\ell(A) - x$ $(\text{mod } 2\pi)$, respectively, where $0 \leq x < \pi$.

This follows directly from Proposition 9.17. See Exercise 10.

Corollary 9.30 *Given any spherical ray $\overset{\hookrightarrow}{AB}$ and a real number x, $0 \leq x < \pi$, there exists a unique point C of $\overset{\hookrightarrow}{AB}$ such that $d(A, C) = x$.*

(See Exercise 11.)

Corollary 9.31 *If C is a point of $\overset{\hookrightarrow}{AB}$ and $C \neq A$, then $\overset{\hookrightarrow}{AB} = \overset{\hookrightarrow}{AC}$.*

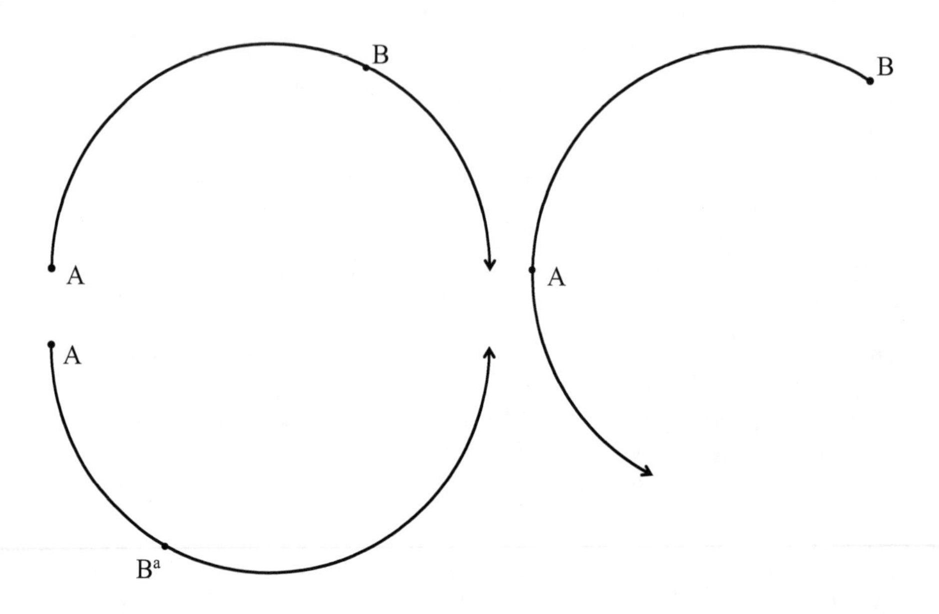

Figure 3.3: Three spherical rays from the same great circle: rays $\overrightarrow{AB}$ and $\overrightarrow{AB^a}$ are opposite; rays $\overrightarrow{AB}$ and $\overrightarrow{BA}$ intersect in $\widehat{AB}$.

(For the proof of Corollary 9.31, see Exercise 12.)

Proposition 9.32 *The intersection of the rays $\overrightarrow{AB}$ and $\overrightarrow{BA}$ is $\widehat{AB}$.*

(See Figure 3.3.) The proof is Exercise 13.

Definition 9.33 *Spherical rays $\overrightarrow{AB}$ and $\overrightarrow{AC}$ are said to be opposite if they are not the same ray but A, B, and C lie on a single great circle.*

(See Figure 3.3.)

Definition 9.34 *Let A be a point, and $\overrightarrow{AB}$ a spherical ray with vertex A. Then the* great semicircle $A\widehat{B}A^a$ *is the union of the ray $\overrightarrow{AB}$ and the point A^a.*

Proposition 9.35 *The great semicircle ABA^a is the union of the arcs $\widehat{AB}$ and $\widehat{A^aB}$.*

See Exercise 15.

Note that a spherical ray contains no pair of antipodal points. The set of points consisting of antipodes of the points on the spherical ray is another spherical ray on the same great circle whose vertex is the antipode of the vertex of the given ray. Every point on the great circle of that ray is either on the ray or its antipode is on the ray, but not both. (See Exercise 24.)

Definition 9.36 *The unique point of a spherical arc whose spherical distance to the end points is the same is called the* midpoint *of the arc.*

That a midpoint exists and is unique is a consequence of Proposition 9.30.

Proposition 9.37 *A spherical ray and a great circle intersect in exactly one point unless the ray is contained in the great circle.*

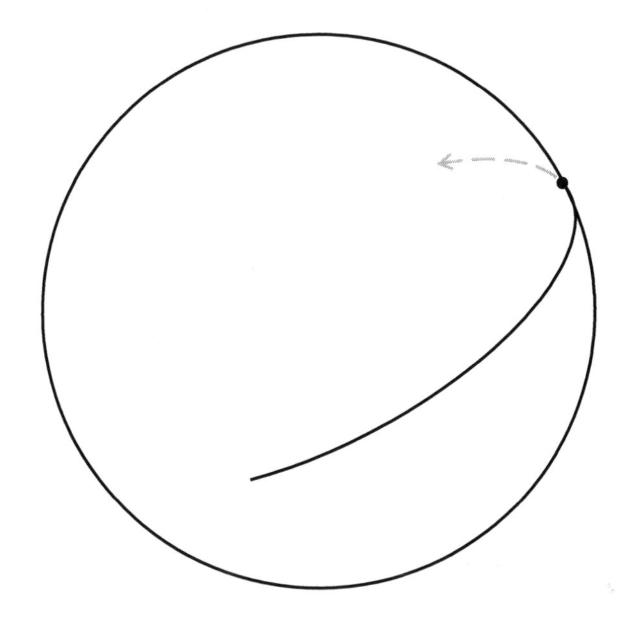

Figure 3.4: Intersection of a spherical ray and a great circle.

Proof. Suppose the great circle (Γ_1) does not contain the ray ($\overset{\hookrightarrow}{r}$). Then a second great circle (Γ_2) which does contain $\overset{\hookrightarrow}{r}$ meets Γ_1 in two points which divide Γ_2 into two great semicircles. Because those two points are antipodal, exactly one of them lies on $\overset{\hookrightarrow}{r}$. $\diamondsuit$

Definition 9.38 *The set of all points lying less than a quarter circle from a given point is called a* hemisphere. *The given point is said to be the* center *of the hemisphere.*

If A is a point, then we leave it to Exercise 19 to check that every point on the sphere is either in the hemisphere centered at A, the hemisphere centered at A^a, or on the polar of A. Also, a hemisphere contains no pair of antipodal points, and the antipode of a point in a hemisphere centered at A is in the hemisphere centered at A^a.

Definition 9.39 *Given a great circle with poles A, A^a, the hemispheres with centers at A and A^a are said to be the* sides *of the great circle. The polar circle* *of A is said to be the* edge *of each hemisphere. Two points lying in the same hemisphere are said to be on the* same side *of the great circle. If one lies in the hemisphere centered at A and the other lies in the hemisphere centered at A^a, the points are said to be on* opposite sides *of the great circle.*

Definition 9.40 *A given subset of the sphere is said to be spherically convex if for every pair of non-antipodal points, the unique arc between them is contained in the given subset.*

Proposition 9.41 *A great circle and a spherical ray are spherically convex.*

Proof. Given two non-antipodal points on a great circle, the spherical arc between them is a subset of the great circle by definition, so the great circle is spherically convex.

Suppose we are given two points C and D on a spherical ray $\overset{\hookrightarrow}{AB}$. By Proposition 9.29, if $\ell(B)-\ell(A) = d(A,B) (\text{mod}\, 2\pi)$, then the points C and D of $\overset{\hookrightarrow}{AB}$ satisfy $\ell(C) = \ell(A)+d(A,C)(\text{mod}\, 2\pi)$ and $\ell(D) = \ell(A)+d(A,D)(\text{mod}\, 2\pi)$ where we may assume without loss of generality that $d(A,C) < d(A,D)$ so that $0 \leq d(A,C) < d(A,D) < \pi$. Suppose E is between C and D. Then since either $C = A$ or C is between A and D, E must also be between A and D (by Proposition 9.18). Since $\ell(D) = \ell(A) + d(A,D)(\text{mod}\, 2\pi)$ then by Proposition 9.17, $\ell(E) = \ell(A) + d(A,E)(\text{mod}\, 2\pi)$ for $0 < d(A,E) < d(A,D) < \pi$. Then by Proposition 9.29, E is on $\overset{\hookrightarrow}{AD} = \overset{\hookrightarrow}{AB}$.

If $\ell(A) - \ell(B) = d(A,B)(\text{mod}\, 2\pi)$ then the argument is analogous. ◊

In Exercise 18, we will also see that open spherical rays and spherical arcs are spherically convex.

Axiom 9.42 (A-7) *A hemisphere is spherically convex.*

See Figure 2.6. In §13, Exercises 11 and 12, we will see that the interior of small circles are also spherically convex. We leave it to Exercise 17 to show that the intersection of spherically convex sets is also spherically convex, so that using hemispheres and interiors of small circles we may create many other spherically convex sets.

Proposition 9.43 *If two points are on opposite sides of a great circle and are not antipodal, the arc between them meets the great circle.*

Proof. Let A and B be the points and Γ the given great circle. Since A and B are on opposite sides of Γ, A and B^a are on the same side of Γ (see Exercise 19). Thus the arc $\overset{\frown}{AB^a}$ does not meet Γ. But $\overset{\hookrightarrow}{B^aA}$ meets Γ in a unique point, which thus must be between A and B, as desired. ◊

The following consequence of the convexity of a hemisphere is a useful tool in the proof of a number of theorems, including Propositions 10.8, 10.9, 10.11, 13.1, and Theorem 15.3.

Proposition 9.44 *If $\bigcirc AB$ is a great circle, and C is not on that great circle, then the points of $\overset{\hookrightarrow}{AC}$ other than A are on the same side of $\bigcirc AB$ as C.*

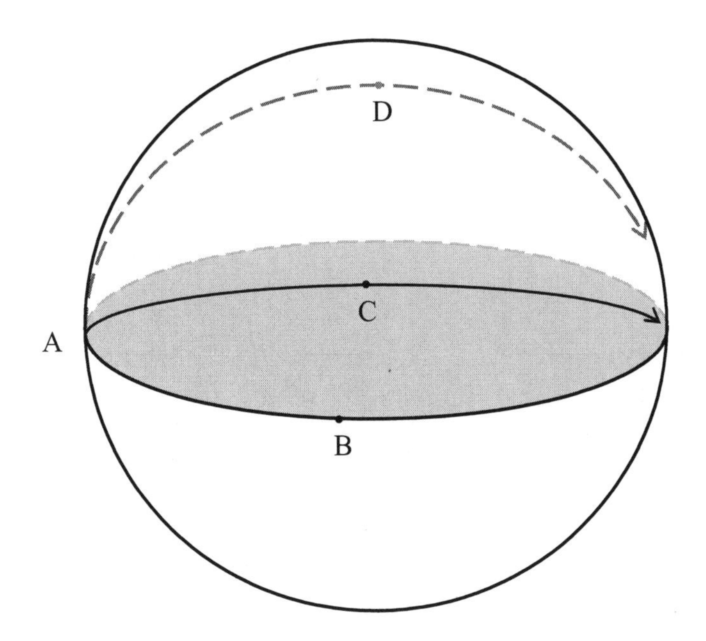

Figure 3.5: The points of $\overset{\hookrightarrow}{AC}$ and $\overset{\hookrightarrow}{AD}$ lie on the same side of $\bigcirc AB$, except for A.

Proof. (See Figure 3.5.) Suppose $X \neq A$ is on $\overset{\hookrightarrow}{AC}$ and is on $\bigcirc AB$. Then $\overset{\hookrightarrow}{AX}=\overset{\hookrightarrow}{AC}$ is contained in $\bigcirc AB$, so C is on $\bigcirc AB$, a contradiction.

Suppose X is on $\overset{\hookrightarrow}{AC}$ and belongs to the side of $\bigcirc AB$ opposite to C. By Proposition 9.43 above, $\overset{\frown}{CX}$ meets Γ in some point Y. Since $\overset{\hookrightarrow}{AC}$ is spherically convex, point Y must be on $\overset{\hookrightarrow}{AC}$. Since $\overset{\hookrightarrow}{AC}$ meets Γ in only one point, and A is such a point, $A = Y$. But an open ray is spherically convex (Exercise 18), and points $C \neq A$ and $X \neq A$ are on the ray $\overset{\hookrightarrow}{AC}$, so Y is also different from A, a contradiction. Thus the assumption was false: X must be on the same side of Γ as C. $\diamondsuit$

Definition 9.45 *We say that two (great or small) circles are* tangent *if they meet in a single point. This point is called the point of tangency. A* chord *of a small circle is a spherical arc between two points of the small circle.*

Since a pair of great circles must meet in two points, we only have tangent circles if one of the circles is a small circle. We will characterize tangency in later sections.

Exercises §9

1. Explain why the antipode of the antipode of a point is the given point.
2. Prove Proposition 9.5.
3. Prove Proposition 9.9.
4. Prove Proposition 9.10.
5. Justify the statement that $d(B, A) = d(A, B)$.
6. Prove that for any pair of points A, B on a sphere, $d(A^a, B^a) = d(A, B)$.
7. Prove Proposition 9.14.
8. Prove that there are infinitely many great circles.
9. Prove Proposition 9.22.
10. Prove Proposition 9.29.
11. Prove Corollary 9.30.
12. Prove Corollary 9.31.
13. Prove Proposition 9.32.
14. If A is a point, $B \neq A$, and $B \neq A^a$ then $m\, \overset{\frown}{AB} + m\, \overset{\frown}{A^aB} = \pi$.
15. Prove Proposition 9.35.
16. Given a small circle with center P, prove that the points of the circle are also equidistant from the antipode of P.
17. Prove that the intersection of spherically convex sets is also spherically convex.
18. Prove that open spherical rays and spherical arcs are spherically convex.
19. If A is a point, then every point on the sphere is either in the hemisphere centered at A, the hemisphere centered at A^a, or on the polar of A. Also, a hemisphere contains no pair of antipodal points, and the antipode of a point in a hemisphere centered at A is in the hemisphere centered at A^a.
20. Suppose A is between B and C. Prove that $\overset{\hookrightarrow}{AB}$ and $\overset{\hookrightarrow}{AC}$ are well-defined opposite rays.
21. Suppose A, B, and C are three (distinct) points on a great circle no two of which is antipodal. Prove that one of them is between the other two or between the antipodes of the other two.

22. Prove that rays $\overset{\hookrightarrow}{AB}$ and $\overset{\hookrightarrow}{AC}$ are opposite if and only if one of them consists of the set of points X on $\bigcirc AB$ such that $\ell(X) = \ell(A) + x (\bmod 2\pi)$, $0 \leq x < \pi$ and the other consists of the set of points X on $\bigcirc AB$ such that $\ell(X) = \ell(A) - x (\bmod 2\pi)$, $0 \leq x < \pi$. (Here ℓ is the one-one correspondence for $\bigcirc AB$.)

23. If a great circle has a pole lying on a second great circle, the poles of the second great circle lie on the first great circle. Conclude that a set of great circles all pass through a particular point if and only if their poles all lie on a single great circle.

24. Verify the assertions made earlier: a spherical ray contains no pair of antipodal points. The set of points consisting of antipodes of the points on the spherical ray is another spherical ray on the same great circle whose vertex is the antipode of the vertex of the given ray. Every point on the great circle of that ray is either on the ray or its antipode is on the ray, but not both.

25. Suppose that we are given two spherical rays whose endpoints are neither the same nor antipodal. Suppose that the points of the rays (endpoints excepted) lie on the same side of the great circle passing through those endpoints. Then the rays meet in exactly one point.

10 Angles

We now discuss angles on the sphere and their measurement.

Definition 10.1 *Let A, B, and C be points of a sphere which do not lie on a single great circle. Then the* (spherical) angle *$\sphericalangle ABC$ with* vertex *B is the union of the spherical rays $\overset{\hookrightarrow}{BA}$ and $\overset{\hookrightarrow}{BC}$. These rays are known as the* sides *of the angle. The* interior *of the angle is the intersection of two hemispheres: the side of $\bigcirc AB$ containing C and the side of $\bigcirc BC$ containing A.*

See Figure 3.7. Note that the assumption that the points A, B, C are not on the same great circle means that $\overset{\hookrightarrow}{BA}$ and $\overset{\hookrightarrow}{BC}$ are not the same and not opposite rays.

In the plane, the measure of angles comes from the desire to measure how far apart the side rays are. The process of assigning measure comes about by placing a protractor centered at the vertex of an angle. A protractor is essentially a circle whose perimeter has been labeled, typically with degrees. The measure of an angle is determined by looking at the two points where the sides of the angle intersect the protractor circle. The measure of the angle is the measure of the arc on the protractor circle between those two points.

The motivation on the sphere is the same.

Definition 10.2 *Let $\prec ABC$ be a spherical angle. Choose points D and E on $\overrightarrow{BA}$ and $\overrightarrow{BC}$, respectively, which are both at a quarter circle from B. Then the* measure *of $\prec ABC$ is defined to be $m\ \widehat{DE}$. (See Figure 3.6.)*

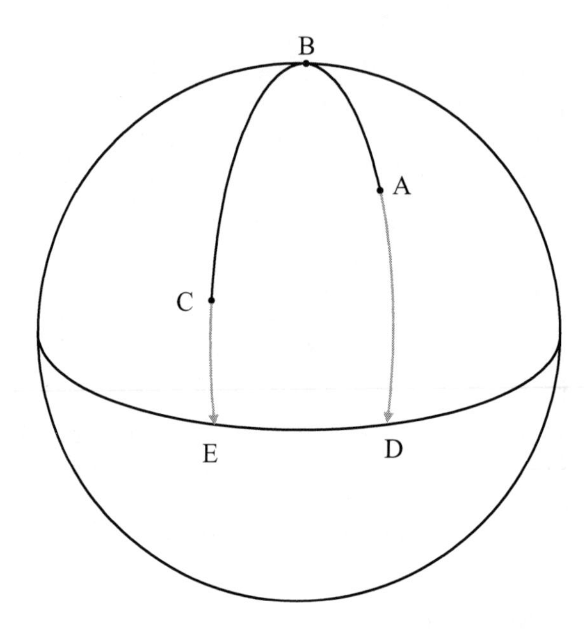

Figure 3.6: Definition 10.2: $m \prec ABC = m\ \widehat{DE}$.

Here we may regard the protractor circle as the polar circle of B ($\bigcirc DE$) and the vertex B as the center of the protractor. We say that $\prec ABC$ is *acute*, *right*, or *obtuse* if its measure is, respectively, less than, equal to, or greater than $\frac{\pi}{2}(90°)$.

Recall that in §5 (see Figure 2.4) we noted several different equivalent ways of understanding the measure of a spherical angle in space. Only one of those was intrinsic to the sphere, and this is the understanding of Definition 10.2.

Definition 10.3 *Two spherical angles are said to be* congruent *if their measures are the same. If the angles are $\prec ABC$ and $\prec DEF$ then we write $\prec ABC \cong \prec DEF$. The angles are said to be* supplementary *if their measures add up to π radians (180°). The angles are said to be* complementary *if their measures add up to $\frac{\pi}{2}$ radians (90°).*

Proposition 10.4 *Suppose $\prec ABC$ and $\prec DBC$ are such that $\overrightarrow{BA}$ is opposite to $\overrightarrow{BD}$. Then $\prec ABC$ and $\prec DBC$ are supplementary.*

Proof. Choose points E, F, and G on $\overrightarrow{BA}$, $\overrightarrow{BC}$, and $\overrightarrow{BD}$ so that $m\ \widehat{BE} = m\ \widehat{BF} = m\ \widehat{BG} = \pi/2$. Since $\overrightarrow{BA}$ and $\overrightarrow{BD}$ are opposite, E and G are a semicircle apart on the same great circle, so $E = G^a$. All of E, F, and G lie on the polar of B. F cannot be the same as either E or G since if either were the case, neither $\prec ABC$ nor $\prec DBC$ would be an angle. Then EFG forms a great semicircle such that (by §9 Exercise 14) $m\ \overset{\frown}{EF} + m\ \overset{\frown}{FG} = \pi$ so $m \prec ABC + m \prec DBC = \pi$, so $\prec ABC$ and $\prec DBC$ are supplementary by definition. ♢

Definition 10.5 *Suppose we are given $\prec ABC$ and $\prec DBE$ such that $\overset{\hookrightarrow}{BA}$ is opposite to $\overset{\hookrightarrow}{BD}$ and $\overset{\hookrightarrow}{BC}$ is opposite to $\overset{\hookrightarrow}{BE}$. Then $\prec ABC$ and $\prec DBE$ are said to be* vertical *angles.*

Proposition 10.6 *Vertical angles are congruent.*

Proof. If $\prec ABC$ and $\prec DBE$ are vertical as in Definition 10.5, then $\prec ABC$ and $\prec DBE$ are both supplementary to $\prec CBD$, so must be congruent. ♢

Definition 10.7 *A* lune *is the union of an angle with the antipode of its vertex. The* measure *of a lune is the measure of its associated angle. If the angle is $\prec ABC$ then we use the notation BCB^aAB to refer to this lune, where B^a is the antipode of B. The points B and B^a are the* vertices *of the lune.*

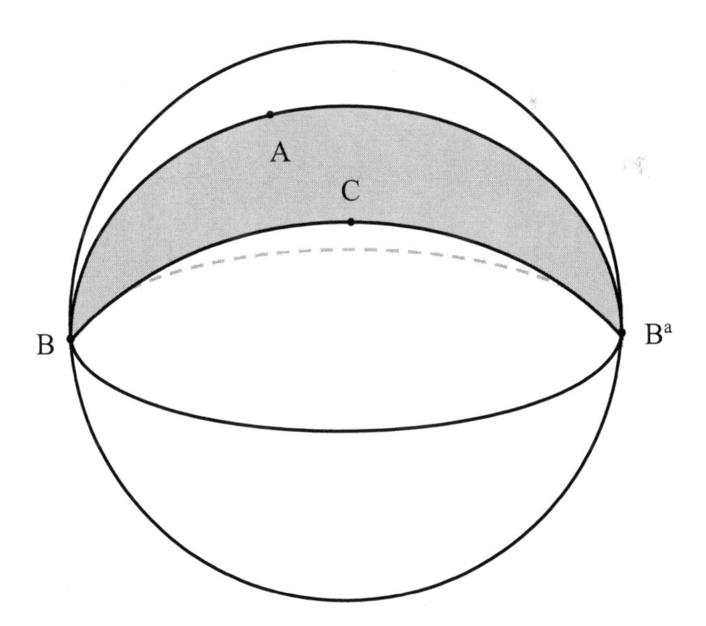

Figure 3.7: Angle $\prec ABC$, its interior, and the associated lune BAB^aCB.

Proposition 10.8 (Protractor theorem) *Suppose that we are given (1) a spherical ray* $r = \overrightarrow{PQ}$ *with* $\widehat{PQ}$ *a quarter circle and (2) a hemisphere* h *on one side of* r*. Then there exists a one to one to one correspondence between the set of real numbers* $-\pi < \theta \leq \pi$*, the points of the polar of* P *and the set of spherical rays which have the same endpoint as* r *such that (1)* $\theta = 0$ *is associated with* r *and* Q*, (2)* $\theta = \pi$ *is associated with the ray opposite to* r *and* Q^a *(3)* $0 < \theta < \pi$ *is associated with point* R *on the polar of* P *and ray* $\overrightarrow{PR}$ *such that* R *is in* h*,* $m\ \widehat{QR} = \theta$*, and* $\overrightarrow{QR}$ *is the same ray for all such* θ*, (4)* $-\pi < \theta < 0$ *is associated with point* S *on the polar of* P *and ray* $\overrightarrow{PS}$*, such that* S *is in the hemisphere opposite to* h*,* $m\ \widehat{QS} = -\theta$*, and* $\overrightarrow{QS}$ *is the same ray for all such* θ*, (5) if* r_1 *and* r_2 *are any two such rays forming an angle with associated real numbers* θ_1 *and* θ_2 *then the angle formed has measure found by reducing* $(\theta_1 - \theta_2)$ *modulo* 2π *to a value in* $(-\pi, \pi]$ *and taking the absolute value.*

Thus there exists a unique spherical ray r' *which (6) has the same endpoint as* r*, (7) passes through the hemisphere* h *and (8) forms an angle with* r *of measure* $\theta > 0$*. In fact, the spherical ray* r' *lies entirely in* h *except for its endpoint.*

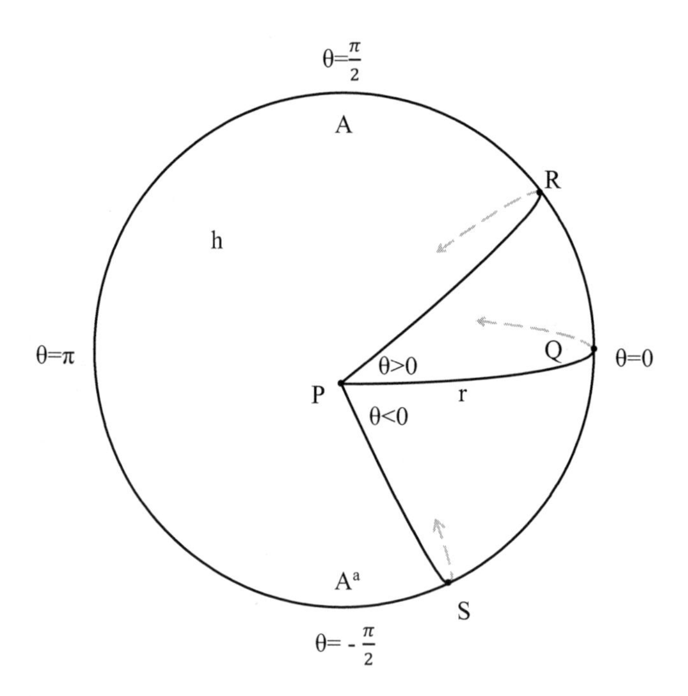

Figure 3.8: Proposition 10.8.

Proof. (See Figure 3.8.) Let A be the center of h, so A^a is the center of the opposite hemisphere. (So A, A^a are the poles of r.) Then both A and Q are a quarter circle from P, so $\bigcirc AQ$ is the polar of P (by Proposition 9.26). Points A and Q are neither the same nor antipodal because Q and Q^a are on the great circle of r but A is not). Let ℓ be the one-one correspondence coordinate for the polar of P such that $\ell(Q) = 0$ and $\ell(A) = \frac{\pi}{2}$. Then for $0 < \theta \leq \pi$, θ is associated with the point R on $\overrightarrow{QA}$ such that $m\,\widehat{QR} = \theta$. For $-\pi < \theta < 0$, θ is associated with the point S on $\overrightarrow{QA^a}$ such that $m\,\widehat{QS} = -\theta$. Next we can see that the points on the polar of P are in one-one correspondence with the rays emanating from P. Given a point X on the polar of P, we associate it with the ray $\overrightarrow{PX}$ emanating from P. If X_1 and X_2 are distinct, $\overrightarrow{PX_1}$ and $\overrightarrow{PX_2}$ are distinct because if $\overrightarrow{PX_1} = \overrightarrow{PX_2}$ then X_1 and X_2 are both the unique point on the ray at a quarter circle from P. Furthermore, given any ray s with endpoint P, it meets the polar of P at a point X other than P by Proposition 9.37: then $\overrightarrow{PX} = s$. Thus $X \leftrightarrow \overrightarrow{PX} = s$ is a one-one correspondence between points of the polar of P and the spherical rays emanating from P. This point X either lies on $\bigcirc PQ$, in h or in the hemisphere opposite to h. It then is associated with a value θ as in (1)-(4) above.

For $0 < \theta < \pi$ (corresponding to $\overrightarrow{PR}$), the ray $\overrightarrow{QR}$ is $\overrightarrow{QA}$ for all R, which is in h (except for Q) by Proposition 9.44. Similarly, for $-\pi < \theta < 0$, the ray $\overrightarrow{QS}$ is $\overrightarrow{QA^a}$ which lies in the hemisphere opposite to h (except for Q).

We see that property (5) holds as follows. Values θ_1, θ_2 correspond to rays $\overrightarrow{PX_1}$ and $\overrightarrow{PX_2}$. Then $m \measuredangle X_1PX_2 = m\,\widehat{X_1X_2}$ whose measure is the absolute value of $(\theta_1 - \theta_2)$ reduced mod 2π by definition of arc measure (Definition 9.13).

Further details are left to Exercise 3. $\diamondsuit$

Proposition 10.9 *Let $\measuredangle ABC$ be a spherical angle and let P be a point on the sphere. Then the following conditions are equivalent:*

(1) P is in the interior of $\measuredangle ABC$.

(2) P is on the same side of $\bigcirc BC$ as A and $m \measuredangle PBC < m \measuredangle ABC$.

(3) P is not on either of $\bigcirc AB$ or $\bigcirc BC$, and $m \measuredangle ABC = m \measuredangle ABP + m \measuredangle PBC$.

(4) P does not lie on either side of the angle, but for any Q on $\overrightarrow{BA}$ other than B, there is a point R on $\overrightarrow{BC}$ different from B such that P is between Q and R.

Proof. (See Figures 3.9 and 3.10.) For P to satisfy any of these four properties, P cannot be on $\bigcirc AB$ or $\bigcirc BC$, so we assume this is the case throughout. Our line of argument rests substantially on obtaining statements equivalent to (1) through (3) on the polar circle of B. Let points D, E, and F be on $\overrightarrow{BA}$, $\overrightarrow{BP}$, and $\overrightarrow{BC}$, respectively, at a quarter circle from B. Since P is not on $\bigcirc AB$ or $\bigcirc BC$, E is different from D and F (Proposition 10.8).

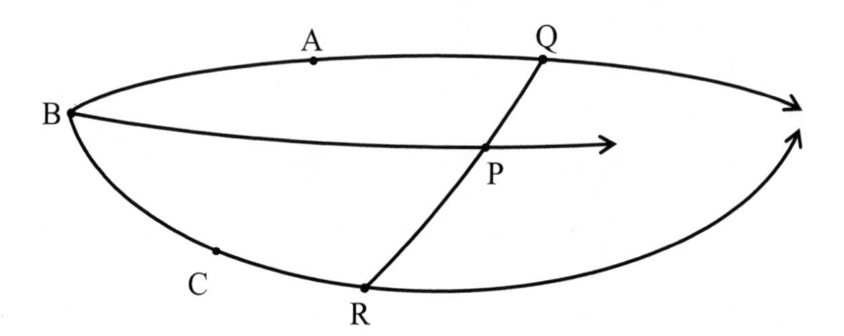

Figure 3.9: Proposition 10.9.

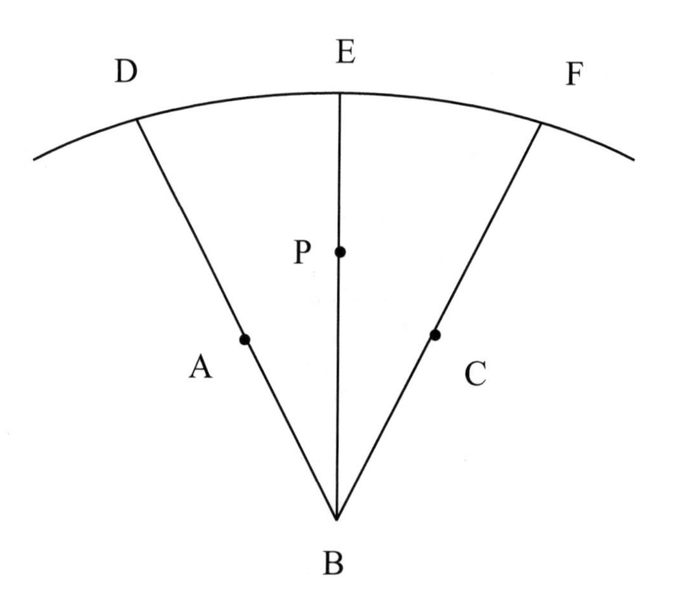

Figure 3.10: Proposition 10.9.

If (1) holds, then P and A are on the same side of $\bigcirc BC$, so (by Proposition 9.44) D and E are on the same side of $\bigcirc BC$. Thus $\overset{\hookrightarrow}{FE}$ and $\overset{\hookrightarrow}{FD}$ are the same ray (Proposition 10.8, where r is $\overset{\hookrightarrow}{BC}$). Similarly $\overset{\hookrightarrow}{DE}$ and $\overset{\hookrightarrow}{DF}$ are the same. Since the intersection of $\overset{\hookrightarrow}{DF}$ and $\overset{\hookrightarrow}{FD}$ is $\overset{\frown}{DF}$ (Proposition 9.32) E is on $\overset{\frown}{DF}$. Since E is different from D and F E is between D and F, so $m\ \overset{\frown}{DF}= m\ \overset{\frown}{DE}$ $+m\ \overset{\frown}{EF}$, so $m \prec ABC = m \prec ABP + m \prec PBC$, which gives (3).

If (3) holds, then E is different from D and F and $m\ \overset{\frown}{DF}= m\ \overset{\frown}{DE} +m\ \overset{\frown}{EF}$, so E is between D and F and $m\ \overset{\frown}{FE}< m\ \overset{\frown}{FD}$ (so $\overset{\hookrightarrow}{FE}=\overset{\hookrightarrow}{FD}$), so D and E are on the same side of $\bigcirc BC$ and ($m\ \overset{\frown}{FE}=$)$m \prec PBC < m \prec ABC(= m\ \overset{\frown}{DF})$, and finally P and A are on the same side of $\bigcirc BC$ (applying Proposition 9.44 to $\overset{\hookrightarrow}{BE}$ and $\overset{\hookrightarrow}{BD}$). Thus (2) holds.

If (2) holds, D and E are on the same side of $\bigcirc BC$ (applying Proposition 9.44 to $\overset{\hookrightarrow}{BE}$ and $\overset{\hookrightarrow}{BD}$). Applying Proposition 10.8 where r is $\overset{\hookrightarrow}{BC}$, $\overset{\hookrightarrow}{FE}=\overset{\hookrightarrow}{FD}$. Since $m \prec PBC < m \prec ABC$, $m\ \overset{\frown}{FE}< m\ \overset{\frown}{FD}$. Since $\overset{\hookrightarrow}{FE}=\overset{\hookrightarrow}{FD}$ and $m\ \overset{\frown}{FE}<$ $m\ \overset{\frown}{FD}$, E can only be between D and F. Thus $\overset{\hookrightarrow}{DE}$ is the same as $\overset{\hookrightarrow}{DF}$ so (by Proposition 10.8 applied where r is $\overset{\hookrightarrow}{BA}$) E and F are on the same side of $\bigcirc AB$, so P and C are on the same side of $\bigcirc AB$ (Proposition 9.44 applied to $\overset{\hookrightarrow}{BE}$ and $\overset{\hookrightarrow}{BF}$). Thus (1) holds.

Thus the first three conditions are equivalent. We now prove (1) is equivalent to (4).

If (4) holds, then P is on rays $\overset{\hookrightarrow}{RQ}$ and $\overset{\hookrightarrow}{QR}$. Since Q is on $\overset{\hookrightarrow}{BA}$, Q is on the same side of $\bigcirc BC$ as A. Since R is on $\bigcirc BC$, the points of $\overset{\hookrightarrow}{RQ}$ (except for R) are on the same side of $\bigcirc BC$ as Q, so on the same side of $\bigcirc BC$ as A. So P is on the same side of $\bigcirc BC$ as A. A similar argument shows that P is on the same side of $\bigcirc AB$ as C, so (1) holds.

Suppose that (1) holds, and Q is any point on $\overset{\hookrightarrow}{BA}$ other than B. Then Q is not on $\bigcirc BC$, so $\overset{\hookrightarrow}{QP}$ is not contained in $\bigcirc BC$. Thus $\overset{\hookrightarrow}{QP}$ meets $\bigcirc BC$ in a unique point R. R must be different from Q because Q is not on $\bigcirc BC$ but R is on $\bigcirc BC$. Every point of $\overset{\hookrightarrow}{QP}$ (except Q) is on the same side of $\bigcirc AB$ as P, so on the same side of $\bigcirc AB$ as C. Since R is different from Q, R is on the same side of $\bigcirc AB$ as C (and P). In particular R cannot be B or B^a. If R were on the ray opposite to $\overset{\hookrightarrow}{BC}$ ($\overset{\hookrightarrow}{BC^a}$), R and P would be on opposite sides of $\bigcirc AB$ (a contradiction). Thus R is a point on $\overset{\hookrightarrow}{BC}$ other than B, as desired. Then P is on $\overset{\hookrightarrow}{QR}$ and cannot be Q or R since P is not on $\bigcirc AB$ or $\bigcirc BC$. If R were between P and Q then P would be on the opposite side of $\bigcirc BC$ from Q, hence opposite from A, a contradiction of the assumption that P is in the interior of $\prec ABC$. So by definition of $\overset{\hookrightarrow}{QR}$, P is between Q and R. So (4) holds. This shows that (1) and (4) are equivalent, so all four conditions are equivalent. ◇

We need the above theorem in Proposition 10.13 and Proposition 11.12. Note that property (4) is too strong to be true in plane geometry, as in the plane it is not possible to let Q be any point on $\overset{\hookrightarrow}{BA}$. See Exercise 4.

Definition 10.10 *Let $\overset{\hookrightarrow}{BA}$ and $\overset{\hookrightarrow}{BC}$ form an angle $\prec ABC$ and let $\overset{\hookrightarrow}{BP}$ be a spherical ray. Then we say $\overset{\hookrightarrow}{BP}$ is* between *$\overset{\hookrightarrow}{BA}$ and $\overset{\hookrightarrow}{BC}$ if and only if P is in the interior of $\prec ABC$.*

Proposition 10.11 (Crossbar theorem) *Suppose that we are given angle $\prec ABC$ and D is a point in the interior of $\prec ABC$. Then the spherical ray $\overset{\hookrightarrow}{BD}$ intersects arc $\overset{\frown}{AC}$ in a point between A and C.*

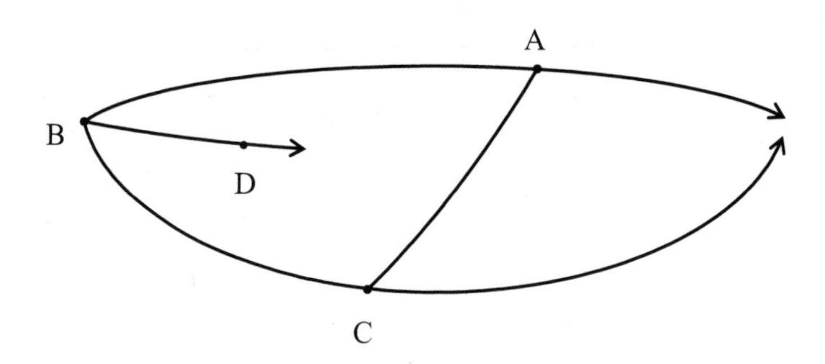

Figure 3.11: Proposition 10.11.

Proof. (See Figure 3.11.) Since D is in the interior of $\prec ABC$, it is on the same side of $\bigcirc BC$ as A. By Proposition 9.44, the points of spherical ray $\overset{\hookrightarrow}{BD}$ (except for B) are all on the same side of $\bigcirc BC$ as A. Similarly, we can argue that the points of spherical ray $\overset{\hookrightarrow}{BD}$ (except for B) are all on the same side of $\bigcirc AB$ as C. By definition, the points of $\overset{\hookrightarrow}{BD}$ (except for B) are all in the interior of $\prec ABC$. We know from Proposition 10.9 that the points of $\overset{\frown}{AC}$ are in the interior of $\prec ABC$, except for A and C. Now $\overset{\hookrightarrow}{BD}$ intersects $\bigcirc AC$ in a single point we call E (Proposition 9.37). Now E cannot be B since B is not on $\bigcirc AC$ so E must be in the interior of $\prec ABC$ since it is a point of $\overset{\hookrightarrow}{BD}$. So E is on the same side of $\bigcirc BC$ as A. Thus E must be on $\overset{\hookrightarrow}{CA}$. (If E were on the opposite ray $\overset{\hookrightarrow}{CA}^a$, it would be on the opposite side of $\bigcirc BC$ from A by Proposition 9.44.) Since E is on the same side of $\bigcirc AB$ as C, E must be on

$\overset{\hookrightarrow}{AC}$. Since E is on both $\overset{\hookrightarrow}{AC}$ and $\overset{\hookrightarrow}{CA}$, it must be on $\overset{\frown}{AC}$, and in fact must be between A and C since it is in the interior of $\prec ABC$. ♢

The notion of "measure of the angle between two lines" in the plane is not well-defined because when two lines meet, four angles are formed, and two (usually distinct) supplementary measures arise from these four angles. The same situation holds on the sphere when two great circles meet. Which of the two supplementary values one chooses depends on the situation. However, when the angles are all right angles, we may say that the great circles are perpendicular.

Proposition 10.12 *If two great circles are perpendicular then each of them passes through the poles of the other. Conversely, if one great circle passes through the poles of another, the two great circles are perpendicular.*

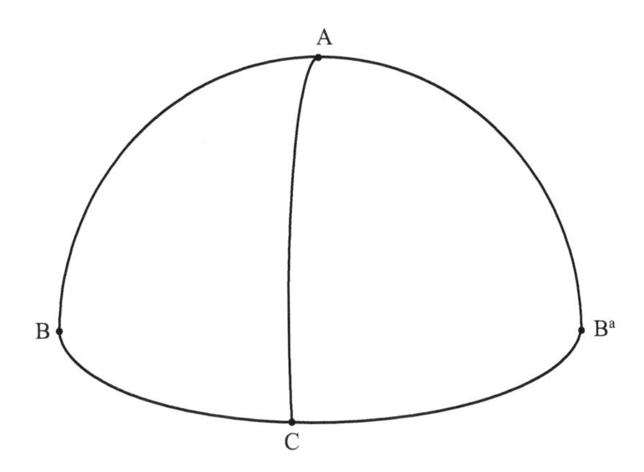

Figure 3.12: Proposition 10.12.

Proof. (See Figure 3.12.) If one of the circles passes through the pole A of the second circle, they are distinct circles, so meet in antipodal points B, B^a. Choose point C of the second circle such that C is a quarter circle from B. By definition of pole, $\overset{\frown}{AB}$ is a quarter circle. By definition, $m \prec ABC = m\ \overset{\frown}{AC}$. But $m\ \overset{\frown}{AC}= \frac{\pi}{2}$ since A is a pole of $\bigcirc BC$. Thus $\bigcirc AB$ is perpendicular to $\bigcirc BC$, as desired.

If the two circles are perpendicular then they are distinct, so meet at antipodal points B, B^a. Choose A on one circle and C on the other which are each a quarter circle from B. Then $m\ \overset{\frown}{AC}= m \prec ABC$ by definition; the latter is $\frac{\pi}{2}$ since the circles are perpendicular. Since $m\ \overset{\frown}{AB}= m\ \overset{\frown}{AC}= \frac{\pi}{2}$, A is a pole of $\bigcirc BC$, by Proposition 9.26. Similarly, C is a pole of $\bigcirc AB$. ♢

A useful fact in geometry is that if one desires to measure an angle between two lines (or planes) one may do so by measuring the angle between their perpendiculars at the point of intersection. The following result reflects this idea in spherical geometry.

Proposition 10.13 *Let A, B and C be points of a sphere which do not lie on a single great circle. Let A' be the pole of great circle $\bigcirc BC$ which is on the same side of $\bigcirc BC$ as A. Let C' be the pole of great circle $\bigcirc AB$ which is on the same side of $\bigcirc AB$ as C. Then $m\ \overset{\frown}{A'C'} = m \prec A'BC' = \pi - m \prec ABC$.*

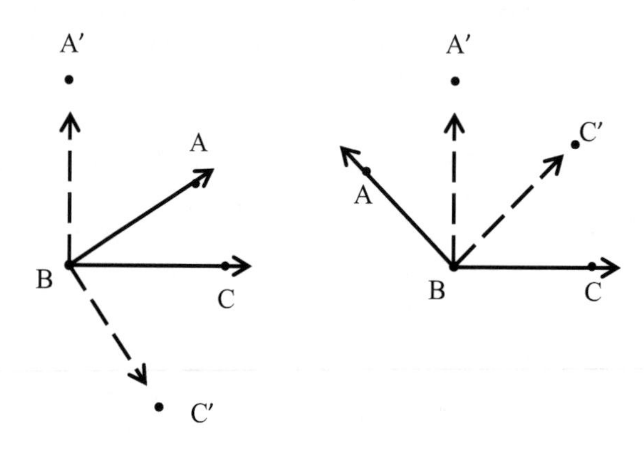

Figure 3.13: Proposition 10.13.

Proof. (See Figure 3.13.) By Proposition 10.12, $\bigcirc A'B$ is perpendicular to $\bigcirc BC$ and $\bigcirc C'B$ is perpendicular to $\bigcirc AB$. There are three cases to consider: $\prec ABC$ is acute, right, or obtuse. If $\prec ABC$ is acute, then (by Proposition 10.9) A is in the interior of $\prec A'BC$ and C is in the interior of $\prec ABC'$. Then[2] $m \prec A'BC' = m \prec A'BA + m \prec ABC + m \prec CBC' = \frac{\pi}{2} - m \prec ABC + m \prec ABC + \frac{\pi}{2} - m \prec ABC$, which is $\pi - m \prec ABC$, as desired. If $\prec ABC$ is obtuse, then A' and C' are both in the interior of $\prec ABC$. Then $m \prec ABC = m \prec ABA' + m \prec A'BC' + m \prec C'BC = \frac{\pi}{2} - m \prec A'BC' + m \prec A'BC' + \frac{\pi}{2} - m \prec A'BC'$, which is $\pi - m \prec A'BC'$. Thus $m \prec ABC = \pi - m \prec A'BC'$, as desired. If $\prec ABC$ is a right angle, then C' is on $\overrightarrow{BA}$ and A' is on $\overrightarrow{BC}$, so $\prec A'BC'$ is also a right angle and hence is supplementary to $\prec ABC$. $\diamondsuit$

Definition 10.14 *Suppose two small circles with centers A and B meet at point C. Then we define the measure of the angle between the circles to be $m \prec ACB$. (See Figure 3.14.)*

We occasionally will need to speak of the angle between a small circle and a great circle. We may follow the same trick used above: Let A be the center

[2] A detail is omitted in justifying the first equality here. It can be justified by using the one-one correspondence ℓ between rays emanating from B and the values $-\pi < \theta \leq \pi$. If ℓ is 0 on $\overset{\hookrightarrow}{BC}$ and $\frac{\pi}{2}$ on $\overset{\hookrightarrow}{BA'}$ then $\ell = m \prec ABC$ on $\overset{\hookrightarrow}{BA}$ since A, A' are on the same side of $\bigcirc BC$. Then since C, C' are on the same side of $\bigcirc AB$, the value of ℓ on $\overset{\hookrightarrow}{BC'}$ is $m \prec ABC - \frac{\pi}{2}$. The measures of the angles may be obtained from these protractor coordinates.

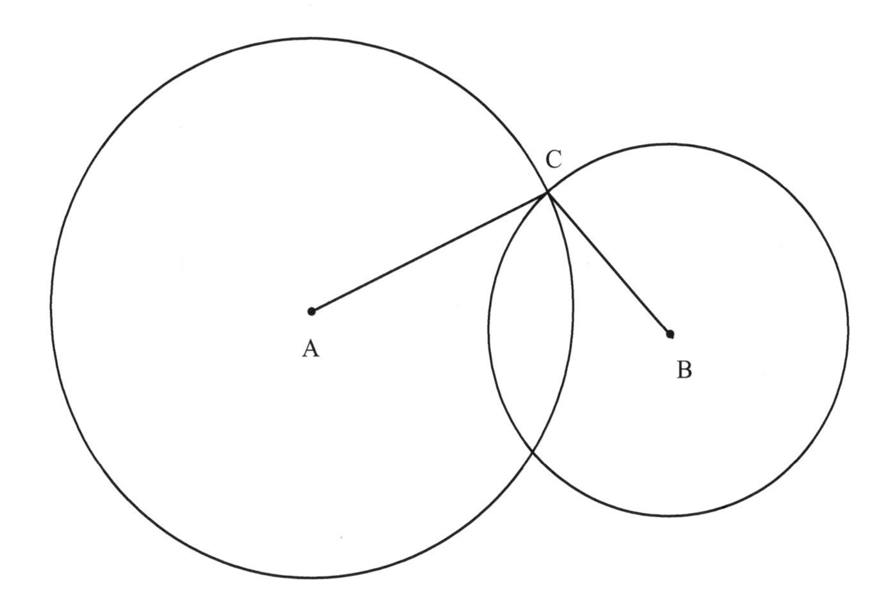

Figure 3.14: Definition 10.14.

of the small circle, B a pole of the great circle, and C a point of intersection. Then the measure of the angle between them is $m \prec ACB$. However, this cannot be made precise because a great circle has two poles, and we get two supplementary angle measures depending on which pole we use.[3] The ambiguity can be resolved by deciding a case by case basis which angle matters. As an example, in astronomy a star's daily path in the sky is a small circle but the horizon is a great circle, so the angle at which the star rises or sets is the angle between a small and great circle. One might choose the acute angle between the two circles as the desired angle measure. See §23, Exercise 10.

Proposition 10.15 *Let E be a great circle, let P be a pole of E, and let X be a point which is not a pole of E. Let M be the great circle passing through P and X, and let Y and Z be the (antipodal) points where E and M meet. Then the spherical distance from X to Y is different from the spherical distance from X to Z.*

Proof. (See Figure 3.15.) Point X lies on a great semicircle with endpoints Y and Z. If X is equidistant from Y and Z, then both distances are equal to $\frac{\pi}{2}$, which means $X = P$ or $X = P^a$, a contradiction. $\diamondsuit$

Definition 10.16 *Let E be a great circle, let P be a pole of E and let X be a point which is not a pole of E. Let M be the great circle passing through*

[3]In fact, since small circles also have two poles, the choice of poles A and B made in Definition 10.14 (instead of their antipodes) is somewhat artificial.

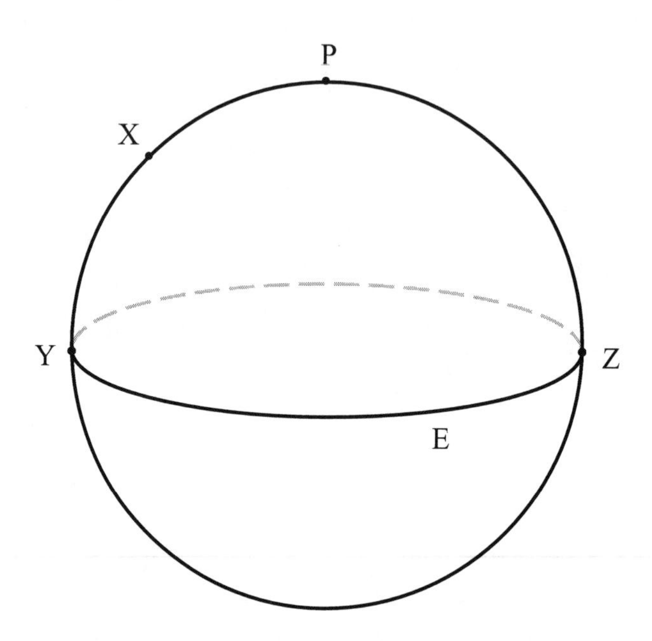

Figure 3.15: Proposition 10.15 and Definition 10.16.

P and X, and let Y and Z be the (antipodal) points where E and M meet. Suppose X is closer to Y than Z. Then the spherical distance from X to Y is said to be the distance *from X to E. If* $X \neq Y$, *then* $\widehat{XY}$ *is said to be the* shorter perpendicular *or* shorter altitude *from X to E,* $\widehat{XZ}$ *is the* longer perpendicular *or* longer altitude *from X to E, Y is the* foot *of the shorter perpendicular, Z is the* foot *of the longer perpendicular. The* distance *from P to E is defined to be* $\frac{\pi}{2}$ *radians (90°). (See Figure 3.15.)*

Note that, by Proposition 9.23, the distance to a great circle from one of its poles is the same as the distance between the pole and any point on the great circle. It is left to Exercise 8 to prove the following proposition.

Proposition 10.17 *Suppose that E is a great circle with a pole P and X is not on E and not one of the poles of E. If W is a point of E such that* $\bigcirc XW$ *is perpendicular to E, then W must be one of the two points where* $\bigcirc PX$ *meets E. (See Figure 3.15.)*

The last proposition of this section will be critical to relating spherical distance and angle measure, and be central to proving congruence of triangles. We shall accept it as an axiom. The reader can either accept it as self-evident for a sphere in space, or use solid geometry to justify it. (See Exercise 7.)

Axiom 10.18 (A-8) *Suppose that in spherical* $\prec A_1B_1C_1$ *and* $\prec A_2B_2C_2$, *we have* $\widehat{A_1B_1} \cong \widehat{A_2B_2}$ *and* $\widehat{B_1C_1} \cong \widehat{B_2C_2}$. *Then* $\prec A_1B_1C_1 \cong \prec A_2B_2C_2$ *if and only* *if* $\overset{\frown}{A_1C_1} \cong \overset{\frown}{A_2C_2}$.

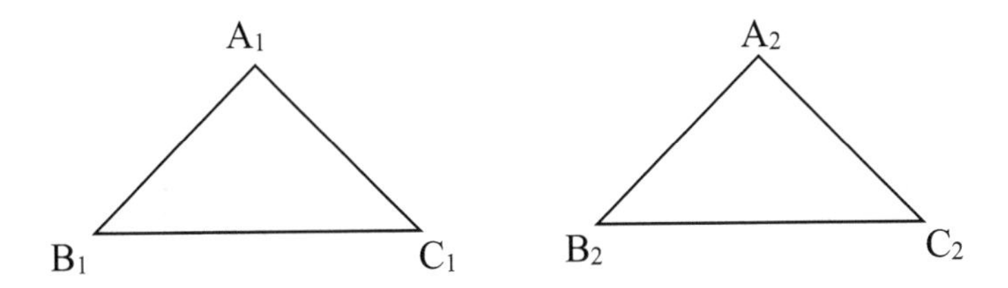

Figure 3.16: Axiom 10.18.

Exercises §10.

1. Given a spherical angle $\prec ABC$, prove that $\prec A^aB^aC^a$ is a well-defined spherical angle with the same measure.
2. In lune BCB^aAB, prove that $m \prec ABC = m \prec AB^aC$.
3. Fill in the details of the proof of Proposition 10.8.
4. Consider the analogue of property (4) of Proposition 10.9 in plane Euclidean geometry. For what Q does such an R exist?
5. Suppose that $\prec ABC$ is a right angle and $\overset{\frown}{AB}$ is a quarter circle. Prove that $\overset{\frown}{AC}$ is a quarter circle and $\prec ACB$ is a right angle.
6. Suppose that $\overset{\hookrightarrow}{BA}$, $\overset{\hookrightarrow}{BP}$, and $\overset{\hookrightarrow}{BC}$ are spherical rays. Using the one-one correspondence of Proposition 10.8, let D, E, and F be the corresponding points on the polar of B. Prove that $\overset{\hookrightarrow}{BP}$ is between $\overset{\hookrightarrow}{BA}$ and $\overset{\hookrightarrow}{BC}$ if and only if E is between D and F.
7. Use solid geometry to justify Axiom 10.18.
8. Prove Proposition 10.17.
9. Prove that two distinct great circles have a unique third great circle perpendicular to both.
10. Let X, Y, Z, and E be as in Definition 10.16. Let F be the great circle whose poles are Y and Z. Prove that F meets E in two points which are at distance $\frac{\pi}{2}$ from X, Y and Z.
11. Suppose that E is a great circle, P is one of the poles of E, and X is a point on E or on the same side of E as P. Prove that if h is the distance from X to E, $h = \frac{\pi}{2} - d(P, X)$. What happens if X is on the opposite side of E from P?

Historical notes. Let X-Y denote Book X, Proposition Y of Menelaus' *Sphaerica*. Then Axiom 10.18 is I-4.

11 Triangles

In plane geometry, a triangle arises by taking three points which do not lie on a line, and connecting them with the unique segments between pairs of these points. Thus three noncollinear points determine a unique triangle. We will do the same on a sphere (where instead the points will not lie on a single great circle, and we connect them with arcs of great circles). But before we do so, one problem will arise. If three points not on a single great circle are to determine a unique spherical triangle, is it possible that some pair of them are antipodal — which would mean that there is not a unique arc of a great circle between them? The answer, fortunately, is no:

Proposition 11.1 *If three points A, B, C on a sphere do not lie on a single great circle then no two of the points A, B, C are antipodal, so that the spherical arcs $\overset{\frown}{AB}, \overset{\frown}{BC}, \overset{\frown}{AC}$ are defined as in Definition 9.19.*

Proof. We proceed by contradiction. If one pair were antipodal (say A, B) then (the three points being distinct) C would not be antipodal to either A or B. Thus the great circles $\bigcirc AC$ and $\bigcirc BC$ are defined. Since B is antipodal to A, B lies on $\bigcirc AC$ (by Proposition 9.5), so A, B, and C lie on a single great circle, a contradiction. So no pair of A, B, or C is antipodal. $\diamondsuit$

Definition 11.2 *Let A, B, C be three points on a sphere which do not lie on a single great circle. Then the* spherical triangle *$\triangle^s ABC$ is the union of the three arcs $\overset{\frown}{AB}, \overset{\frown}{BC}, \overset{\frown}{AC}$. Each of the three points A, B, and C is called a* vertex *(plural:* vertices*) of $\triangle^s ABC$. The arcs $\overset{\frown}{AB}, \overset{\frown}{BC}, \overset{\frown}{AC}$ are called the* sides *of $\triangle^s ABC$. The spherical angles $\prec CAB$, $\prec ABC$ and $\prec BCA$ (also denoted by $\prec A$, $\prec B$, and $\prec C$) are the* angles *of $\triangle^s ABC$.*

By Proposition 11.1, no two of the points A, B, and C are antipodal, so the three spherical arcs in Definition 11.2 are well-defined. Also by Definition 9.19, each of the sides of the triangle is the shorter of the two great circle arcs between the vertices, so the measure of each of the three sides is less than π. We may also conclude from Proposition 10.9 that every side of a spherical triangle is in the interior of the opposite angle (except for the endpoints).

We define several types of special triangles.

Definition 11.3 *Two triangles are said to be colunar if they have two vertices in common and one pair of vertices which are antipodal.*

That is, the triangles have the form $\triangle^s ABC$ and $\triangle^s A^a BC$ where A and A^a are antipodal.

Proposition 11.4 *Given $\triangle^s ABC$ and the antipode A^a of A, the three points A^a, B, and C form a spherical triangle such that $\overset{\frown}{A^aB}$, $\overset{\frown}{A^aC}$, $\prec A^aBC$, and $\prec A^aCB$ are supplementary to $\overset{\frown}{AB}$, $\overset{\frown}{AC}$, $\prec ABC$, and $\prec ACB$, respectively.*

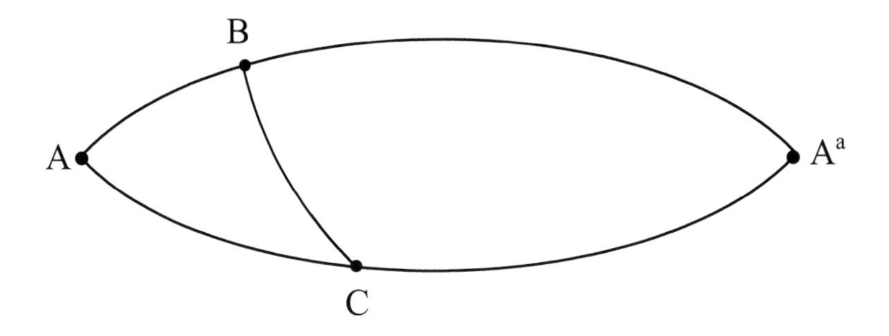

Figure 3.17: Colunar triangles $\triangle^s ABC$ and $\triangle^s A^a BC$.

Proof. See Exercise 1. ♢

Definition 11.5 *Two triangles are said to be antipodal if they can be expressed as* $\triangle^s ABC$ *and* $\triangle^s A^a B^a C^a$*, where* A^a*,* B^a*, and* C^a *are antipodal to* A*,* B*, and* C*, respectively.*

Proposition 11.6 *Given* $\triangle^s ABC$*, the three antipodes* A^a*,* B^a*, and* C^a *form a spherical triangle whose sides and angles are congruent to the corresponding sides and angles in* $\triangle^s ABC$*.*

Proof. See Exercise 2. ♢

Definition 11.7 *A spherical triangle is said to be a* right triangle *if at least one of its angles is a right angle. A side of a triangle which is opposite a right angle is said to be a* hypotenuse *of the right triangle. A side of the triangle which is* not *a hypotenuse is said to be a* leg *of the triangle.*

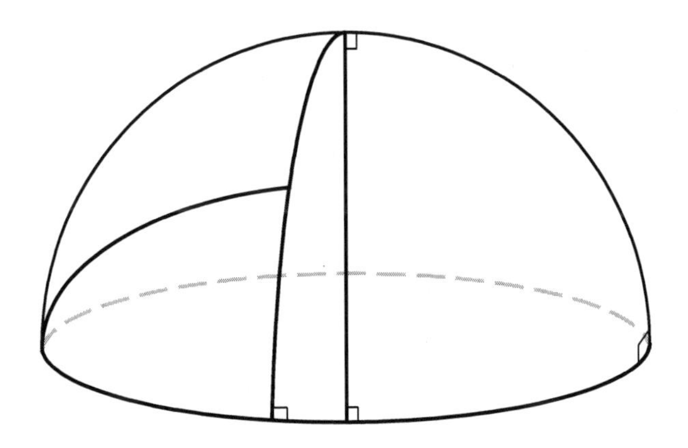

Figure 3.18: Triangles with one, two, and three right angles.

Definition 11.8 *If $\triangle^s ABC$ satisifes the property that one of its sides is a quarter circle, then we say that $\triangle^s ABC$ is* quadrantal. *A side which is a quarter circle is said to be a* right side.

Definition 11.9 *A triangle $\triangle^s ABC$ is said to be* isosceles *with* vertex *A if $\overset{\frown}{AB} \cong \overset{\frown}{AC}$. The angles at B and C are the* base angles *of the triangle. The triangle is said to be* equilateral *if all of its sides are congruent.*

Theorem 11.10 *In an isosceles spherical triangle, the angles opposite the congruent sides are also congruent.*

Proof. Suppose that in $\triangle^s ABC$, $\overset{\frown}{AB} \cong \overset{\frown}{AC}$. Then compare $\prec ABC$ and $\prec ACB$: $\overset{\frown}{AB} \cong \overset{\frown}{AC}$, $\overset{\frown}{BC} \cong \overset{\frown}{CB}$, and $\overset{\frown}{AC} \cong \overset{\frown}{AB}$. By Axiom 10.18, $\prec ABC \cong \prec ACB$, as desired. ◇

The converse of this is also true; this is §12, Exercise 1.

Let us now prove a few propositions in spherical geometry which demonstrate how the world of the sphere will be different from that of the plane. The reader should be used to the fact that in plane geometry, triangles may have only one right angle. This is not the case on the sphere, and the next proposition explains how this happens.

Proposition 11.11 *A pair of angles in a triangle are right angles if and only if the opposite sides are right sides.*

Proof. Suppose that in $\triangle^s ABC$, the angles $\prec B$ and $\prec C$ are both right. Then great circles $\bigcirc AB$ and $\bigcirc AC$ are both perpendicular to $\bigcirc BC$, so by Proposition 10.12, each passes through the poles of $\bigcirc BC$. Since $\bigcirc AB$ and $\bigcirc AC$ are distinct, they meet in only two points (which thus must be the poles of $\bigcirc BC$). Since A is one of these points, it is a pole of $\bigcirc BC$, so $\overset{\frown}{AB}$ and $\overset{\frown}{AC}$ are both right sides, as every point of a great circle lies a quarter circle from each pole.

Conversely, suppose that both $\overset{\frown}{AB}$ and $\overset{\frown}{AC}$ are right sides. By Proposition 9.26, A is a pole of $\bigcirc BC$. By Proposition 10.12, $\bigcirc AB$ and $\bigcirc AC$ are both perpendicular to $\bigcirc BC$, so $\triangle^s ABC$ has right angles at B and C. ◇

In plane geometry, if a triangle has one right angle, the other angles are acute. On a sphere, this also turns out differently.

Proposition 11.12 *Suppose $\triangle^s ABC$ has a right angle at B. Then one of the other angles of $\triangle^s ABC$ is acute, right, or obtuse, if and only if its opposite side is acute, right, or obtuse, respectively.*

Proof. By Proposition 10.12, $\bigcirc AB$ passes through the poles of $\bigcirc BC$. Let A' be the pole of $\bigcirc BC$ on the same side of $\bigcirc BC$ as A. Then $\overset{\hookrightarrow}{BA}$ and

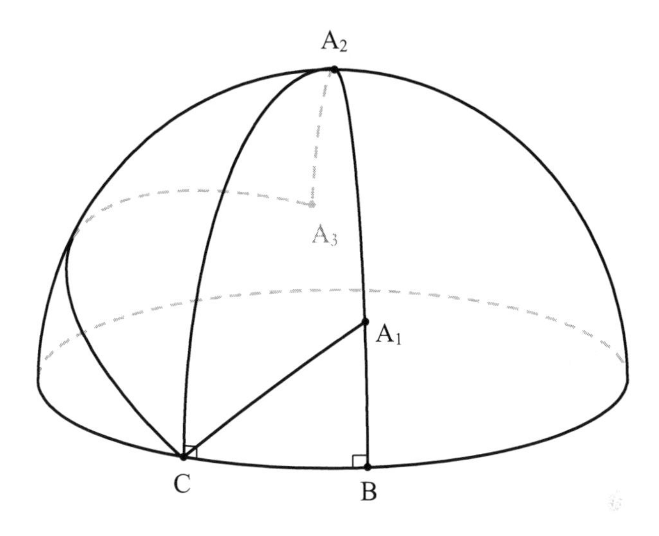

Figure 3.19: Proposition 11.12, cases $A = A_1, A_2, A_3$.

$\overset{\hookrightarrow}{BA'}$ are both perpendicular to $\bigcirc BC$ at B, and A, A' are on the same side of $\bigcirc BC$. By Proposition 10.8 $\overset{\hookrightarrow}{BA}=\overset{\hookrightarrow}{BA'}$.

If side $\overset{\frown}{BA}$ is right, then $A = A'$ (Corollary 9.30). Then by Proposition 10.12, $\bigcirc AC$ is perpendicular to $\bigcirc BC$, so $\prec C$ is a right angle, as desired. If side $\overset{\frown}{BA}$ is acute, then A is between B and A' by Proposition 9.17, since $m\ \overset{\frown}{BA} < m\ \overset{\frown}{BA'} = \frac{\pi}{2}$. From Proposition 10.9 we conclude that A is in the interior of $\prec A'CB$. By Proposition 10.9, $m \prec ACB < m \prec A'CB = \frac{\pi}{2}$, so $\prec ACB$ is acute, as desired.

If side $\overset{\frown}{BA}$ is obtuse, then by Proposition 9.17, A' is between B and A(since $m\ \overset{\frown}{BA} > m\ \overset{\frown}{BA'} = \frac{\pi}{2}$.) From Proposition 10.9 we conclude that A' is in the interior of $\prec ACB$. By Proposition 10.9, $m \prec ACB > m \prec A'CB = \frac{\pi}{2}$, so $\prec ACB$ is obtuse, as desired. $\diamondsuit$

If the assumption that $\prec B$ is right is relaxed to the assumption that $\prec B$ is the angle in the triangle closest to being right, then the same conclusion holds. See §16, Exercise 14 and §19, Exercise 17.

An arc between a vertex of a triangle and a point of the opposite side is called a *cevian* of the triangle from the vertex. If the point on the opposite side is the midpoint of that side, the cevian is called the *median* of the triangle from the vertex. The cevian which bisects the angle at the vertex is called an *angle bisector* from the vertex. An arc between a vertex of a triangle and the great circle containing the opposite side which is perpendicular to the great circle is called an *altitude* of the triangle from that vertex.

Recall that in a plane triangle, an altitude from one vertex of the triangle meets the interior of the opposite side provided that the other two angles are acute. A variation of this holds on the sphere.

Proposition 11.13 *Suppose that in $\triangle^s ABC$, $\prec A$ and $\prec B$ are not right angles. Then C is not a pole of $\bigcirc AB$. Let D and E be the feet of the shorter and longer altitudes from C to $\bigcirc AB$, respectively. Then we may determine the placement of D and E as follows:*

(1) D is between A and B if and only if $\prec A$ and $\prec B$ are both acute.

(2) E is between A and B if and only if $\prec A$ and $\prec B$ are both obtuse.

(3) neither D nor E is between A and B if and only if one of $\prec A$ and $\prec B$ is acute and the other is obtuse.

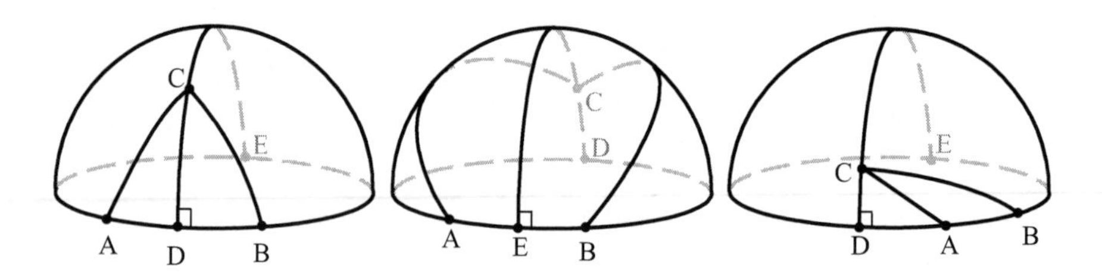

Figure 3.20: Proposition 11.13.

Proof. If C were a pole of $\bigcirc AB$, then by Proposition 9.23, $\overset{\frown}{CA}$ and $\overset{\frown}{CB}$ are right sides, so by Proposition 11.11, $\prec A$ and $\prec B$ would be right angles, a contradiction of our assumptions. So C is not a pole of $\bigcirc AB$. If neither $\prec A$ nor $\prec B$ are right angles then both D and E are distinct from A and B. (If D were equal to A, for example, then $\prec A$ would be right.) Note that D and E cannot be simultaneously between A and B because D and E are antipodal. So either (1) D is between A and B and E is not; (2) E is between A and B and D is not; or (3) neither D nor E is between A and B.

(1) Suppose D is between A and B. Then $\prec CAB$ is the same as $\prec CAD$ and $\prec CBA$ is the same as $\prec CBD$. Applying Proposition 11.12 to $\triangle^s CAD$ and $\triangle^s CBD$, we find that since $\overset{\frown}{CD}$ is acute, so $\prec A$ and $\prec B$ are also.

(2) Suppose E is between A and B. Then we apply the proof of (1), replacing D by E, and "acute" by "obtuse" to conclude that $\prec A$ and $\prec B$ are obtuse.

(3) Suppose neither D nor E is between A and B. We apply Proposition 11.12 again to $\triangle^s CAD$ and $\triangle^s CBD$ to conclude that $\prec CAD$ and $\prec CBD$ are acute. Then A and B are on the same side of the great circle $\bigcirc CD = \bigcirc CE$ because were they on opposite sides, the arc $\overset{\frown}{AB}$ would cross this great circle somewhere (by Proposition 9.43) and this could only occur at the intersection of $\bigcirc AB$ and $\bigcirc CD$ — that is, at either D or E — and this was ruled out by assumption. By Proposition 10.8, $\overset{\hookrightarrow}{DA}=\overset{\hookrightarrow}{DB}$. If A is between D and B then $\overset{\hookrightarrow}{AD}$ and $\overset{\hookrightarrow}{AB}$ are opposite (§9, Exercise 20) so $\prec CAB$ is supplementary to $\prec CAD$.

Furthermore (again assuming A is between D and B), $\overset{\hookrightarrow}{BA}=\overset{\hookrightarrow}{BD}$, so $\prec CBA$ is equal to $\prec CBD$. Then $\prec CAB$ is obtuse and $\prec CBA$ is acute. If B is between D and A then $\prec CAB$ is equal to $\prec CAD$ and $\prec CBA$ is supplementary to $\prec CBD$. Then $\prec CBA$ is obtuse and $\prec CAB$ is acute.

As to the opposite direction, suppose that $\prec A$ and $\prec B$ are both acute. If E is between A and B then from what we just proved, $\prec A$ and $\prec B$ are both obtuse, a contradiction. If neither D nor E is between A and B then from what we just proved, one of $\prec A$ or $\prec B$ is obtuse, a contradiction. So the only possibility remaining is that D is between A and B.

A similar argument works if $\prec A$ and $\prec B$ are both obtuse, or if one is obtuse and the other acute. ♢

Proposition 11.14 *If a triangle has two sides of measure a and the two opposite angles have measure A, then a and A are both acute, both right or both obtuse.*

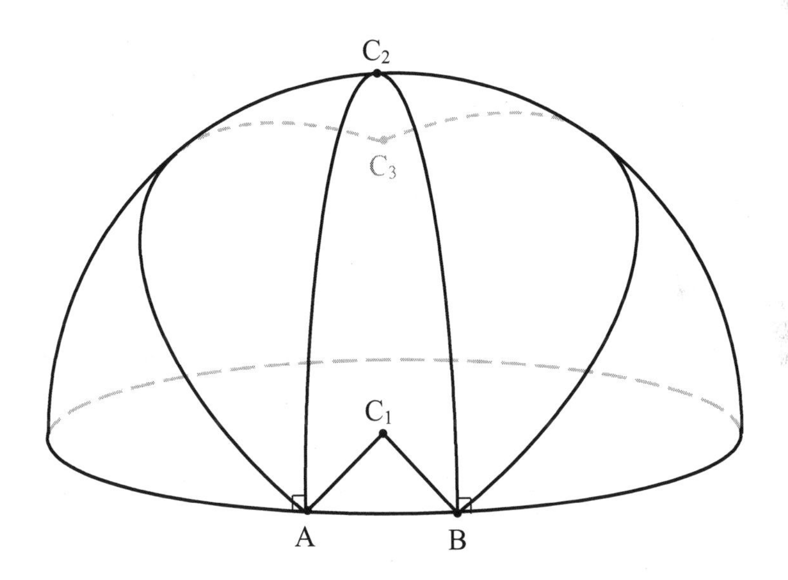

Figure 3.21: Proposition 11.14, cases $C = C_1, C_2, C_3$.

Proof. Suppose that the triangle is $\triangle^s ABC$, the sides mentioned are $\overset{\frown}{BC}$ and $\overset{\frown}{AC}$, and the angles are $\prec A$ and $\prec B$. We have dealt with the case of right angles/right sides in Theorem 11.11. So we may assume here that a, A are each either acute or obtuse (not necessarily the same). Thus C is not a pole of $\bigcirc AB$. Let D and E be the feet of the shorter and longer perpendiculars from C to $\bigcirc AB$, respectively. By Proposition 11.13, because the angles at A and B are congruent, they must be both acute or both obtuse, so either D or E is between A and B (but not both, since D and E are antipodal). (Note that D and E are both different from A and B because $\prec A$ and $\prec B$ are not right angles.)

If a is acute then A and B belong to the hemisphere centered at C, so by Axiom 9.42, so do all points on $\overset{\frown}{AB}$. This would include either D or E, but it cannot be E because $\overset{\frown}{CE}$ is obtuse. Thus $\prec CAB = \prec CAD$. Applying Proposition 11.12 to $\triangle^s CDA$ we conclude that since $\overset{\frown}{CD}$ is acute by definition the opposite angle $\prec CAD$ (with measure A) is acute. (So $\prec CBD$ is also acute, having the same measure as $\prec CAD$.)

If a is obtuse, then A and B belong to the hemisphere centered at the antipode of C, so the points of $\overset{\frown}{AB}$ are also, by Axiom 9.42. This must include either D or E, and since $\overset{\frown}{CD}$ is acute, it could only be E. Thus $\prec A = \prec CAE$. Applying Proposition 11.12 to $\triangle^s CAE$ we conclude that the opposite angle $\prec A$ (with measure A) is obtuse since $\overset{\frown}{CE}$ is obtuse by definition. (So $\prec B$ is also obtuse, having the same measure A.) $\diamondsuit$

Definition 11.15 *The* interior *of a triangle is the intersection of the interiors of its angles. The* exterior *of a triangle is the set of points which are neither on a triangle nor in its interior.*

We now introduce a notion that has no equivalent in plane Euclidean geometry, and which will be helpful in understanding many relationships in spherical triangles which do not occur in plane triangles. We note that if $\triangle^s ABC$ is a spherical triangle, then by Definition 11.2 the point A is not on the great circle $\bigcirc BC$. Thus A must lie in one of the two hemispheres into which the sphere is divided by $\bigcirc BC$, and hence lies on a particular side of $\bigcirc BC$. Similarly, B lies on a particular side of $\bigcirc AC$ and C lies on a particular side of $\bigcirc AB$.

Definition 11.16 *Let $\triangle^s ABC$ be a spherical triangle. We define the* polar triangle *$\triangle^s A'B'C'$ of $\triangle^s ABC$ (also sometimes called the* supplemental triangle *or the* dual triangle*) as follows. (See Figure 3.22.) We let A' be the pole of $\bigcirc BC$ which lies on the same side of $\bigcirc BC$ as A. We define B' and C' analogously: B' is the pole of $\bigcirc AC$ on the same side of $\bigcirc AC$ as B, and C' is the pole of $\bigcirc AB$ on the same side of $\bigcirc AB$ as C.*

We need to check that $\triangle^s A'B'C'$ is a well-defined triangle.

Theorem 11.17 *If $\triangle^s ABC$ is a well-defined spherical triangle, $\triangle^s A'B'C'$ is also; i.e., the points A', B', C' do not lie on a single great circle.*

Proof. Suppose A', B', C' all lie on a single great circle Γ with poles N, S. By Proposition 9.23 the points on $\bigcirc BC$ are the set of all points on Γ at a quarter circle from A'. By the same proposition, since N and S are a quarter circle from A', N and S lie on $\bigcirc BC$. Similarly we conclude that N and S belong to $\bigcirc AB$ and $\bigcirc AC$. Since A, B, C do not all lie on the same great circle, $\bigcirc AB$

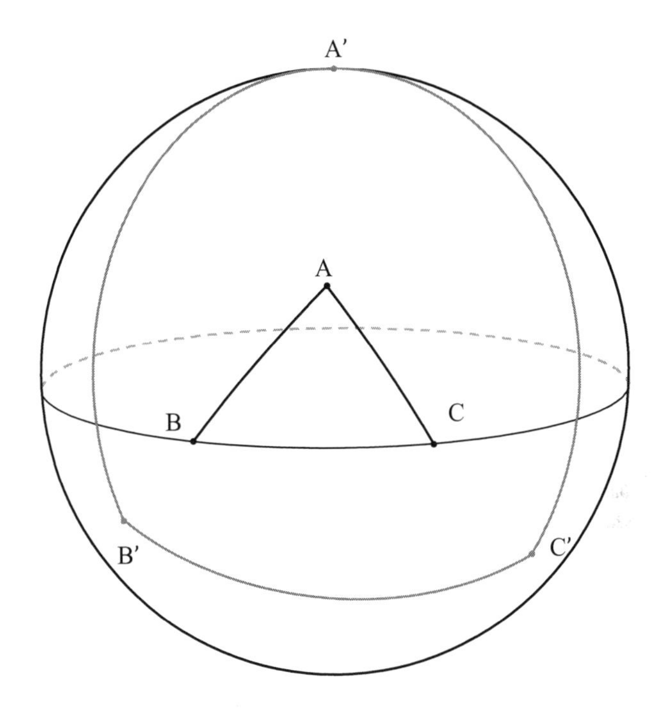

Figure 3.22: $\triangle^s ABC$ and its polar triangle $\triangle^s A'B'C'$.

and $\bigcirc BC$ must be different. Thus these two great circles meet in only two points (by Proposition 9.12), which we already know are N and S. But $\bigcirc AB$ and $\bigcirc BC$ also have the point B in common, so B must be the same as N or S. A similar argument shows A is the same as N or S, and C is the same as N or S. But then A, B, C all lie on a single great circle (any great circle passing through N and S contains all of A, B, C). This is a contradiction, so the assumption is false, and A', B', C' cannot all lie on a single great circle.$\diamondsuit$

The following theorem is one of the reasons the polar triangle is sometimes referred to as the dual triangle. (See [Je1994].[4])

Theorem 11.18 *The polar triangle of* $\triangle^s A'B'C'$ *is* $\triangle^s ABC$.

Proof. Since C and C' are on the same side of $\bigcirc AB$, and C' is a pole of $\bigcirc AB$, the distance between C and C' is strictly less than $\pi/2$. Since A' is a pole of $\bigcirc BC$, A' is at distance $\pi/2$ from C. Since B' is a pole of $\bigcirc AC$, B' is at distance $\pi/2$ from C. Since A' and B' are vertices of a spherical triangle, they are neither identical nor antipodal. Thus we conclude (from Proposition 11.1) that $\bigcirc A'B'$ is well-defined. Since A' and B' are both at distance $\frac{\pi}{2}$ from C, by Propositions 9.26 and 9.23 every point on $\bigcirc A'B'$ is at distance $\pi/2$

[4]Jennings defines the dual triangle to behave well with respect to the vector (cross) product, so that C' and $A \times B$ point in the same direction.

from C and C is one of the poles of $\bigcirc A'B'$. Since C and C' are at distance less than $\pi/2$ from each other, and C is a pole of $\bigcirc A'B'$, C' must be on the same side of $\bigcirc A'B'$ as C. Thus C is the pole of $\bigcirc A'B'$ on the same side of $A'B'$ as C'. In a similar manner, we can show B is the pole of $\bigcirc A'C'$ on the same side of $\bigcirc A'C'$ as B', and A is the pole of $\bigcirc B'C'$ on the same side of $\bigcirc B'C'$ as A'. By definition, this proves that $\triangle^s ABC$ is the polar triangle of $\triangle^s A'B'C'$.$\diamondsuit$

The following theorem is the justification for sometimes referring to the polar triangle as the supplemental triangle.

Theorem 11.19 *If $\triangle^s ABC$ is a spherical triangle and $\triangle^s A'B'C'$ is the polar triangle of $\triangle^s ABC$, then*

$$m \prec A' = \pi - m(\overset{\frown}{BC}) \tag{3.2}$$
$$m \prec B' = \pi - m(\overset{\frown}{AC}) \tag{3.3}$$
$$m \prec C' = \pi - m(\overset{\frown}{AB}) \tag{3.4}$$
$$m(\overset{\frown}{A'B'}) = \pi - m \prec C \tag{3.5}$$
$$m(\overset{\frown}{A'C'}) = \pi - m \prec B \tag{3.6}$$
$$m(\overset{\frown}{B'C'}) = \pi - m \prec A \tag{3.7}$$

Proof. The equation (3.6) follows immediately from Proposition 10.13. Equations (3.5) and (3.7) follow by permuting the vertices of the triangle.

To obtain the first three equations, we apply what we have just proven to $\triangle^s A'B'C'$, using the fact now known from Theorem 11.18 that the polar triangle of $\triangle^s A'B'C'$ is $\triangle^s ABC$. This gives us (3.2), (3.3), and (3.4). $\diamondsuit$

The foregoing theorems are valuable in spherical geometry because they show how statements involving sides and angles in one triangle may immediately be changed into statements involving corresponding angles and sides in another triangle (i.e., the polar triangle).

As in the plane, we sometimes consider objects with more than three sides.

Definition 11.20 *A spherical* quadrilateral *is the union of four spherical arcs of the form $\overset{\frown}{AB}$, $\overset{\frown}{BC}$, $\overset{\frown}{CD}$, and $\overset{\frown}{DA}$, where A, B, C, and D are chosen so that no three of the points lie on a great circle, and the four arcs meet only at their endpoints. The quadrilateral is said to be spherically convex if for each of the four arcs of the quadrilateral, the points of the quadrilateral not on that arc lie on only one side of the great circle containing that arc. Each of the points A, B, C, and D is said to be a* vertex *of the quadrilateral.*

That is, the points of the quadrilateral not on $\overset{\frown}{AB}$ lie on the same side of $\bigcirc AB$, the points of the quadrilateral not on $\overset{\frown}{BC}$ lie on the same side of $\bigcirc BC$, and so on. We will refer to $\overset{\frown}{AB}$, $\overset{\frown}{BC}$, $\overset{\frown}{CD}$, and $\overset{\frown}{DA}$ as the "sides" of the quadrilateral, but this use of the word "side" must not be confused with the use of the word "side" used in the definition. No pair of the points A, B, C, and D is antipodal, so the arcs are well-defined (see Exercise 23). We use the notation $\diamond ABCD$ for the quadrilateral. A similar definition may be made for spherical *pentagon*, *hexagon*, or *n-gon*.

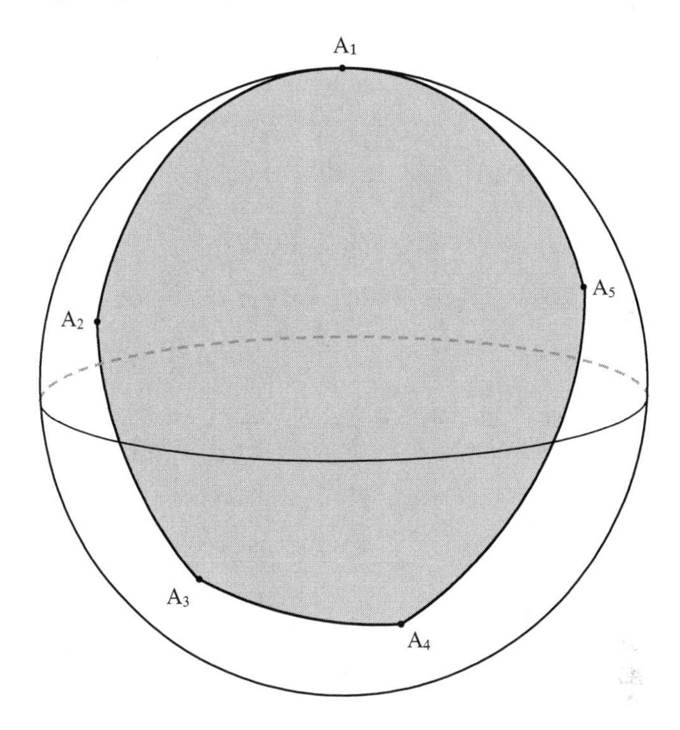

Figure 3.23: A spherically convex polygon $A_1A_2A_3A_4A_5$ and its interior.

Definition 11.21 *A spherical polygon of n sides, or n-gon, is the union of n spherical arcs $\overset{\frown}{A_1A_2}, \overset{\frown}{A_2A_3}, \ldots, \overset{\frown}{A_{n-1}A_n}, \overset{\frown}{A_nA_1}$ which meet only at their endpoints and such that of the n points $A_1, A_2, \ldots, A_n$, no three lie on a great circle. The n-gon is said to be* spherically convex *if for each of the n arcs of the n-gon, the points of the n-gon not on that arc lie on only one side of the great circle containing that arc. The n points are said to be the* vertices *of the n-gon. (See Figure 3.23.)*

A spherically convex n-gon is closely related to the notion of spherically convex set defined earlier. Given a great circle which contains a side of a spherically convex n-gon, one of the sides of that great circle contains the rest of the n-gon. Thus we obtain n hemispheres, one for each side of the n-gon. A hemisphere is a spherically convex set. Thus the intersection of the n hemispheres we obtain this way (one from each side of the n-gon) is a spherically convex set called the *interior* of the n-gon. The union of a spherically convex n-gon and its interior is also spherically convex; this is left to Exercise 24.

Exercises §11

1. Prove Proposition 11.4.

2. Prove Proposition 11.6.

3. Prove the "obtuse" case of Proposition 11.12 from the "acute" case using colunar triangles.

4. Prove the "obtuse" case of Proposition 11.14 from the "acute" case using colunar triangles.

5. Suppose that $\measuredangle ABC$ is acute (respectively, obtuse). Prove that the foot of the shorter (longer) perpendicular from A to $\bigcirc BC$ is a point on $\overset{\hookrightarrow}{BC}$ other than B.

6. Suppose that $\triangle^s ABC$ has a right angle at C. Suppose that the side opposite angle C has measure $\frac{\pi}{2}$. Use Exercise 3 of §2 to prove that $\triangle^s ABC$ must have a right angle at either A or B.

7. Suppose that $\triangle^s ABC$ has a right angle at C. Suppose that the side opposite angle C has measure $\frac{\pi}{2}$. Using only propositions from spherical geometry (as opposed to using the solid geometry of Exercise 6), prove that $\triangle^s ABC$ must have a right angle at either A or B.

8. Suppose that $\triangle^s ABC$ has a right angle at C. Suppose that the side opposite angle C has measure less than $\frac{\pi}{2}$. Using only propositions from spherical geometry (as opposed to using solid geometry), prove that in $\triangle^s ABC$, sides $\overset{\frown}{AC}$ and $\overset{\frown}{BC}$ are both acute or both obtuse.

9. Suppose that $\triangle^s ABC$ has a right angle at C. Suppose that the side opposite angle C has measure greater than $\frac{\pi}{2}$. Using only propositions from spherical geometry (as opposed to using solid geometry), prove that of $\overset{\frown}{AC}$ and $\overset{\frown}{BC}$ one is acute and the other is obtuse.

10. Prove that in any spherical triangle, a cevian is always a well-defined spherical arc (that is, the vertex and a point of the opposite side are not antipodal). Then show that the points of the cevian are in the interior of the triangle, except for the endpoints.

11. Prove that in any spherical triangle, there is a well-defined angle bisector from each vertex which (except for the endpoints) lies in the interior of the triangle.

12. Show that the intersection of the interiors of only two of the angles of a spherical triangle must be the interior of the triangle.

13. Suppose that two points are in the interior of a spherical triangle. Prove that the spherical arc between them lies in the interior of the triangle.

14. Suppose that given a spherical arc, one endpoint is in the interior of a triangle and the other is in the exterior. Prove that the arc contains at least one point of the triangle.

15. In a spherical $\triangle^s ABC$, suppose D is a point of $\overset{\frown}{BC}$ different from B and C. Suppose that $m\,\overset{\frown}{AB} < \frac{\pi}{2}$ and $m\,\overset{\frown}{AC} < \frac{\pi}{2}$. Prove that $m\,\overset{\frown}{AD} < \frac{\pi}{2}$. What happens if the lengths of the two sides given are not less than $\frac{\pi}{2}$?

16. Suppose that a great circle meets a side of a spherical triangle but does not contain any of the vertices. Prove that the great circle contains at least one other point of the triangle not on that side. (This is the spherical analogue of a theorem in the plane known as Pasch's theorem.)

17. Suppose that the sides of a spherical triangle all have measure less than $\frac{\pi}{2}$. Prove that the sides of the polar triangle do not meet any sides of the original triangle.

18. Suppose that the angles of a spherical triangle all are obtuse. Prove that the sides of the polar triangle do not meet any sides of the original triangle.

19. Suppose that the sides of a spherical triangle all have measure greater than $\frac{\pi}{2}$. Prove that the sides of the polar triangle do not meet any sides of the original triangle.

20. Suppose that the angles of a spherical triangle all are acute. Prove that the sides of the polar triangle do not meet any sides of the original triangle.

21. Suppose $\triangle^s ABC$ has a right side $\overset{\frown}{AC}$. Then one of the other sides of $\triangle^s ABC$ is acute, right or obtuse, if and only if its opposite angle is acute, right or obtuse, respectively. (Hint: look at the polar triangle.)

22. Suppose an altitude of a spherical triangle is acute (respectively, obtuse). Prove that it is contained in (respectively, contains) an altitude of the polar triangle and these altitudes are supplementary. If an altitude is right, then it coincides with an altitude of the polar triangle.

23. Prove that in a spherical $\diamond ABCD$, the "sides" $\overset{\frown}{AB}$, $\overset{\frown}{BC}$, $\overset{\frown}{CD}$, and $\overset{\frown}{DA}$ and "diagonals" $\overset{\frown}{AC}, \overset{\frown}{BD}$ are well-defined in the sense that no pair of the points is antipodal. Then prove that the diagonals meet in a single point if and only if the quadrilateral is spherically convex.

24. Prove that the union of a spherically convex n-gon and its interior is also spherically convex.

25. Suppose that $A_1A_2 \ldots A_n$ is a spherically convex n-gon. Let $A_{n+1} = A_1$. For $i = 1, 2, \ldots n$ let A_i' be the pole of $\bigcirc A_iA_{i+1}$ on the same side of $\bigcirc A_iA_{i+1}$ as the other vertices of the n-gon (which are all on the same side by definition of spherically convex n-gon). Prove that $A_1'A_2' \ldots A_n'$ is also a spherically convex n-gon. (We call $A_1'A_2' \ldots A_n'$ the polar n-gon of $A_1A_2 \ldots A_n$.)

26. Prove that if $A_1'A_2' \ldots A_n'$ is the polar n-gon of the spherically convex n-gon $A_1A_2 \ldots A_n$, then $A_1A_2 \ldots A_n$ is the polar n-gon of $A_1'A_2' \ldots A_n'$.

27. Let $A_1A_2 \ldots A_n$ be a spherically convex n-gon and $A_1'A_2' \ldots A_n'$ its polar n-gon. Let $A_{n+1} = A_1$ and $A_{n+1}' = A_1'$, $a_i = m\ \overset{\frown}{A_iA_{i+1}}$, and $a_i' = m\ \overset{\frown}{A_{i-1}'A_i'}$. Prove that for all i, $a_i + m \prec A_i' = \pi$ and $a_i' + m \prec A_i = \pi$.

Historical notes. We let X-Y denote Book X, Proposition Y of Menelaus' *Sphaerica*. Then Theorem 11.10 is I-2.

12 Congruence

In this section we discuss what conditions are needed on sides and angles of two triangles for the two triangles to be congruent. We first state the definition, which is the same as for plane geometry.

Definition 12.1 *Two spherical triangles $\triangle^s ABC$ and $\triangle^s DEF$ are said to be* congruent *if $\overset{\frown}{AB} \cong \overset{\frown}{DE}$, $\overset{\frown}{AC} \cong \overset{\frown}{DF}$, $\overset{\frown}{BC} \cong \overset{\frown}{EF}$, $\prec A \cong \prec D$, $\prec B \cong \prec E$, and $\prec C \cong \prec F$. We say that* corresponding sides and angles *are all congruent.*

The corresponding sides and angles in Definition 12.1 are determined by the specific one-one correspondence of the vertices in each triangle: $A \leftrightarrow D$, $B \leftrightarrow E$, $C \leftrightarrow F$. Thus to say that $\triangle^s ABC \cong \triangle^s DEF$ is different from saying that $\triangle^s ACB \cong \triangle^s DEF$.

The reader should be familiar with the triangle congruence propositions of plane geometry, often abbreviated by SAS, SSS, and ASA. The first of these indicates that if two triangles have two pairs of congruent corresponding sides and the corresponding angles included between those sides are congruent, then the triangles are congruent, i.e., all of the other corresponding sides and angles are congruent. The SSS and ASA propositions are similar. These three propositions will turn out to hold for spherical triangles as well. In order to avoid confusion, we shall sometimes write "planar SSS" to refer to the SSS congruence proposition for planar triangles, and "spherical SSS" to refer to the corresponding proposition for spherical triangles. We will use similar language with SAS and ASA.

In high school plane geometry courses, these congruence propositions are usually assumed as axioms. Logically this is not necessary: one need only assume one of them (typically SAS) and the others can be proven.

Here all our congruence properties for spherical triangles will be established as theorems. The main vehicle for proving them will be Axiom 10.18.

Theorem 12.2 (SSS Congruence) *Suppose that in the spherical triangles $\triangle^s A_1B_1C_1$ and $\triangle^s A_2B_2C_2$, we have:*

$$\overset{\frown}{A_1B_1} \cong \overset{\frown}{A_2B_2}, \overset{\frown}{B_1C_1} \cong \overset{\frown}{B_2C_2}, \overset{\frown}{A_1C_1} \cong \overset{\frown}{A_2C_2}\ .$$

Then $\triangle^s A_1B_1C_1 \cong \triangle^s A_2B_2C_2$.

Proof. We apply Proposition 10.18 to conclude that $\prec A_1B_1C_1 \cong \prec A_2B_2C_2$. By permuting the letters A, B, and C, we can conclude the same for the other corresponding angles. Thus $\triangle^s A_1B_1C_1 \cong \triangle^s A_2B_2C_2$. ♢

Theorem 12.3 (SAS Congruence) *Suppose that in the spherical triangles $\triangle^s A_1B_1C_1$ and $\triangle^s A_2B_2C_2$, we have:*

$$\overset{\frown}{A_1B_1} \cong \overset{\frown}{A_2B_2}, \overset{\frown}{B_1C_1} \cong \overset{\frown}{B_2C_2}, \prec B_1 \cong \prec B_2.$$

Then $\triangle^s A_1B_1C_1 \cong \triangle^s A_2B_2C_2$.

Proof. We apply Proposition 10.18 to conclude that the sides opposite the congruent pair of angles are congruent. But then we have an SSS correspondence between the triangles, and Theorem 12.2 implies that the triangles are congruent. ♢

As with plane geometry, we also have an ASA congruence theorem. However, we present a proof of it which makes use of the magic of polar triangles.

Theorem 12.4 (ASA Congruence) *Suppose that in the spherical triangles $\triangle^s A_1B_1C_1$ and $\triangle^s A_2B_2C_2$, we have:*

$$\overset{\frown}{B_1C_1} \cong \overset{\frown}{B_2C_2}, \prec B_1 \cong \prec B_2, \prec C_1 \cong \prec C_2.$$

Then $\triangle^s A_1B_1C_1 \cong \triangle^s A_2B_2C_2$.

Proof. We let $\triangle^s A_i'B_i'C_i'$ be the polar triangle of $\triangle^s A_iB_iC_i$ for $i = 1, 2$. By Theorem 11.19, $m \prec A_1' = \pi - m\ \overset{\frown}{B_1C_1} = \pi - m\ \overset{\frown}{B_2C_2} = m \prec A_2'$ (so $\prec A_1' \cong \prec A_2'$,) $m\ \overset{\frown}{A_1'B_1'} = \pi - m \prec C_1 = \pi - m \prec C_2 = m\ \overset{\frown}{A_2'B_2'}$ (so $\overset{\frown}{A_1'B_1'} \cong \overset{\frown}{A_2'B_2'}$), and $m\ \overset{\frown}{A_1'C_1'} = \pi - m \prec B_1 = \pi - m \prec B_2 = m\ \overset{\frown}{A_2'C_2'}$ (so $\overset{\frown}{A_1'C_1'} \cong \overset{\frown}{A_2'C_2'}$). Thus we obtain an SAS correspondence between $\triangle^s A_1'B_1'C_1'$ and $\triangle^s A_2'B_2'C_2'$. By the spherical SAS congruence theorem, $\triangle^s A_1'B_1'C_1' \cong \triangle^s A_2'B_2'C_2'$. Thus all corresponding sides and angles of $\triangle^s A_1'B_1'C_1'$ and $\triangle^s A_2'B_2'C_2'$ have the same measure. By using Theorem 11.19 again in the same way in the other direction, we can show that corresponding sides and angles of spherical $\triangle^s A_1B_1C_1$ and $\triangle^s A_2B_2C_2$ are congruent, so $\triangle^s A_1B_1C_1 \cong \triangle^s A_2B_2C_2$, as desired. ♢

In another moment's reflection we realize that by applying the same polar triangle trick to SSS congruence, we obtain a much more surprising theorem, one that does not hold in plane geometry.

Theorem 12.5 (AAA Congruence) *Suppose that in the spherical triangles* $\triangle^s A_1B_1C_1$ *and* $\triangle^s A_2B_2C_2$*, we have:*

$$\prec A_1 \cong \prec A_2, \prec B_1 \cong \prec B_2, \prec C_1 \cong \prec C_2.$$

Then $\triangle^s A_1B_1C_1 \cong \triangle^s A_2B_2C_2$*.*

Proof. We let $\triangle^s A_i'B_i'C_i'$ be the polar triangle of $\triangle^s A_iB_iC_i$ for $i = 1, 2$. By Theorem 11.19, $m\ \overset{\frown}{B_1'C_1'} = \pi - m \prec A_1 = \pi - m \prec A_2 = m\ \overset{\frown}{B_2'C_2'}$, so $\overset{\frown}{B_1'C_1'} \cong \overset{\frown}{B_2'C_2'}$. A similar argument shows that $\overset{\frown}{A_1'B_1'} \cong \overset{\frown}{A_2'B_2'}$ and $\overset{\frown}{A_1'C_1'} \cong \overset{\frown}{A_2'C_2'}$. Thus $\triangle^s A_1'B_1'C_1' \cong \triangle^s A_2'B_2'C_2'$ by the spherical SSS congruence (Theorem 12.2). Thus corresponding angles and sides of triangles $\triangle^s A_i'B_i'C_i'$ are congruent. Using Theorem 11.19 again to go back to the original triangles $\triangle^s A_iB_iC_i$, we conclude similarly that their corresponding sides and angles all have the same measure (so are congruent). By definition, $\triangle^s A_1B_1C_1 \cong \triangle^s A_2B_2C_2$, as desired. $\diamondsuit$

There are two other cases to consider: can we conclude the congruence of spherical triangles in the event of an SSA or SAA congruence of corresponding sides and angles?

In plane geometry, let us recall that an SAA correspondence does guarantee congruence of triangles. One way to see this is that if two pairs of corresponding angles of a triangle are congruent, so is the third pair because in both triangles, the sum of the measures of the angles is π. Then we have an ASA correspondence between the triangles, from which congruence of the triangles follows. This argument will turn out to fail on the sphere because by Theorem 13.7 the sum of the measures of the angles is greater than π, and this sum is not generally the same from one triangle to another.

However, the angle sum theorem in the plane depends on the parallel postulate, and in fact the SAA congruence theorem in the plane does not depend on the parallel postulate. The reader will find a proof of SAA congruence in the plane in [Mo1963] (or [MD1982]) which makes use of the exterior angle theorem in the plane (which states that an exterior angle of a triangle has measure greater than the measures of either of the opposite interior angles). The spherical analogue of this theorem also turns out to be false.

In plane geometry the SSA correspondence does not guarantee congruence of triangles, although we might say that it almost does. Knowledge of the measures of two sides and an angle that is not included leads to the so-called "ambiguous case" in determining the other sides of the triangle. The other sides are uniquely determined if the angle is a right angle, but otherwise there are either two possibilities for the other sides and angles, or the triangle cannot be constructed at all.

We can make an immediate observation for spherical geometry: either SSA and SAA both guarantee congruence of triangles or neither does. The reason is that if one guarantees congruence, then we could prove that the other does as well by employing the same argument with polar triangles used in the proofs of ASA and AAA congruence. For example, suppose there were an SAA congruence theorem. To prove that an SSA correspondence guarantees congruence, we would consider two triangles which have an SSA correspondence. Their polar triangles would have an SAA correspondence by Theorem 11.19, and hence would be congruent. The congruence of the pair of polar triangles would in turn guarantee (by Theorem 11.19 again) that the original triangles be congruent. A similar line of reasoning would allow us to use an SSA congruence theorem to prove an SAA congruence theorem.

It turns out that on a sphere neither SSA nor SAA correspondences guarantee congruence of triangles in general. A single counterexample will dispense with both. Let $A_1 = A_2$ be the pole of any great circle. Choose points B_1, B_2, C_1, and C_2 on the great circle such that $m\,\overset{\frown}{B_1C_1} \neq m\,\overset{\frown}{B_2C_2}$. Then by definition of pole, $\overset{\frown}{A_1B_1}$, $\overset{\frown}{A_1C_1}$, $\overset{\frown}{A_2B_2}$, and $\overset{\frown}{A_2C_2}$ are all quarter circles. Furthermore, $\prec A_1B_1C_1$, $\prec A_1C_1B_1$, $\prec A_2B_2C_2$, and $\prec A_2C_2B_2$ are all right angles by Proposition 10.12. Then triangles $\triangle^s A_1B_1C_1$ and $\triangle^s A_2B_2C_2$ are in both SSA and SAA correspondences, but are not congruent because $m\,\overset{\frown}{B_1C_1} \neq m\,\overset{\frown}{B_2C_2}$.

The counterexample above is somewhat extreme in that if the triangles involved have no more than one right angle, the situation is more like that in the plane. If there is exactly one right angle, we do obtain congruence theorems. If there are no right angles, then there are at most two possibilities for a triangle where two sides and an angle opposite one of them (or two angles and a side opposite one of them) are known.

Theorem 12.6 (SSAA Congruence) *(1)Suppose we have that* $\triangle^s ABC$ *and* $\triangle^s DEF$ *satisfy* $\overset{\frown}{AB} \cong \overset{\frown}{DE}$, $\overset{\frown}{AC} \cong \overset{\frown}{DF}$, $\prec B \cong \prec E$, *and* $\prec C$ *is not supplementary to* $\prec F$. *Then* $\triangle^s ABC \cong \triangle^s DEF$. *(2)Suppose that* $\triangle^s ABC$ *and* $\triangle^s DEF$ *satisfy* $\overset{\frown}{AB} \cong \overset{\frown}{DE}$, $\prec C \cong \prec F$, $\prec B \cong \prec E$, *and* $\overset{\frown}{AC}$ *is not supplementary to* $\overset{\frown}{DF}$. *Then* $\triangle^s ABC \cong \triangle^s DEF$.

Proof. We prove (1) and leave (2) to Exercise 8. If $\overset{\frown}{BC} \cong \overset{\frown}{EF}$ then $\triangle^s ABC \cong$

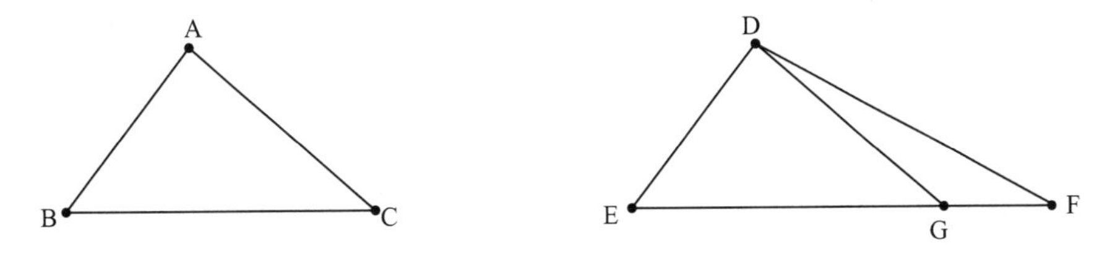

Figure 3.24: Theorem 12.6, part (1).

$\triangle^s DEF$ by SSS congruence. If this is not the case then one of the two arcs $\overset{\frown}{BC}$ and $\overset{\frown}{EF}$ is longer. Assume without loss of generality that $\overset{\frown}{EF}$ is the longer arc. Then there exists a point G on $\overset{\frown}{EF}$ not equal to E or F such that $\overset{\frown}{BC} \cong \overset{\frown}{EG}$.

Then $\triangle^s ABC \cong \triangle^s DEG$ by SAS congruence. Then $\prec DGE$ is congruent to $\prec ACB$ and $\overset{\frown}{AC} \cong \overset{\frown}{DG}$ (as corresponding parts). Since $\overset{\frown}{AC} \cong \overset{\frown}{DF}$ we also have $\overset{\frown}{DG} \cong \overset{\frown}{DF}$. Then in isosceles $\triangle^s DGF$ the angles opposite sides $\overset{\frown}{DG}$ and $\overset{\frown}{DF}$ are congruent, i.e., $\prec DGF \cong \prec DFE$. But $\prec DGF$ is supplementary to $\prec DGE \cong \prec ACB$, so $\prec DFE$ is supplementary to $\prec ACB$, a contradiction of the assumptions. Thus indeed we must have had $\overset{\frown}{BC} \cong \overset{\frown}{EF}$ as above. This concludes the proof of (1).◇

This theorem suggests that if one is given two pair of sides in a spherical triangle and the opposite angles — then the other side and angle can generally be determined. This is simple to do in plane trigonometry (since the sum of the angle measures is known) but not so simple on the sphere. In fact, if the triangles are not isosceles it can be done with Napier's analogies (see §18, Exercise 6 and §19, Exercise 11). If the triangles are isosceles but the known angles are not right angles it can be done using right triangle trigonometry (see §17, Exercise 21).

Definition 12.7 *Two spherical triangles $\triangle^s ABC$ and $\triangle^s DEF$ are said to have a* hypotenuse-leg correspondence *if they have (corresponding) right angles at B and E, the opposite hypotenuses are congruent, and they have a pair of corresponding congruent (non-right) legs.*

Corollary 12.8 (Hypotenuse-leg Theorem) *Two triangles which have a hypotenuse-leg correspondence must be congruent.*

Proof. See Exercise 25. The reader should note that it is important that a leg is not a hypotenuse of the triangle. ◇

Definition 12.9 *Two spherical triangles $\triangle^s ABC$ and $\triangle^s DEF$ are said to have a* hypotenuse-angle correspondence *if they have (corresponding) right angles at B and E, the opposite hypotenuses are congruent, and they have a pair of corresponding congruent* non-right angles.

Corollary 12.10 (Hypotenuse-angle Theorem) *Two spherical triangles which have a hypotenuse-angle correspondence must be congruent.*

Proof. See Exercise 31. The assumption that the second angles are not right angles is important. ◇

Exercises §12

1. Prove the converse of Theorem 11.10: that if two angles in a spherical triangle are congruent, the opposite sides are congruent.

2. Suppose that $\triangle^s ABC$ is in an SAA correspondence with $\triangle^s DEF$ in the sense that $\prec A \cong \prec D$, $\prec B \cong \prec E$, and $\overset{\frown}{BC} \cong \overset{\frown}{EF}$. Suppose that $\triangle^s ABC$ is not congruent to $\triangle^s DEF$. Let A^a be the antipode of A, so that $\triangle^s A^a BC$ is colunar with $\triangle^s ABC$. Prove that there exists an SSA correspondence between $\triangle^s DEF$ and $\triangle^s A^a BC$.

3. Suppose that in spherical $\triangle^s ABC$, $m\ \overset{\frown}{AB} = m\ \overset{\frown}{AC}$ and D is the midpoint of $\overset{\frown}{BC}$. Prove that $\bigcirc AD \perp \bigcirc BC$.

4. Suppose that in spherical $\triangle^s ABC$, $m\ \overset{\frown}{AB} = m\ \overset{\frown}{AC}$ and D is the midpoint of $\overset{\frown}{BC}$. Prove that $\prec BAD \cong \prec CAD$.

5. Suppose that in spherical $\triangle^s ABC$, $m\ \overset{\frown}{AB} = m\ \overset{\frown}{AC} \neq \frac{\pi}{2}$ and D is between B and C such that $\overset{\frown}{AD} \perp \overset{\frown}{BC}$. Prove that D is the midpoint of $\overset{\frown}{BC}$. Is it important that $\overset{\frown}{AB}$ and $\overset{\frown}{AC}$ be not right?

6. Suppose that a median of a triangle is perpendicular to the opposite side. Prove that the triangle is isosceles.

7. Suppose we are given two chords of a small circle. Show that the two chords have the same measure if and only if the perpendiculars from the center to each chord have the same measure. (Note that by Exercise 5, the perpendicular is the arc between the center and the midpoint of the chord.)

8. Prove part (2) of Theorem 12.6.

9. Prove that two sides of a spherical triangle are supplementary if and only if the opposite angles are supplementary.

10. Prove that the set of all points on a sphere which are at the same spherical distance from two given distinct points of the sphere must be a great circle which is perpendicular to a spherical arc between the two given points at its midpoint. (If the two given points are not antipodal, this great circle is called the *perpendicular bisector* of the spherical arc between the two given points.)

11. Prove that if three points do not lie on a single great circle, then there exists a unique small circle passing through them whose center is on each of the perpendicular bisectors of the arcs between the pairs of points. (This circle is called the *circumscribed circle* or *circumcircle* of the triangle of the three points and its center is the *circumcenter*.)

12. Let $\prec ABC$ be an angle. Let $\overset{\hookrightarrow}{r}$ be a spherical ray with vertex B on the same side of $\overset{\hookrightarrow}{BC}$ as A such that $\overset{\hookrightarrow}{r}$ forms an angle with $\overset{\hookrightarrow}{BC}$ of

measure half of $m \prec ABC$. Show that $\overleftrightarrow{r}$ also forms an angle of measure $\frac{1}{2}m \prec ABC$ with $\overleftrightarrow{BA}$. Then $\overleftrightarrow{r}$ is called the *angle bisector* of $\prec ABC$. Show that a point in the interior of a spherical angle is on the angle bisector of the angle if and only if its spherical distances to the sides of the angle are the same. Also, under these circumstances the distance to the sides is less than $\frac{\pi}{2}$.

13. Suppose that an angle bisector of an angle of a triangle is perpendicular to the opposite side. Prove that the triangle is isosceles.

14. Suppose that in $\triangle^s ABC$, $m\ \overset{\frown}{AB} < m\ \overset{\frown}{AC}$ and D is on $\overset{\frown}{BC}$ so that $\overleftrightarrow{AD}$ bisects $\prec BAC$. Prove that $m \prec ADB < m \prec ADC$ (so $\prec ADB$ is acute and $\prec ADC$ is obtuse).

15. Suppose that in $\triangle^s ABC$, D is on $\overset{\frown}{BC}$ so that $\overleftrightarrow{AD}$ bisects $\prec BAC$ or D bisects $\overset{\frown}{BC}$. Prove that if $m\ \overset{\frown}{AD} = \frac{\pi}{2}$ then $m\ \overset{\frown}{AB} + m\ \overset{\frown}{AC} = \pi$.

16. Suppose that in $\triangle^s ABC$, $m\ \overset{\frown}{AB} + m\ \overset{\frown}{AC} = \pi$. Suppose that D is on $\overset{\frown}{BC}$. Prove that $\overleftrightarrow{AD}$ bisects $\prec BAC$ if and only if D is the midpoint of $\overset{\frown}{BC}$ and in either case $m\ \overset{\frown}{AD} = \frac{\pi}{2}$.

17. Suppose that in $\triangle^s ABC$, $m\ \overset{\frown}{AB} \neq m\ \overset{\frown}{AC}$ and D is on $\overset{\frown}{BC}$ so that $\overleftrightarrow{AD}$ bisects $\prec BAC$ and D bisects $\overset{\frown}{BC}$. Prove that $m\ \overset{\frown}{AB} + m\ \overset{\frown}{AC} = \pi$. Can you remove the condition that $m\ \overset{\frown}{AB} \neq m\ \overset{\frown}{AC}$?

18. Suppose that base $\overset{\frown}{BC}$ of a spherical triangle $\triangle^s ABC$ is known, and the sum of the measures of the angles at B and C is known. Prove that regardless of the measures of the angles at B and C, the bisector of the angle at A passes through a fixed point.

19. Prove that in any $\triangle^s ABC$, the three spherical rays which bisect the three angles must intersect in a single point. (This point is called the *incenter* of the triangle.) Prove that the incenter of a triangle is at the same spherical distance $r < \frac{\pi}{2}$ from the great circles containing the sides of the triangle, and the feet of the three shorter perpendiculars from I to these three great circles lie on the sides of the triangle between the endpoints. Conclude that the small circle with spherical radius r and center I is tangent to each of $\bigcirc AB$, $\bigcirc BC$, and $\bigcirc AC$. (The circle is called the *inscribed circle* of $\triangle^s ABC$, its center is the *incenter*, and the number r is the *inradius* of $\triangle^s ABC$.

20. Suppose that $\triangle^s ABC$ has incenter I. Prove that I is the circumcenter of the polar triangle of $\triangle^s ABC$ and the inradius of $\triangle^s ABC$ is complementary to the circumradius of the polar triangle.

21. Suppose that in $\Delta^s ABC$, $m\ \overset{\frown}{AB} \neq m\ \overset{\frown}{AC}$ and $m\ \overset{\frown}{AB} + m\ \overset{\frown}{AC} = \pi$. Suppose D and E are on $\overset{\frown}{BC}$ so that $m\ \overset{\frown}{AD} + m\ \overset{\frown}{AE} = \pi$. Prove that $\overset{\frown}{BD} \cong \overset{\frown}{CE}$ and $\prec BAD \cong \prec CAE$. Can you remove the condition that $m\ \overset{\frown}{AB} \neq m\ \overset{\frown}{AC}$?

22. Following [Ca1889], we define a triangle to be *diametrical* if its vertices are at the same spherical distance from the midpoint of one of its sides. Prove that a triangle is diametrical if and only if the measure of one of its angles equals the sum of the measures of the other two.

23. Prove that in a diametrical triangle, two of its colunar triangles are also diametrical, and in the third the sum of the measures of the angles[5] must be 2π.

24. Let $\overset{\frown}{BC}$ be an arc and let k be a real number. Determine the set of all points A such that A, B, and C form a spherical triangle where $B + C - A = k$. What are the possible values of k? (Hint: see Exercise 22.)

25. Prove Theorem 12.8. Explain (by giving a counterexample) why it is important that the "leg" of the theorem is not a hypotenuse.

26. Suppose that a great circle is tangent to a small circle. Prove that the radius of the small circle to the point of tangency is perpendicular to the great circle.

27. Suppose that two great circles passing through a point A are tangent to a small circle at points B and C of the small circle (see Exercise 26). Prove that $\overset{\frown}{AB} \cong \overset{\frown}{AC}$.

28. Suppose that in a spherical right triangle, the two angles other than the right angle are acute. Prove that the hypotenuse is longer than either of the legs.

29. Suppose that in a spherical right triangle, the hypotenuse is acute and one of the non-right angles is acute. Prove that in this triangle, the measure of that angle is greater than or equal to the measure of the opposite side.

    What happens if the hypotenuse is obtuse? Right?

30. Suppose that an isosceles spherical triangle has legs which are acute. Prove that the measure of the base is less than the measure of the vertex angle. (Hint: see Exercise 29.)

31. Prove Theorem 12.10. Explain (by giving a counterexample) why it is important that the second angle not be a right angle.

---

[5] See [Ca1889].

32. Two spherical triangles $\triangle^s ABC$ and $\triangle^s DEF$ are said to have a *leg-angle correspondence* if they have (corresponding) right angles at B and E, a pair of corresponding legs is congruent, and a pair of corresponding non-right angles is congruent. Are two such triangles always congruent? If not, what conditions might be added to guarantee congruence?

33. Suppose that in a convex spherical quadrilateral the opposite sides are congruent. Prove that the opposite angles are congruent and the diagonals bisect each other.

34. Suppose that in a convex spherical quadrilateral the opposite angles are congruent. Prove that the opposite sides are congruent.

35. Suppose that in a convex spherical quadrilateral all the sides are congruent. Prove that the diagonals are perpendicular and bisect the angles. Do the diagonals have to be congruent?

36. Suppose that in a convex spherical quadrilateral all the angles are congruent. Prove that the diagonals are congruent. Do all the sides have to be congruent?

Historical notes. Most theorems in this section appear in Menelaus' *Sphaerica*, although most proofs are different since *Sphaerica* does not make use of the notion of polar triangle. We let X-Y denote Book X, Proposition Y of *Sphaerica*. Then Exercise 1 is I-3, Theorem 12.10 is I-12, Theorem 12.6 part (1) is I-13, Theorem 12.4 is I-14 and I-15, Theorem 12.6 part (2) is I-17, Theorem 12.5 is I-18, Exercise 16 is I-29 and I-30, and Exercise 21 is I-31.

13 Inequalities

Having considered many situations where objects in spherical geometry are equal or congruent, we now consider situations where inequalities occur. Recall that in the plane the sum of the measures of the angles in a triangle is π radians (180°). This is untrue for spherical triangles, and is probably the most important distinction between triangles in the plane and those on the sphere. We first examine this question in the case of right triangles.

Proposition 13.1 *In a right spherical triangle, the sum of the measures of the angles is larger than π (180°).*

Proof. Suppose without loss of generality that $\triangle^s ABC$ has a right angle at B. If either of the angles at A or C is right or obtuse, we are done. So we may assume that the angles at A and C are both acute. By Proposition 11.12, the opposite sides $\overset{\frown}{AB}$ and $\overset{\frown}{BC}$ are both acute also. Let A' be the pole of $\bigcirc BC$ on the same side of $\bigcirc BC$ as A. Then since $\overset{\frown}{AB}$ is acute, A is between A' and B. Then $\prec A'AC$ is obtuse, since its measure is the supplement of the

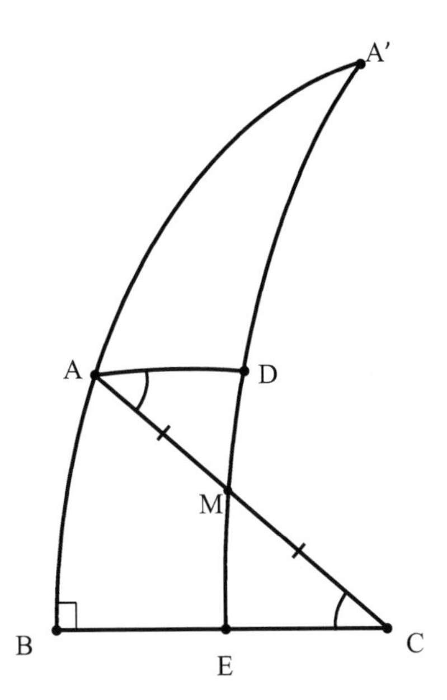

Figure 3.25: Figure for Proposition 13.1.

measure of $\prec A$. Since $\prec C$ is acute, $m \prec C < m \prec A'AC$. Then by Proposition 10.8, there exists a ray $\overleftrightarrow{r}$ emanating from A making an angle with $\overleftrightarrow{AC}$ whose measure is the same as that of $\prec C$, and such that $\overleftrightarrow{r}$ is on the same side of $\bigcirc AC$ as A'. The points of $\overleftrightarrow{r}$ are in the interior of $\prec A'AC$ (except for A). Let M be the midpoint of $\frown{AC}$. By Proposition 10.11, $\overleftrightarrow{r}$ meets $\frown{A'M}$ at a point we call D which lies between A' and M. Since A and A' are on the same side of $\bigcirc BC$, by Proposition 9.44, the points of $\overleftrightarrow{CA}$ (except for C) lie on the same side of $\bigcirc BC$ as A (so on the same side of $\bigcirc BC$ as A'); thus M is on the same side of $\bigcirc BC$ as A', so $\frown{A'M}$ is acute. Since D is between M and A', $\frown{A'D}$ is also acute. Since M is between A and C, M is in the interior of $\prec AA'C$. Thus $\overleftrightarrow{A'M}$ is also in the interior of $\prec AA'C$ (except for A'), and meets $\frown{BC}$ at a point E between B and C. Since $\overleftrightarrow{A'M}$ passes through the pole A' of $\frown{BC}$, it meets $\frown{BC}$ at a right angle. Thus $\prec MEC$ is right. Since $\frown{AM} \cong \frown{MC}$ (definition of M), $\prec AMD \cong \prec CME$ (vertical angles) and $\prec MAD \cong \prec MCE$ (by definition of D), we obtain that $\triangle^s AMD \cong \triangle^s CME$ by spherical ASA congruence. Then since $\triangle^s CME$ has a right angle at E, $\triangle^s AMD$ has a right angle at D. Then $\triangle^s A'DA$ has a right angle at D also. Since $\frown{A'D}$ is acute, by Proposition 11.12 applied to $\triangle^s A'AD$, $\prec A'AD$ is acute. But then $\prec BAD$ is obtuse. Since $\prec BAC$ is acute, $m \prec BAC < m \prec BAD$. Now all of ray

$\overset{\hookrightarrow}{A'M}$ is on the same side of $\bigcirc AB$ except for A', so D and M are on the same side of $\bigcirc AB$. By Proposition 10.9, M is in the interior of $\prec BAD$. Thus $m \prec BAD = m \prec BAC + m \prec CAD = m \prec A + m \prec C$. So the sum of the measures of $\prec A$ and $\prec C$ is greater than $\frac{\pi}{2}$ (90°). Since $\prec B$ is a right angle, the sum of the measures of $\prec A$, $\prec B$ and $\prec C$ is at least π (180°). ♢

Theorem 13.2 *The sum of the measures of the angles in a spherical triangle is greater than π radians (180°).*

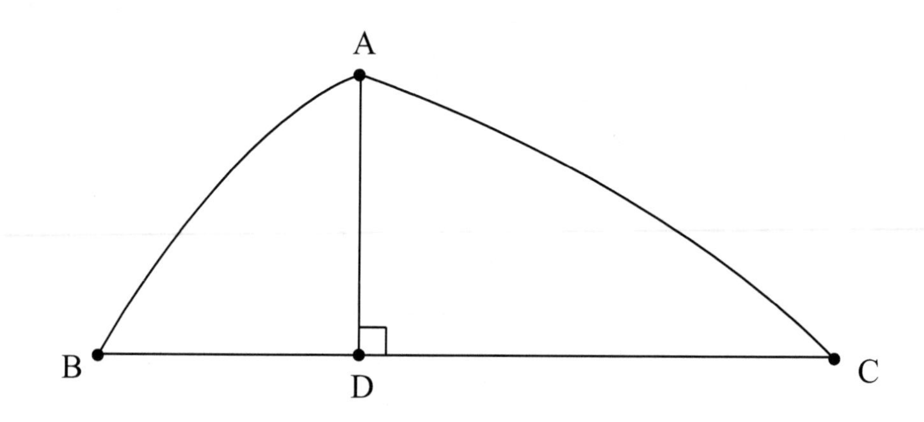

Figure 3.26: Figure for Proposition 13.2.

Proof. If the given triangle is right, we apply Proposition 13.1. If the given triangle has only one acute angle, then the sum of the measures of the other two (both obtuse) must be larger than π, so we are done.

Now suppose the triangle ($\triangle^s ABC$) has two acute angles (say at B and C). Then the foot of the shorter perpendicular from A to $\bigcirc BC$ is a point D between B and C. The sum of the measures of the non-right angles in $\triangle^s ADB$ is greater than $\frac{\pi}{2}$. The same is true in $\triangle^s ADC$. But the sum of the measures of the angles in $\triangle^s ABC$ is $m \prec A + m \prec B + m \prec C = m \prec BAD + m \prec CAD + m \prec B + m \prec C$ since D is between B and C. Then this sum is

$$(m \prec B + m \prec BAD) + (m \prec C + m \prec CAD),$$

which (by Proposition 13.1) is greater than $\frac{\pi}{2} + \frac{\pi}{2} = \pi$, as desired. ♢

Theorem 13.2 has a whole series of consequences that we obtain simply by making use of colunar and polar triangles on the sphere.

Theorem 13.3 *In any spherical triangle, the sum of the measures of the sides is less than 2π.*

Proof. Let the given triangle be $\triangle^s ABC$, and let its polar triangle be $\triangle^s A'B'C'$. By Theorem 13.2, $m \prec A' + m \prec B' + m \prec C' > \pi$. But by Theorem

11.19, $m \prec A' = \pi - m\ \widehat{BC}$, $m \prec B' = \pi - m\ \widehat{AC}$, and $m \prec C' = \pi - m\ \widehat{AB}$. Substituting these into the above inequality,

$$(\pi - m\ \widehat{BC}) + (\pi - m\ \widehat{AB}) + (\pi - m\ \widehat{AC}) > \pi,$$

or $m\ \widehat{AB} + m\ \widehat{AC} + m\ \widehat{BC} < 2\pi$, as desired. ◇

Theorem 13.4 *The sum of the measures of any two sides of a spherical triangle is greater than the measure of the third.*

No doubt this theorem, commonly known as the *triangle inequality* on the sphere, is intuitively obvious to most readers.

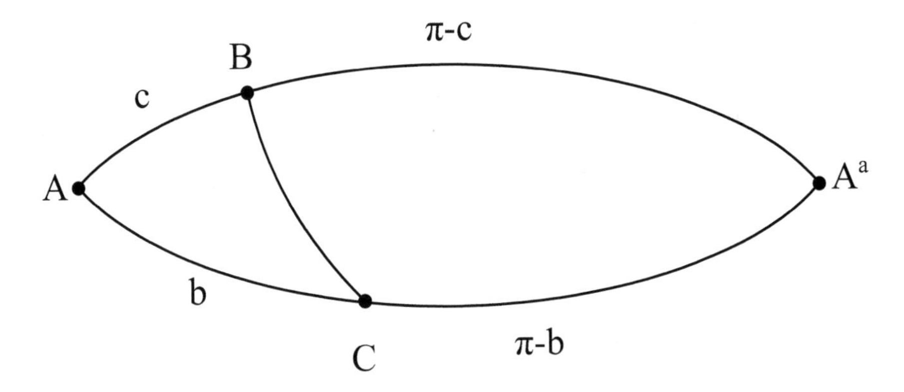

Figure 3.27: Figure for Proposition 13.4.

Proof. Suppose the triangle is $\triangle^s ABC$. Let $\triangle^s A^a BC$ be a triangle colunar with $\triangle^s ABC$. Then $m\ \widehat{A^a B} = \pi - m\ \widehat{AB}$ and $m\ \widehat{A^a C} = \pi - m\ \widehat{AC}$. Applying Theorem 13.3 to $\triangle^s A^a BC$, we obtain $(\pi - m\ \widehat{AB}) + (\pi - m\ \widehat{AC}) + m\ \widehat{BC} < 2\pi$, or $m\ \widehat{BC} < m\ \widehat{AB} + m\ \widehat{AC}$, as desired. The other two inequalities are obtained by permuting the vertices. ◇

Theorem 13.5 *If A, B, and C are three (distinct) points on a sphere, then $d(A, C) \leq d(A, B) + d(B, C)$, where equality occurs if and only if B is between A and C or $A = C^a$.*

Proof. If the points A, B, and C do not lie on a single great circle, they form a spherical triangle and the conclusion is immediate from Theorem 13.4 (where B cannot be between A and C because the three points are not on a great circle, and A cannot be antipodal to C by Proposition 11.1). If the points do lie on a great circle, then we leave the details to Exercise 20. ◇

Corollary 13.6 *Let $A_1, A_2, A_3, \ldots, A_n$ be distinct points on a sphere. Then*

$$d(A_1, A_n) \leq d(A_1, A_2) + d(A_2, A_3) + d(A_3, A_4) + \ldots + d(A_{n-1}, A_n). \quad (3.8)$$

Proof. See Exercise 21. ♢

Corollary 13.6 shows that if a sequence of points is chosen on a sphere, then the sum of the spherical distances traveled from each point to the next is always greater than or equal to the distance from the first point of the sequence to the last point in the sequence. This is a version of the principle stated earlier that the length of the shortest spherical path between two points on a sphere travels the route of a great circle arc of shortest length between the points. Note that such a great circle arc is not necessarily unique: if two points are antipodal, then any of the infinitely many great semicircles between them is a great circle arc of shortest length.

Of course, Corollary 13.6 is not the final word on the subject, since it only considers paths between two points on the sphere which consist of pieces which are great circle arcs. What about other paths?

In order to consider the last question, we need to use the integral and differential calculus — even to define what is meant by the length of a curve. The problem is unusually difficult because a minimum value is being sought for a function (length) which is defined on a set (the set of reasonably smooth curves) which cannot be parametrized with finitely many variables. Thus this minimum value problem does not belong in a typical course in calculus; instead it belongs to the field of *calculus of variations.* The interested reader will find a discussion of this problem in Exercises 54a and 54b.

We now use the notion of polar triangle to prove two more theorems which allow us to find relations among the measures of the angles of a spherical triangle.

Theorem 13.7 *In any spherical* $\triangle^s ABC$*, we have*

$$m \prec A + m \prec B < \pi + m \prec C \tag{3.9}$$

$$m \prec B + m \prec C < \pi + m \prec A \tag{3.10}$$

$$m \prec A + m \prec C < \pi + m \prec B \tag{3.11}$$

$$\pi < m \prec A + m \prec B + m \prec C < 3\pi \tag{3.12}$$

Proof. Let $\triangle^s A'B'C'$ be the polar triangle of $\triangle^s ABC$. By Theorem 13.4, we have $m\ \overset{\frown}{B'C'} + m\ \overset{\frown}{A'C'} > m\ \overset{\frown}{A'B'}$. By Theorem 11.19, we have (3.5), (3.6), and (3.7). Substituting these into the above inequality, we find that $(\pi - m \prec A) + (\pi - m \prec B) > \pi - m \prec C$, which gives us (3.9). We can permute the variables to obtain (3.10) and (3.11).

The first inequality in (3.12) is merely Theorem 13.2. The second inequality in (3.12) results from the fact that no angle can have measure greater than π, so the sum of the measures of three angles is less than 3π. ♢

To summarize Theorem 13.3 and 13.7: *in any spherical triangle, the sum of the measures of the sides is between* 0 *and* 2π *and the sum of the measures of the angles is between* π *and* 3π.

One might wonder whether we can improve on these statements, or whether there exist triangles whose side sums and angle sums approach these extremes. In fact, there are triangles which approach these extremes as closely as we would like, but being precise about their existence is a little tricky. We indicate informally how such extreme triangles might be found. A triangle with all three sides extremely small will clearly have the sum of the measures of its sides close to zero (but still slightly positive). Since such a triangle is close to being planar, the sum of the measures of its angles will be very close to π (but still slightly larger). For the other extremes, choose any great circle on the sphere with pole P, and then choose three points A, B, and C on the same side of that great circle but very close to it, all approximately equally spaced. Then the sides of the triangle will all be very close to the given great circle. As a result, the sum of the measures of the sides will be very close to 2π (the measure of the great circle). Since the great circles determined by the sides are all very close to the given great circle, the angles between the sides will all have measures close to π, hence the sums of the measures of the angles will be very close to 3π. For more details, see Exercises 46-49.

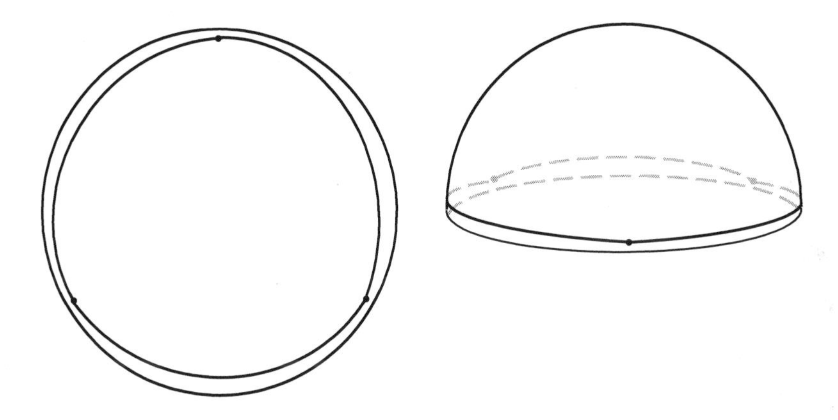

Figure 3.28: A triangle whose angle sum is near 3π and sums of whose sides is near 2π — top and side view.

One might further wonder if it is possible to have other combinations. For example, could one have a triangle, the sum of the measures of whose sides is very close to 2π, but the sum of the measures of whose angles is very close to π? This kind of question is explored in Exercises 50 and 51.

Definition 13.8 *Let $\triangle^s ABC$ be a spherical triangle. Let A^a, B^a, and C^a be the antipodes of A, B, and C, respectively. Then the* exterior angles *at A are the angles $\prec BAC^a$ and $\prec CAB^a$. The* opposite interior *angles (or* remote interior *angles) to $\prec BAC^a$ and $\prec CAB^a$ are the angles $\prec ABC$ and $\prec ACB$.*

Theorem 13.9 (Spherical Exterior Angle Theorem) *The measure of any exterior angle of a spherical triangle is less than the sum of the measures of the opposite interior angles and greater than the (absolute value of the) difference between those measures.*

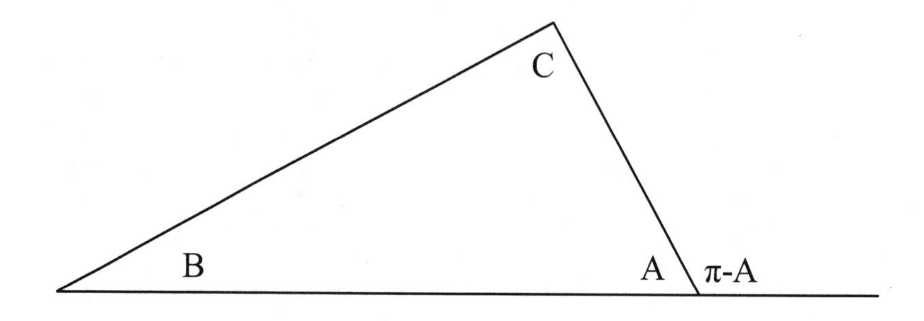

Figure 3.29: Figure for Theorem 13.9: $|B - C| < \pi - A < B + C$.

Proof. We let the triangle be $\triangle^s ABC$ and consider the exterior angle at A. By Theorem 13.2, $\pi < m \prec A + m \prec B + m \prec C$, so $\pi - m \prec A < m \prec B + m \prec C$. Since the exterior angle at A has measure $\pi - m \prec A$, this shows that the exterior angle at A has measure less than the sum of the measures of the angles $\prec B$ and $\prec C$. From equations (3.9) and (3.11) we find $\pi - m \prec A > m \prec B - m \prec C$ and $\pi - m \prec A > m \prec C - m \prec B$, respectively, so $\pi - m \prec A$ is greater than $|m \prec B - m \prec C|$, the absolute value of the difference between the measures of angles B and C. This is what we wanted. Inequalities for angles exterior at B and C are similar. ◊

Theorem 13.10 *Given two sides of a triangle whose measures are unequal and the angles opposite them, then the angles are unequal in measure and the larger angle is opposite the longer side. Similarly, given two angles of a triangle whose measures are unequal, the opposite sides have unequal measures and the longer side is opposite the larger angle.*

Proof. Suppose the triangle is $\triangle^s ABC$, and $m \overset{\frown}{AB} > m \overset{\frown}{BC}$. Choose point D on $\overset{\frown}{AB}$ such that $m \overset{\frown}{BD} = m \overset{\frown}{BC}$. By Theorem 11.10, $m \prec BDC = m \prec BCD$. Now $\prec BDC$ is exterior to $\triangle^s ADC$, so by Theorem 13.9,

$$m \prec BDC > |m \prec BAC - m \prec DCA| \geq m \prec BAC - m \prec DCA.$$

Then $m \prec BDC + m \prec DCA > m \prec BAC$. Since $m \prec BDC = m \prec BCD$, $m \prec BCD + m \prec DCA > m \prec BAC$. By Proposition 10.9, $m \prec BCD + m \prec DCA = m \prec BCA$, so $m \prec BCA > m \prec BAC$, as desired. The second sentence of the proof follows from the first and Theorem 11.10: if the opposite sides were equal in measure, the angles would also be (by Theorem 11.10).

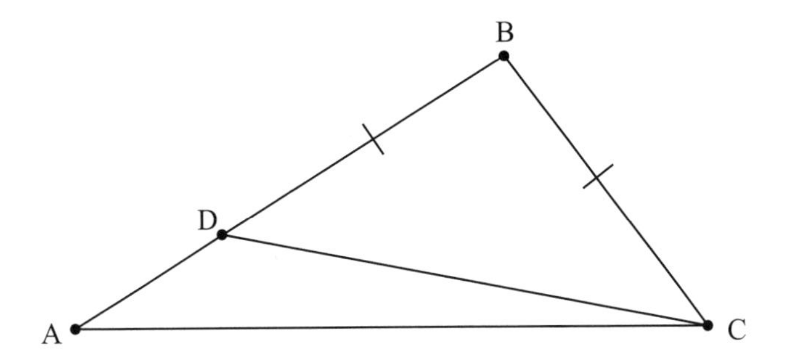

Figure 3.30: Figure for Theorem 13.10.

Thus the opposite sides are different in measure, and by the first sentence, the larger angle is opposite the longer side. ◇

Another approach to Theorem 13.10, using the triangle inequality, can be found in Exercise 3.

Proposition 13.11 *Suppose that $\triangle^s ABC$ has a right angle at C. Then side $\overset{\frown}{BC}$ is shorter than, congruent to, or longer than hypotenuse $\overset{\frown}{AB}$ if and only if $\overset{\frown}{BC}$ is acute, right, or obtuse, respectively. Furthermore, if $\overset{\frown}{BC}$ is not right, $m\ \overset{\frown}{AB}$ is between $m\ \overset{\frown}{BC}$ and $\pi - m\ \overset{\frown}{BC}$.*

Proof. We know from Proposition 11.12 (see Figure 3.19) that $\overset{\frown}{BC}$ is acute, right or obtuse if and only if $\prec BAC$ is acute, right or obtuse, respectively. By Theorem 13.10, Theorem 11.10 and §12, Exercise 1, this occurs if and only if the side opposite $\prec BAC$ is shorter than, congruent to, or longer than the side opposite $\prec ACB$, respectively — that is, if $\overset{\frown}{BC}$ is shorter than, congruent to, or longer than the hypotenuse $\overset{\frown}{AB}$.

Consider $\triangle^s A^a BC$. If $\overset{\frown}{BC}$ is acute, $\prec BAC$ is acute, so $\prec BA^aC$ is acute. So $m\ \overset{\frown}{A^aB} > m\ \overset{\frown}{BC}$, so $\pi - m\ \overset{\frown}{AB} > m\ \overset{\frown}{BC}$ and $m\ \overset{\frown}{AB} < \pi - m\ \overset{\frown}{BC}$. Since we already have $m\ \overset{\frown}{AB} > m\ \overset{\frown}{BC}$ from above, we conclude $m\ \overset{\frown}{BC} < m\ \overset{\frown}{AB} < \pi - m\ \overset{\frown}{BC}$. The case where $\overset{\frown}{BC}$ is obtuse is similar. ◇

It is worth making some observations about the order of logic in the proofs of the theorems in this section. The proof of Theorem 13.2 took considerable effort. Once this was done, the proofs of Theorems 13.3, 13.4, 13.7, 13.9, and 13.10 (as well as Exercise 9) follow rather quickly through the careful use of colunar and polar triangles. In fact, once we prove any of these, the others all follow fairly quickly in a similar manner. The real difficulty seems to be in proving one of them. In Exercise 27 the reader is encouraged to find a way to prove Theorem 13.3 as the first in the group (from which all the others would follow).

We conclude with the Hinge theorem on the sphere.

Proposition 13.12 (Spherical Hinge Theorem) *Suppose that $\triangle^s ABC$ and $\triangle^s DEF$ are spherical triangles with $\overset{\frown}{AB} \cong \overset{\frown}{DE}$ and $\overset{\frown}{BC} \cong \overset{\frown}{EF}$. Then $m \measuredangle ABC < m \measuredangle DEF$ if and only if $m\ \overset{\frown}{AC} < m\ \overset{\frown}{DF}$.*

Theorem 13.12 may be proven using the same method as used to prove the plane version, and we refer the reader to Exercise 28 for the argument.

Exercises §13

1. Prove that if spherical $\triangle^s ABC$ has a right angle at C, then $m \measuredangle A$ and $m \measuredangle B$ differ by no more than $\frac{\pi}{2}$ and their sum lies between $\frac{\pi}{2}$ and $\frac{3\pi}{2}$.

2. Suppose that in $\triangle^s ABC$, $m\ \overset{\frown}{AB} < m\ \overset{\frown}{AC}$ and let D be the midpoint of $\overset{\frown}{BC}$. Prove that $m \measuredangle ADB < m \measuredangle ADC$.

3. Suppose that in a spherical triangle one angle is larger than another. By constructing inside the given triangle another triangle with two angles equal, prove that, of the two opposite sides, the larger angle is opposite the larger side.

4. Suppose that in $\triangle^s ABC$, D is chosen in the interior of the triangle. Prove that $m\ \overset{\frown}{DB} + m\ \overset{\frown}{DC} < m\ \overset{\frown}{AB} + m\ \overset{\frown}{AC}$.

5. Suppose that D is chosen in the interior of $\triangle^s ABC$ so that $m \measuredangle BAC + m \measuredangle BDC < \pi$. Prove that $m\ \overset{\frown}{AC} > m\ \overset{\frown}{DC}$ and $m\ \overset{\frown}{AB} > m\ \overset{\frown}{DB}$.

6. Suppose that Γ is a great circle and X is not one of its poles. Let Y and Z be the feet of the shorter and longer perpendiculars, respectively, from X to Γ (see Definition 10.16). Using Proposition 13.11, prove that if W is any point of Γ other than Y and Z, $m\ \overset{\frown}{XY} < m\ \overset{\frown}{XW} < m\ \overset{\frown}{XZ}$. (Another approach to this problem is found in §17, Exercise 22.)

7. Let X, Y, Z and Γ be as in Exercise 6. If U and V are points of Γ other than Y and Z then: U is closer to Y than V if and only if $m\ \overset{\frown}{XU} < m\ \overset{\frown}{XV}$.

8. Suppose that a great circle meets a small circle at a point where the great circle is perpendicular to the radius of the small circle to the point of intersection. Prove that the great and small circles are tangent at the point of intersection. (That is, they meet at only one point.)

9. Prove that in any spherical triangle, the sum of the measures of two sides of a spherical triangle is (less than, equal to, greater than, respectively) π if and only if the sum of the measures of the opposite angles is (less than, equal to, greater than, respectively) π. Conclude that the measure of the exterior angle at one vertex of a triangle is greater than, equal to, or less than, the measure of the (interior) angle at a second vertex if the sum of the measure of the two sides not between the two given vertices

is, respectively, less than, equal to, or greater than π. (This is Book I, Proposition 10 of Menelaus' *Sphaerica*; it shows to what extent the planar version of the weak exterior angle theorem holds on the sphere.)

10. Let X, Y, Z, and Γ be as in Exercise 6. Let P be a pole of $\bigcirc XY$. If U and V are two points on Γ such that $m\,\overset{\frown}{YU} < m\,\overset{\frown}{YV} \leq m\,\overset{\frown}{YP} = \frac{\pi}{2}$, prove that $m \prec XUY > m \prec XVY \geq m \prec XPY = m\,\overset{\frown}{XY}$. Conclude that for W in Γ, $m \prec XWY$ is least when W is a pole of $\bigcirc XY$.

11. Suppose that in $\triangle^s ABC$, $m\,\overset{\frown}{AB} + m\,\overset{\frown}{AC} < \pi$ and $m\,\overset{\frown}{AB} \leq m\,\overset{\frown}{AC}$. Let D be a point between B and C. Prove that $m\,\overset{\frown}{AD} < m\,\overset{\frown}{AC}$.

12. Use Exercise 11 to conclude that the interior of a small circle is spherically convex. (Recall that a small circle by definition has radius less than $\frac{\pi}{2}$.) In fact, given two points on the small circle itself, or a point on the small circle and a point in the interior, then the points between them lie in the interior of the small circle.

13. Suppose that in $\triangle^s ABC$, $m\,\overset{\frown}{AB} + m\,\overset{\frown}{AC} < \pi$. Suppose D is between B and C and E is chosen on the same side of $\bigcirc BC$ as A so that $m \prec EDC = m \prec ABC$. If $m\,\overset{\frown}{AB} \leq \frac{\pi}{2}$, prove that $\overset{\hookrightarrow}{DE}$ meets $\overset{\frown}{AC}$ between A and C.

14. Suppose that in $\triangle^s ABC$, $a < \frac{\pi}{2}$, $b < \frac{\pi}{2}$, $a + c < \pi$ and $b + c < \pi$. Let D be a point in the interior of $\triangle^s ABC$. Prove that $m \prec ADB > m \prec ACB$. (See §16, Exercise 28 to see a partial converse of this statement.)

15. Suppose that in $\triangle^s ABC$, D lies on $\overset{\frown}{BC}$ so that $\overset{\hookrightarrow}{AD}$ bisects $\prec A$ (that is, $\prec BAD \cong \prec CAD$). Also assume $m\,\overset{\frown}{AB} + m\,\overset{\frown}{AC} < \pi$. If $m\,\overset{\frown}{AB} < m\,\overset{\frown}{AC}$, prove that $m\,\overset{\frown}{DB} < m\,\overset{\frown}{DC}$. What happens if $m\,\overset{\frown}{AB} + m\,\overset{\frown}{AC} \geq \pi$?

16. Suppose that in $\triangle^s ABC$, $m\,\overset{\frown}{AB} + m\,\overset{\frown}{AC} < \pi$. Let M be the midpoint of $\overset{\frown}{BC}$ and suppose $m\,\overset{\frown}{AB} < m\,\overset{\frown}{AC}$. Using Exercise 15, show that $m \prec BAM > m \prec CAM$.

17. Suppose that in $\triangle^s ABC$, $m\,\overset{\frown}{AB} < m\,\overset{\frown}{AC} < \frac{\pi}{2}$. Assume that the foot of the shorter perpendicular from A to $\bigcirc BC$ is not on $\overset{\frown}{BC}$. Let D be between B and C. Assume $m \prec BAD / m \prec CAD$ is a rational number. Prove that $m \prec BAD / m \prec CAD > m\,\overset{\frown}{BD} / m\,\overset{\frown}{CD}$.

18. Suppose that in $\triangle^s ABC$, D lies on $\overset{\frown}{BC}$ so that $\overset{\hookrightarrow}{AD}$ bisects $\prec A$ (that is, $\prec BAD \cong \prec CAD$). Also assume $m\,\overset{\frown}{AB} + m\,\overset{\frown}{AC} < \pi$. Let F be a point between A and D. If $m\,\overset{\frown}{AB} < m\,\overset{\frown}{AC}$, prove that $m\,\overset{\frown}{FB} < m\,\overset{\frown}{FC}$, $m \prec FCA < m \prec FBA$ and $m \prec FCB < m \prec FBC$.

19. Suppose that in $\Delta^s ABC$, G is the midpoint of $\overset{\frown}{BC}$. Also assume $m\ \overset{\frown}{AB} + m\ \overset{\frown}{AC} < \pi$. Let H be a point between A and G. If $m\ \overset{\frown}{AB} < m\ \overset{\frown}{AC}$, prove that $m\ \overset{\frown}{HB} < m\ \overset{\frown}{HC}$, $m \prec HCA < m \prec HBA$ and $m \prec HCB < m \prec HBC$.

20. Prove Theorem 13.5.

21. Prove Corollary 13.6.

22. Prove that in any spherical right triangle, if two sides are acute then so is the third.

23. Let $\Delta^s A_iB_iC_i$ be two right triangles with right angles at C_i for $i = 1, 2$. Let the lengths of the sides be a_i, b_i and c_i. If $m \prec A_1B_1C_1 = m \prec A_2B_2C_2 < \frac{\pi}{2}$, and $c_1 < c_2$ prove that $a_1 < a_2$. If also $c_2 < \frac{\pi}{2}$, prove that $b_1 < b_2$.

24. Let $\Delta^s A_iB_iC_i$ be two right triangles with right angles at C_i for $i = 1, 2$. Let the lengths of the sides be a_i, b_i and c_i. Suppose that $c_1 = c_2$ and $m \prec A_1B_1C_1 < m \prec A_2B_2C_2 \leq \frac{\pi}{2}$. Prove that $b_1 < b_2$. If also $c_1 < \frac{\pi}{2}$ prove that $a_1 > a_2$.

25. Suppose that two sides of a spherical triangle are acute and the angle between them is not acute. Prove that the other two angles are acute.

26. In $\Delta^s ABC$, suppose that $C \geq \frac{\pi}{2}$, $a < \frac{\pi}{2}$ and $c < \frac{\pi}{2}$. Prove that $A < \frac{\pi}{2}$, $B < \frac{\pi}{2}$ and $b < \frac{\pi}{2}$.

27. Prove Theorem 13.3 as follows. Given an arbitrary spherical triangle $\Delta^s ABC$, let P be the center of its circumcircle. (See §12, Exercise 11.) Consider the three arcs $\overset{\frown}{PA}$, $\overset{\frown}{PB}$, and $\overset{\frown}{PC}$ and apply §12, Exercise 30.

28. Prove the Spherical Hinge Theorem as follows. First, we may arrange without loss of generality that $B = E$, $C = F$ and A is in the interior of $\prec DEF$. Construct the angle bisector of $\prec DEA$; it must meet $\overset{\frown}{DF}$ at some point G. Show that $\Delta^s DEG \cong \Delta^s AEG$, and then $m\ \overset{\frown}{AF} < m\ \overset{\frown}{FD}$.

29. Consider the following proposition: In $\Delta^s ABC$ and $\Delta^s DEF$, suppose that $\prec B \cong \prec E$ and $\prec C \cong \prec F$. Then $m\ \overset{\frown}{BC} < m\ \overset{\frown}{EF}$ if and only if $m \prec A < m \prec D$. Explain how this proposition is equivalent to the Hinge theorem on the sphere.

30. In $\Delta^s ABC$ and $\Delta^s DEF$, suppose that $\prec B \cong \prec E$, $\prec C \cong \prec F$, and $m \prec B < m \prec C < \pi - m \prec B$. If $m\ \overset{\frown}{BC} < m\ \overset{\frown}{EF}$ prove that $m\ \overset{\frown}{AB} < m\ \overset{\frown}{DE}$.

31. In $\triangle^s ABC$ and $\triangle^s DEF$, suppose that $\prec B \cong \prec E$, $\prec C \cong \prec F$, and $m\ \overset{\frown}{BC} < m\ \overset{\frown}{EF}$. Then prove that $m\ \overset{\frown}{AC} + m\ \overset{\frown}{DF}$ is equal to, less than, or greater than π, respectively, if $m\ \overset{\frown}{AB}$ is equal to, less than, or greater than $m\ \overset{\frown}{DE}$, respectively.

32. Given $\triangle^s ABC$ and $\triangle^s DEF$ so that $\overset{\frown}{AC} \cong \overset{\frown}{DF}$, $m \prec A > m \prec D$, $m \prec C < m \prec F$, and $m \prec B + m \prec E \geq \pi$. Prove that $m\ \overset{\frown}{BC} > m\ \overset{\frown}{EF}$ and $m\ \overset{\frown}{DE} > m\ \overset{\frown}{AB}$.

33. Given $\triangle^s ABC$ and $\triangle^s DEF$ so that $m\ \overset{\frown}{AB} < m\ \overset{\frown}{DE}$, $m\ \overset{\frown}{AC} > m\ \overset{\frown}{DF}$, $\prec A \cong \prec D$ and $m\ \overset{\frown}{BC} + m\ \overset{\frown}{EF} \leq \pi$. Prove that $m \prec B > m \prec E$ and $m \prec C < m \prec F$.

34. By §12 Exercise 11, a spherical triangle has a unique small circle passing through its three vertices called the *circumscribed circle* or *circumcircle.* Given a $\triangle^s ABC$ with circumcircle of center P, prove that (1) P is on the same (respectively, opposite) side of $\bigcirc BC$ as A if and only if (2) $m \prec A < m \prec B + m \prec C$ (respectively, $m \prec A > m \prec B + m \prec C$) if and only if (3) the median from A to $\overset{\frown}{BC}$ has measure greater (respectively, less) than $\frac{1}{2} m\ \overset{\frown}{BC}$. Conclude that the circumcenter is in the interior of the triangle if and only if the measure of any angle of the triangle is less than the sum of the measures of the other two angles. (Compare this problem to §12, Exercise 22.)

35. Prove that in a spherical triangle, the arc connecting the midpoints of two sides (a *midline*) has measure greater than half the measure of the third side of the triangle. (Hint. Suppose that in $\triangle^s ABC$, D, E are the midpoints of $\overset{\frown}{AB}, \overset{\frown}{AC}$, respectively. Extend $\overset{\frown}{DE}$ to F so $\overset{\frown}{DF}$ has twice the measure of $\overset{\frown}{DE}$. Then you must show $m\ \overset{\frown}{BC} < m\ \overset{\frown}{DF}$. Do this by applying the Spherical Hinge Theorem to $\triangle^s BDC$ and $\triangle^s FCD$.)

36. Suppose that in a spherical triangle the sum of the measures of two sides is less than π. Consider an arc from their common vertex to the opposite side. If this arc either bisects the angle or the opposite side, prove that the arc has measure less than $\frac{\pi}{2}$.

37. Extend Exercise 36 to prove that half the sum of the measures of the two given sides is greater than the measure of the arc constructed from the vertex to the opposite side.

38. Assume that Exercise 37 is true. Suppose then that in a triangle where the sum of the measures of two sides is less than π, the sides are unequal. We construct a segment from their common vertex to the opposite side whose measure is half the sum of the measures of those two sides. Prove

that both the vertex angle and base are divided into two unequal parts such that the larger part is adjacent to the shorter side.

39. Suppose that in $\triangle^s ABC$, $m\ \overset{\frown}{AB} < m\ \overset{\frown}{AC}$, $m\ \overset{\frown}{AB} + m\ \overset{\frown}{AC} < \pi$, and M is the midpoint of $\overset{\frown}{BC}$. Suppose that D and E are chosen on $\overset{\frown}{BM}$ and $\overset{\frown}{MC}$, respectively, so that $\overset{\frown}{BD} \cong \overset{\frown}{EC}$. Prove that $m \prec BAD > m \prec EAC$ and $m\ \overset{\frown}{AD} + m\ \overset{\frown}{AE} < m\ \overset{\frown}{AB} + m\ \overset{\frown}{AC}$. Hint. Exercise 36 shows that $m\ \overset{\frown}{AM} < \frac{\pi}{2}$. Choose H on $\overrightarrow{AM}$ so that $m\ \overset{\frown}{AM} = m\ \overset{\frown}{MH}$. Consider $\triangle^s ABH$ and $\triangle^s ADH$.

40. Suppose that in $\triangle^s ABC$, $m\ \overset{\frown}{AB} < m\ \overset{\frown}{AC}$, $m\ \overset{\frown}{AB} + m\ \overset{\frown}{AC} < \pi$, and F is on $\overset{\frown}{BC}$ such that $\overrightarrow{AF}$ bisects $\prec BAC$. Suppose that D and E are chosen on $\overset{\frown}{BF}$ and $\overset{\frown}{FC}$, respectively, so that $\prec BAD \cong \prec EAC$. Prove that $m\ \overset{\frown}{BD} < m\ \overset{\frown}{EC}$ and $m\ \overset{\frown}{AD} + m\ \overset{\frown}{AE} < \pi$.

41. Suppose that a lune has vertices A and A^a, and that B is a point such that $\overrightarrow{AB}$ bisects the angle at A. Suppose that points C_i and D_i, $i = 1, 2$, are on the lune such that B is between C_i and D_i for $i = 1, 2$ and that $m \prec ABC_1 < m \prec ABC_2 \leq \frac{\pi}{2}$. Prove that $m\ \overset{\frown}{C_1D_1} > m\ \overset{\frown}{C_2D_2}$, and conclude that among all arcs with endpoints on the lune passing through B, the shortest is perpendicular to $\overrightarrow{AB}$.

42. Suppose that in $\triangle^s ABC$, $m\ \overset{\frown}{AB} + m\ \overset{\frown}{AC} < \pi$, M is the midpoint of $\overset{\frown}{BC}$ and that D is between A and M. Prove that $m \prec BDC > m \prec BAC$.

43. Suppose that in $\triangle^s ABC$, $m\ \overset{\frown}{AB} + m\ \overset{\frown}{AC} < \pi$ and M is the midpoint of $\overset{\frown}{BC}$. Choose D and E on $\overset{\frown}{BM}$ and $\overset{\frown}{MC}$ so that $\overset{\frown}{BD} \cong \overset{\frown}{EC}$. Prove that $m \prec ADM + m \prec AEM > m \prec ABC + m \prec ACB$.

44. Suppose that in $\triangle^s ABC$, D and E are the midpoints of $\overset{\frown}{AB}$ and $\overset{\frown}{AC}$, respectively. Suppose that $\prec A$ is not acute. Prove that $m \prec ADE < m \prec ABC$ and $m \prec AED < m \prec ACB$.

45. Suppose that in $\triangle^s ABC$, D, E, and F are the midpoints of $\overset{\frown}{AB}, \overset{\frown}{BC}$, and $\overset{\frown}{AC}$, respectively. Suppose that $\prec A$ is not acute. Prove that both $m \prec BDE$ and $m \prec EFC$ are smaller than $m \prec A$.

46. Let δ be a real number between 0 and $\frac{\pi}{2}$. Suppose that a great circle has pole P and points A, B, and C are chosen at (the same) distance greater than $\frac{\pi}{2} - \delta$ from P. Suppose that the measures of angles $\prec APB$, $\prec BPC$, and $\prec APC$ are all $\frac{2\pi}{3}$. Prove that the perimeter of the spherical triangle $\triangle^s ABC$ is greater than or equal to $2\pi - 6\delta$.

47. Let δ, P, A, B, and C be as in Exercise 46. Show that if A, B, and C are far enough from P (i.e., close enough to the polar of P) then $m \prec A + m \prec B + m \prec C > 3\pi - 12\delta$. (Hint: let D and E be the points on $\overset{\hookrightarrow}{PA}$ and $\overset{\hookrightarrow}{PB}$ at distance a quarter circle from P. Let B be chosen so that $m \prec BDE < \delta$.)

48. Using Exercises 46 and 47 obtain a triangle $\triangle^s ABC$ such that both $m\ \widehat{AB} + m\ \widehat{AC} + m\ \widehat{BC} \geq 2\pi - 6\delta$ and $m \prec A + m \prec B + m \prec C > 3\pi - 12\delta$.

49. Using the triangle from Exercise 48, obtain a triangle $\triangle^s DEF$ whose spherical perimeter is less than 12δ and the sum of whose angles is less than or equal to $\pi + 6\delta$.

50. Is it possible for a spherical triangle to have the sum of its side measures close to 2π (360°) but have its angle sum close to π (180°)? If so, where could one place the vertices to produce such a triangle?

51. Is it possible for a spherical triangle to have the sum of its side measures close to 0 (0°) but have its angle sum close to 3π (540°)? If so, where could one place the vertices to produce such a triangle?

52. Let $A_1A_2 \ldots A_n$ be a spherically convex n-gon. Prove that the sum of the measures of the angles of the n-gon is greater than $(n-2)\pi$.

53. Prove that the sum of the measures of the sides of a spherically convex n-gon is less than 2π. (Hint. Use the polar n-gon.)

54. (For readers familiar with calculus)

    (a) Assume that the length of a smooth curve $t \mapsto (x(t), y(t), z(t))$ in space is given by $\int_a^b \sqrt{x'(t)^2 + y'(t)^2 + z'(t)^2}\, dt$. For a curve on a sphere of radius r, show that in spherical coordinates (ϕ, θ) this formula is given by $\int_a^b r\sqrt{\phi'(t)^2 + \sin^2(\phi)\theta'(t)^2}\, dt$. (See Theorem 2.3.)

    (b) Assume that a curve $t \mapsto (r, \phi(t), \theta(t))$ for $a \leq t \leq b$ satisfies $\theta(b) = \theta(a)$ (i.e., the initial and terminal point of the curve lie on the same meridian in spherical coordinates). Show that the length of this curve is greater than or equal to the shortest great circle arc from the initial point to the terminal point.

55. (a) Prove the following theorem from solid geometry: (Euclid's Elements, Book XI, Proposition 20) Given four points O, A, B, C in space which do not all lie in a single plane, $m\angle AOB < m\angle AOC + m\angle BOC$.

    (b) Use part (a) to come up with another proof of Theorem 13.4.

56. (a) Use Exercise 55 and solid geometry to prove the following theorem: (Euclid's Elements, Book XI, Proposition 21) If O, A, B, C are four points in space which do not all lie in a single plane, then $m\angle AOB + m\angle AOC + m\angle BOC < 2\pi$.

    (b) Use part (a) to conclude another proof of Theorem 13.3.

Historical notes. The theorems and exercises of this section are featured extensively in the *Sphaerica* of Menelaus. Let X-Y denote Book X, Proposition Y. Then Theorem 13.4 is I-5, Exercise 4 is I-6, Theorem 13.10 is I-7 and I-9, Proposition 13.12 is I-8, Exercise 9 is I-10, Theorem 13.9 contains I-11, Exercises 29 and 31 are I-19, Exercise 32 is I-22, Exercise 34 contains I-23, Exercise 25 is I-24, Exercise 26 is I-25, Exercise 35 is I-26, Exercise 44 is I-27, Exercise 45 is I-28, Exercise 36 is I-32, Exercise 15 is I-33, Exercise 37 is I-34, Exercise 38 is I-35, Exercise 19 is I-36, Exercise 39 is I-37, Exercise 40 is I-38, Exercise 13 is II-3, and Exercise 11 is used in the proof of II-3 without justification.

14 Area

In §7 we justified the well-known classical formula for the surface area of a sphere and lune of radius r in space. Having done so we now accept these formulas as axioms of our axiomatic system. The main objective of this section is to obtain formulas for the areas enclosed by spherical triangles and polygons on the sphere.

Axiom 14.1 (A-9) *There exists an $r > 0$ such that the area of a sphere is $4\pi r^2$, the area of the region enclosed by a lune whose angle has radian measure θ is $2\theta r^2$, and the area of a spherical arc is 0.*

This may seem like a dramatic proposition to assume as an axiom, but because our system of axioms for spherical geometry does not presuppose a particular radius for the sphere, the exact value is not so important. By assuming that the area of a sphere is $4\pi r^2$ we are establishing the sphere's area as a unit by which all the others will be compared — in the same way that the area of a square is taken as a unit in the plane.

Axiom 14.2 (A-10) *If two triangles are antipodal, their interiors have the same area.*

Axiom 14.3 (A-11) *If two spherical regions have interiors which do not overlap then the area of the union of the regions is the sum of the areas of the regions.*

Because a spherical arc has zero area, the area of a triangle itself is zero, but the area of its interior is positive. Thus the area of the interior of a triangle is equal to the area of the union of the triangle with its interior. We can speak of the "area enclosed by a triangle" to refer to either of these, but it is also common to abuse language and refer to the "area of a triangle" when it is clear that we mean the area of the region enclosed by the triangle.

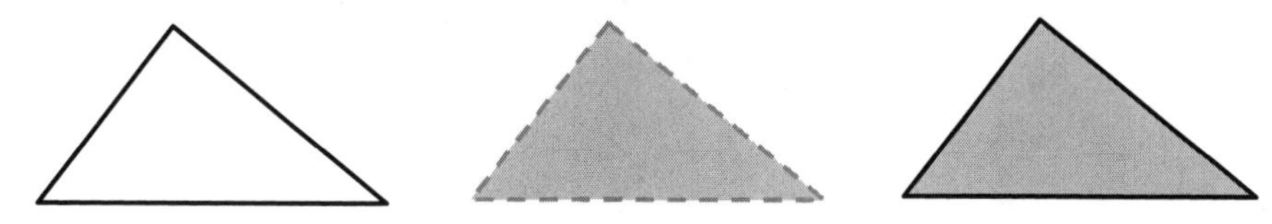

Figure 3.31: A triangle, its interior, and the union of the triangle and its interior.

Theorem 14.4 *The area enclosed by a spherical $\triangle^s ABC$ is given by*

$$area(\triangle^s ABC) = (m \prec A + m \prec B + m \prec C - \pi)r^2. \tag{3.13}$$

This theorem might seem unusual, given that there is no theorem for the area of a plane triangle which depends only on the angles of the triangle. In the plane, two triangles can have the same angles without having the same area: they need only be similar. But after the results of the last section, the result should seem more natural: in spherical geometry, a triangle is uniquely determined (up to congruence) by its angles (by the AAA congruence theorem), so its area is also.

The history of the discovery of this theorem is complex. It is usually attributed to Albert Girard, who published a proof of it in 1629. However, his proof had logical weaknesses and is far more complex than the proof used today. Lagrange commented in 1800 that because of problems with the Girard proof, the theorem should instead be attributed to the Italian mathematician Bonaventura Cavalieri, who published a proof in 1632. The theorem was also apparently found but not published by Thomas Harriot in 1603. The proof given here is basically that given by Euler in 1781. (See [Ro1988], p. 31 and [Pa2014].)

Proof. Let A^a, B^a, and C^a be the antipodes of A, B, and C, respectively. The great circle $\bigcirc AB$ divides the sphere into two hemispheres; C is in one of these hemispheres and C^a is in the other. In this proof we refer informally to the hemisphere with C in it as the "top" hemisphere and the hemisphere with C^a in it as the "bottom" hemisphere. On $\bigcirc AB$, we have points A, B, A^a, B^a in that order. Then the top hemisphere can be written as the union of the regions bounded by four triangles: $\triangle^s ABC$, $\triangle^s BA^aC$, $\triangle^s A^aB^aC$, and $\triangle^s B^aAC$. (We neglect the contribution of the sides of the triangles themselves, which have area zero.) The union of the regions bounded by the first two of these ($\triangle^s ABC$ and $\triangle^s BA^aC$) is the lune ABA^aCA; its angle has measure $m \prec A$,

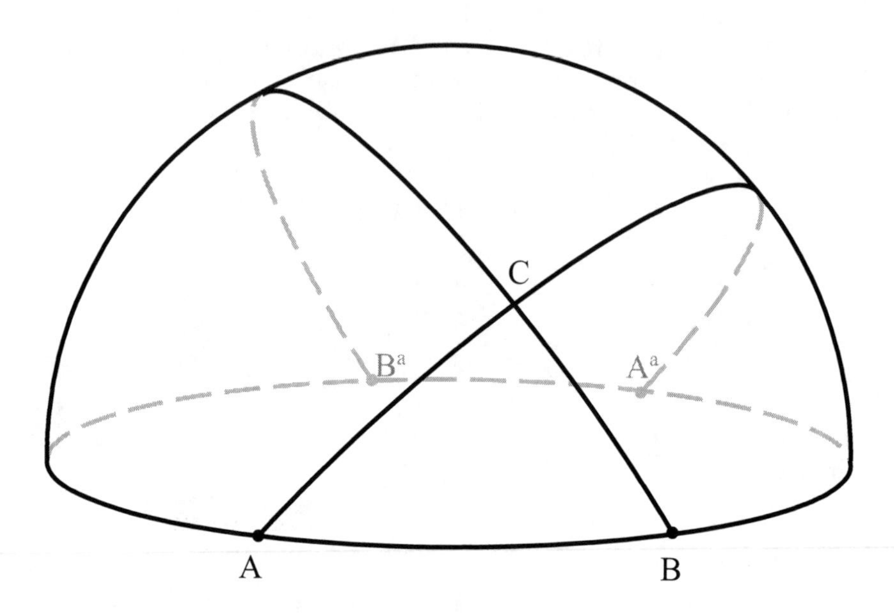

Figure 3.32: Theorem 14.4.

so it has area $2(m \prec A)r^2$. The union of the regions of the first and fourth triangles ($\triangle^s ABC$ and $\triangle^s B^a AC$) is the lune BAB^aCB; its angle has measure $m \prec B$, so it has area $2(m \prec B)r^2$. The lune CAC^aBC has angle whose measure is $m \prec C$ (so its area is $2(m \prec C)r^2$) and is the union of the regions of triangles $\triangle^s ABC$ and $\triangle^s ABC^a$. Having just written three lunes each as the union of two nonoverlapping triangular regions, we may write that the area of each lune is the sum of the areas of the two triangles of which it is the union:

$$2(m \prec A)r^2 = area(\triangle^s ABC) + area(\triangle^s BA^aC) \quad (3.14)$$
$$2(m \prec B)r^2 = area(\triangle^s ABC) + area(\triangle^s B^aAC) \quad (3.15)$$
$$2(m \prec C)r^2 = area(\triangle^s ABC) + area(\triangle^s ABC^a) \quad (3.16)$$

Note that in (3.16), $\triangle^s ABC^a$ is antipodal to $\triangle^s A^aB^aC$, hence has the same area, so we can replace (3.16) with

$$2(m \prec C)r^2 = area(\triangle^s ABC) + area(\triangle^s A^aB^aC). \quad (3.17)$$

Adding together (3.14),(3.15) and (3.17), we get

$$\begin{aligned} & (2m \prec A + 2m \prec B + 2m \prec C)r^2 \quad (3.18) \\ = \; & 3area(\triangle^s ABC) + area(\triangle^s BA^aC) \\ + \; & area(\triangle^s B^aAC) + area(\triangle^s A^aB^aC) \end{aligned}$$

The right side of (3.18) consists of the area of four triangular regions whose nonoverlapping union covers the top hemisphere. Since the area of the top hemisphere is $2\pi r^2$, we may write its area as the sum of the areas of the four triangles:

$$2\pi r^2 = area(\triangle^s ABC) + area(\triangle^s BA^aC) + area(\triangle^s B^aAC) + area(\triangle^s A^aB^aC).$$

Substituting this into (3.18), we obtain

$$(2m \prec\!A + 2m \prec\!B + 2m \prec\!C)r^2 = 2area(\triangle^s ABC) + 2\pi r^2$$

and solving for $area(\triangle^s ABC)$ we obtain (3.13) as desired. ♢

Definition 14.5 *In a spherical triangle* $\triangle^s ABC$*, the* spherical excess *is defined to be the quantity* $2E$*, where* $E = (m \prec\!A + m \prec\!B + m \prec\!C - \pi)/2$.

Later on we will prove a number of formulas involving E; these may be seen as new formulas for the area of a spherical triangle. In particular, we will prove Cagnoli's formula ((4.68) of §19 and Theorem 34.6 of §34), Euler's formula (Theorem 34.5 of §34), and Lhuilier's formula ((8.47) of §34). Also of note are the formulas found in §34, Exercises 12, 13, and 15.

Exercises §14.

1. On a sphere with $r = 10$ find the area of a spherical triangle whose angles have measures
   (a) $A = \frac{\pi}{2}$, $B = \frac{\pi}{3}$, $C = \frac{\pi}{3}$.
   (b) $A = \frac{2\pi}{3}$, $B = \frac{2\pi}{3}$, $C = \frac{\pi}{2}$.
   (c) $A = \frac{4\pi}{5}$, $B = \frac{3\pi}{5}$, $C = \frac{3\pi}{5}$.
   (d) $A = \frac{5\pi}{7}$, $B = \frac{4\pi}{7}$, $C = \frac{3\pi}{7}$.
   (e) $A = \frac{7\pi}{9}$, $B = \frac{5\pi}{6}$, $C = \frac{2\pi}{3}$.

2. On a sphere with $r = 5$ find the area of a spherical triangle whose angles have measures
   (a) $A = 80°$, $B = 70°$, $C = 60°$.
   (b) $A = 110°$, $B = 90°$, $C = 75°$.
   (c) $A = 120°$, $B = 110°$, $C = 100°$.
   (d) $A = 145°$, $B = 125°$, $C = 115°$.
   (e) $A = 175°$, $B = 170°$, $C = 170°$.

3. Assume the earth has a radius of 3963 miles. Find the area in square miles of a triangle on the earth with angles 115°, 100° and 95°.

4. Prove that the area of the region enclosed by a spherically convex n-gon is found by taking the sum of the radian measures of its angles minus $(n-2)\pi$ and multiplying that total by r^2. What happens if the polygon is not convex?

5. Suppose on a sphere with $r = 5$ a spherically convex quadrilateral has four angles of measure $100°$. Find the area of the quadrilateral.

6. Suppose on a sphere with $r = 10$ a spherically convex hexagon has six angles of measure $130°$. Find the area of the hexagon.

7. Given $\overset{\frown}{BC}$ and constant k, determine the set of all points A on the sphere such that A, B, and C form a spherical triangle such that $A+B+C = k$. (Hint: see §12, Exercise 24. The answer to this question is a result known as *Lexell's Theorem.*)

Chapter 4

Trigonometry

15 Spherical Pythagorean theorem and law of sines

Spherical trigonometry is the key to the study of the precise relationship among distances and angles on the sphere. Central to this is the relationship among the distances and angles in a triangle. But what makes these relationships different from the corresponding relationships in the plane?

The central geometric difference between the plane and the sphere is, of course, the fact that the sphere is "curved." In the plane, if we take two rays with the same endpoint, then the rays separate away from the common endpoint. On the sphere, two such rays separate for a while away from the endpoint but then converge back together at the antipode of the endpoint. We first quantify this behavior of rays on the sphere with a proposition.

Proposition 15.1 (A-12) *Suppose two rays form an angle with measure θ. Let x be the spherical distance between two points (one on each ray) at spherical distance ϕ from the vertex. Then* $\sin(\frac{x}{2}) = \sin(\phi)\sin(\frac{\theta}{2})$.

We have seen this proposition already: it is Proposition 6.1 of §6. (See also Figure 2.8.) There we proved it as a theorem for spheres in space, using the properties of three-dimensional geometry. In our axiomatic system we do not have those techniques available and we will assume it as an axiom. However, this is not strictly necessary: if we assume certain properties of trigonometric functions and certain properties of the real numbers from real analysis, we can prove Proposition 15.1 as a theorem. This is sufficiently difficult that it is beyond the scope of this book. However, we refer the interested reader to [Ca1916], p. 136, or [Wo1945], to see how this can be done.[1]

We use this result to determine properties of right triangles on the sphere. We adopt the following notation. In a triangle $\triangle^s ABC$, let A, B, and C

[1] Actually, Carslaw there proves Proposition 15.2, which is very similar.

denote the measures of the angles at vertices A, B, and C, respectively, and let a, b, and c denote the lengths of the sides opposite vertices A, B, and C, respectively. This means that the letters A, B, and C will sometimes denote points and other times they will denote measures of angles, but the notation is common and we hope that context will make confusion unlikely.

Proposition 15.2 *If* $\triangle^s ABC$ *has a right angle at* C, *then* $\sin(A) = \frac{\sin(a)}{\sin(c)}$.

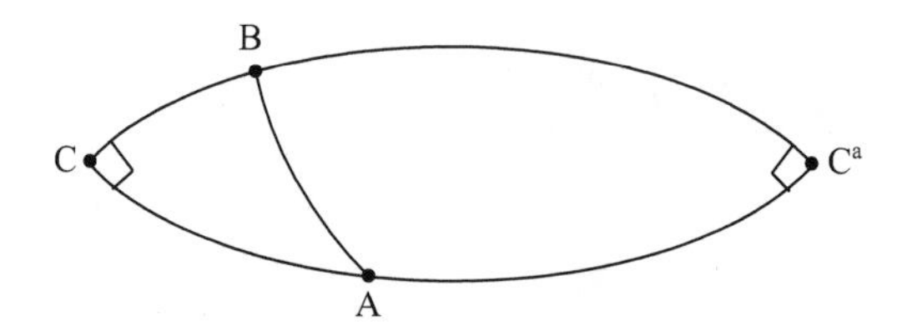

Figure 4.1: Proposition 15.2.

Proof. We first suppose that $\prec A$ is acute. Then $\triangle^s ABC$ may be regarded as half of an isosceles spherical triangle $\triangle^s ABD$ where C is the midpoint of $\overset{\frown}{BD}$. In this case, our result follows immediately from Proposition 15.1.

If $\prec A$ is right, then by Proposition 11.11, so are sides $\overset{\frown}{AB}$ and $\overset{\frown}{BC}$; then the statement to be proven just says $1 = 1$.

If $\prec A$ is obtuse, we consider the colunar triangle $\triangle^s ABC^a$. There $\prec BAC^a$ is acute with measure $\pi - A$, $m\ \overset{\frown}{BC^a} = \pi - a$ so by the acute case just considered, $\sin(\pi - A) = \sin(\pi - a)/\sin(c)$ which gives us what we want by property (1.1) of the sine function. $\diamondsuit$

Proposition 15.2 has the pleasant resemblance to the fact from trigonometry in the plane that the sine of an angle in a right triangle is calculated by taking the quotient "opposite over hypotenuse." On the sphere one must take the sine of the opposite and hypotenuse before taking the quotient.

We now use the setting just given to prove several important relationships among the sides and angles of $\triangle^s ABC$.

Theorem 15.3 *If spherical* $\triangle^s ABC$ *has a right angle at* C, *then we have:*

$$\cos(c) = \cos(a)\cos(b) \tag{4.1}$$

$$\cos(B) = \sin(A)\cos(b) \tag{4.2}$$

Equation (4.1) is called the spherical Pythagorean theorem and equation (4.2) is sometimes called Geber's theorem.

Proof. (See Figure 4.2.) We first suppose for simplicity that $\prec A$ and $\prec B$ are acute (which means sides $\overset{\frown}{AC}$ and $\overset{\frown}{BC}$ are also acute, by Proposition 11.12). We choose points D and E on $\overset{\hookrightarrow}{BA}$ and $\overset{\hookrightarrow}{BC}$ at spherical distance $\frac{\pi}{2}$

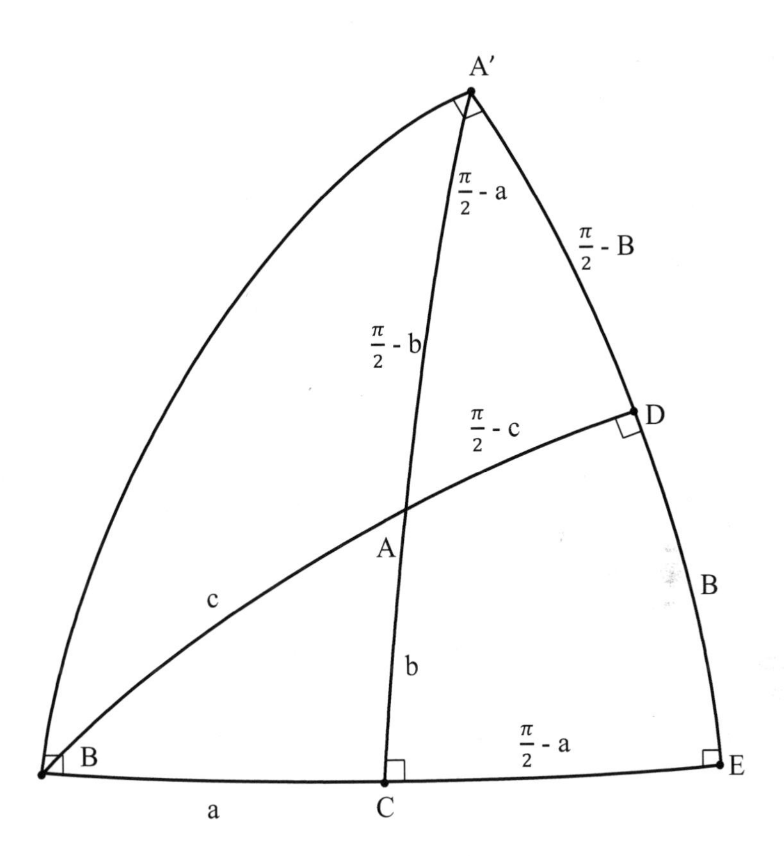

Figure 4.2: Theorem 15.3.

from B. Then $m\ \overset{\frown}{DE}= B$. Let A' be the pole of $\bigcirc BC$ on the same side of $\bigcirc BC$ as A. Then B is the pole of $\bigcirc DE$ and $m\ \overset{\frown}{BD}= m\ \overset{\frown}{BE}= m\ \overset{\frown}{A'C}= m\ \overset{\frown}{A'E}= \frac{\pi}{2}$. Since $\overset{\frown}{AC}$ is acute and $\overset{\frown}{A'C}$ is right, A is between A' and C. Since $m\ \overset{\frown}{BA'}= \frac{\pi}{2}$ and $m\ \overset{\frown}{BC}< \frac{\pi}{2}$, $m\ \overset{\frown}{BA}< \frac{\pi}{2}$ (by applying Proposition 9.44 to $\overset{\hookrightarrow}{A'C}$). We focus on $\triangle^s A'AD$, which has a right angle at D. There we have $m\ \overset{\frown}{A'A}= \frac{\pi}{2}-b$, $m\ \overset{\frown}{AD}= \frac{\pi}{2}-c$ and $m\ \overset{\frown}{A'D}= \frac{\pi}{2}-B$. Furthermore, $m\ \overset{\frown}{CE}= \frac{\pi}{2}-a$ so $m \prec AA'D = \frac{\pi}{2} - a$ by Definition 10.2. We then apply Proposition 15.2 to $\triangle^s A'AD$ twice. First,

$$\sin(\prec AA'D) = \sin(\overset{\frown}{AD})/\sin(\overset{\frown}{A'A}),$$

which gives $\sin(\frac{\pi}{2} - a) = \sin(\frac{\pi}{2} - c)/\sin(\frac{\pi}{2} - b)$ and hence (4.1). Second,

$$\sin(\prec A'AD) = \sin(\overset{\frown}{A'D})/\sin(\overset{\frown}{A'A}),$$

or $\sin(A) = \sin(\frac{\pi}{2} - B)/\sin(\frac{\pi}{2} - b)$, so (4.2) holds.

The cases where $\prec A$ and $\prec B$ are not acute follow by considering colunar triangles.

If $\prec B$ is right, then so are $\overset{\frown}{AB}$ and $\overset{\frown}{AC}$ by Proposition 11.11. So $B = C = b = c = \frac{\pi}{2}$ and both (4.1) and (4.2) merely say $0 = 0$. A similar argument holds if $\prec A$ is right.

Suppose $\prec A$ is acute and $\prec B$ is obtuse. Then by Proposition 11.12, $\overset{\frown}{BC}$ is acute and $\overset{\frown}{AC}$ is obtuse. We consider the colunar right $\triangle^s A^a BC$, where both non-right angles are acute. There we apply (4.1) and (4.2) for the acute case to get $\cos(\pi - c) = \cos(a)\cos(\pi - b)$ and $\sin(A)\cos(\pi - b) = \cos(\pi - B)$ which imply what we want. A similar argument holds if $\prec A$ is obtuse and $\prec B$ is acute.

If $\prec A$ and $\prec B$ are both obtuse, then we consider colunar $\triangle^s ABC^a$ where both non-right angles are acute. There we apply (4.1) and (4.2) for the acute case to get $\cos(c) = \cos(\pi - a)\cos(\pi - b)$ and $\sin(\pi - A)\cos(\pi - b) = \cos(\pi - B)$ which imply what we want. $\diamondsuit$

On a large sphere, a small spherical triangle is nearly planar, so the planar Pythagorean theorem would nearly hold. How can we reconcile the fact that the planar and spherical Pythagorean theorems look so different? The answer lies in looking at the Taylor series for the cosine. We have that $\cos(x) = 1 - x^2/2! + x^4/4! - x^6/6! + \ldots$. If we substitute this into $\cos(c) = \cos(a)\cos(b)$, we find

$$1 - c^2/2! + c^4/4! + \ldots = (1 - a^2/2! + a^4/4! + \ldots)(1 - b^2/2! + b^4/4! + \ldots)$$

which may be thought of as $1 - a^2/2! - b^2/2! +$ higher degree terms, so $c^2 = a^2 + b^2 +$ higher degree terms in a, b, and c. For small values of a, b, and

c, the higher degree terms are much smaller than the squared terms and hence $c^2 = a^2 + b^2$ becomes a good approximation to the original equality $\cos(c) = \cos(a)\cos(b)$.

In the plane we find that the non-right angles in a right triangle are complementary and so the cosine of one equals the sine of the other. Equation (4.2) is reminiscent of this fact. Indeed, if a and b are near 0 then the triangle is nearly planar, $\cos(b)$ is near 1, and (4.2) gives the planar result.

A number of other relations in right spherical triangles can be derived in a similar manner. But because of the hassle associated with the various cases, we leave this to the exercises and pursue other avenues of proving them.

Proposition 15.2 also allows us quickly to obtain an important result for general triangles.

Theorem 15.4 (The spherical law of sines) *In any spherical* $\triangle^s ABC$,

$$\frac{\sin(a)}{\sin(A)} = \frac{\sin(b)}{\sin(B)} = \frac{\sin(c)}{\sin(C)} \tag{4.3}$$

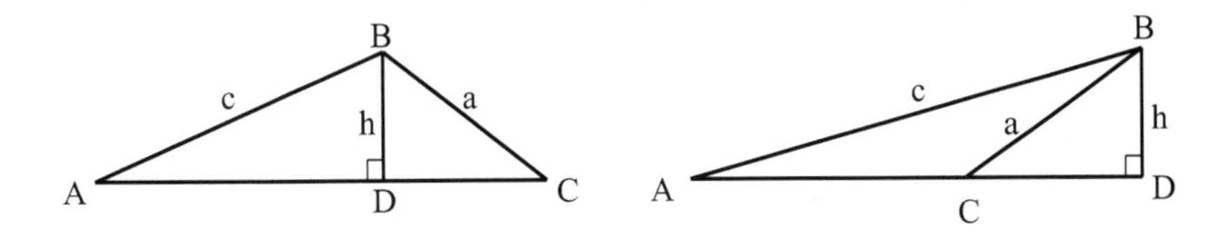

Figure 4.3: Proposition 15.4.

Proof. We will prove

$$\frac{\sin(a)}{\sin(A)} = \frac{\sin(c)}{\sin(C)}. \tag{4.4}$$

Then (4.3) is obtained by permuting the vertices.

If $\triangle^s ABC$ has right angle at C, then $\sin(C) = 1$ and by Proposition 15.2, $\sin(A)/\sin(C) = \sin(A) = \frac{\sin(a)}{\sin(c)}$, which gives us (4.4). A similar observation holds if there is a right angle at A.

If the angles at A and C are not right, then there are two feet of altitudes from B on $\bigcirc AC$. These are antipodal, so one of them (call it D) lies on $\overrightarrow{AC}$. Then D is different from C and A. Let $h = m\,\overset{\frown}{BD}$. Triangles $\triangle^s ADB$ and $\triangle^s CDB$ then have right angles at D, so from Proposition 15.2, $\sin(A) = \sin(h)/\sin(c)$ and $\sin(C) = \sin(h)/\sin(a)$. Dividing these and rearranging factors gives (4.4). ♢

Other approaches to proving the spherical law of sines may be found in Exercises 10 of §16 and Exercise 2 of §18.

Exercises §15.

1. Use trigonometry to do §12, Exercise 9.

2. Suppose that $0 < a, b, A, B < \pi$ and $\frac{\sin(a)}{\sin(b)} = \frac{\sin(A)}{\sin(B)}$. Prove that
$$\tan\frac{a-b}{2}\cot\frac{a+b}{2} = \tan\frac{A-B}{2}\cot\frac{A+B}{2}, \tag{4.5}$$
so if $a + b \neq \pi$ and $A + B \neq \pi$,
$$\frac{\tan\frac{a-b}{2}}{\tan\frac{a+b}{2}} = \frac{\tan\frac{A-B}{2}}{\tan\frac{A+B}{2}} \tag{4.6}$$
This equation, when applied to a spherical triangle, is often known as the *spherical law of tangents.*

3. Suppose that some point on a great circle lies in the interior of a small circle. Prove that the circles meet in exactly two points.

4. Suppose that a small circle contains at least one point on both sides of a great circle. Prove that the two circles meet in exactly two points.

5. Prove that in any spherical triangle, half of the sum of any two angles is acute, right, or obtuse if and only if half the sum of the opposite sides is acute, right, or obtuse, respectively.

6. Suppose that in $\triangle^s ABC$, D is chosen on $\overset{\frown}{BC}$ so that $\overset{\hookrightarrow}{AD}$ bisects $\prec BAC$. Prove that $\sin(b)/\sin(c) = \sin(\overset{\frown}{CD})/\sin(\overset{\frown}{BD})$. Conversely, prove that if the equation holds for some D chosen on $\overset{\frown}{BC}$ then $\overset{\hookrightarrow}{AD}$ bisects $\prec BAC$.

7. Prove that if $0 < x, y < \pi$ and $x + y < \pi$ then $x < y$ if and only if $\sin(x) < \sin(y)$.

8. Suppose that two small circles of spherical radii r and s are tangent to each other and that a great circle is tangent to them both. If the common tangent has measure ℓ, show that $\sin^2(\ell/2) = \tan(r)\tan(s)$.

9. Use trigonometry to prove Exercise 15 of §13.

10. Suppose that A, B, and C do not lie on the same great circle, P is a pole of $\bigcirc AB$, that D, E, and F lie on $\bigcirc AB$, that $\overset{\frown}{PD}$, $\overset{\frown}{PE}$, and $\overset{\frown}{PF}$ meet $\bigcirc BC$ in G, H, and I, respectively, that $0 < m\,\overset{\frown}{PG} < m\,\overset{\frown}{PH} < m\,\overset{\frown}{PI} < \frac{\pi}{2}$, $\frac{\pi}{2} > m\,\overset{\frown}{BG} > m\,\overset{\frown}{BH} > m\,\overset{\frown}{BI} > 0$, and that $m\,\overset{\frown}{GH} = m\,\overset{\frown}{HI}$. Prove that $m\,\overset{\frown}{DE} > m\,\overset{\frown}{EF}$.

11. Suppose that we are given $\triangle^s ABC$ and some great circle intersects $\bigcirc AB$, $\bigcirc BC$, and $\bigcirc AC$ at points X, Y, and Z, respectively. Assume that this great circle does not pass through points A, B, or C. Use the spherical law of sines repeatedly to prove that
$$\frac{\sin(\overset{\frown}{AX})}{\sin(\overset{\frown}{XB})}\frac{\sin(\overset{\frown}{BY})}{\sin(\overset{\frown}{YC})}\frac{\sin(\overset{\frown}{CZ})}{\sin(\overset{\frown}{ZA})} = 1. \tag{4.7}$$

12. Suppose that two small circles of radius r_1 and r_2 meet in two points. Prove that these points are not antipodal and that the great circle passing through them consists of the set of all points whose spherical distances d_1, d_2 to the circle with radius r_1, r_2, respectively, satisfy $\cos(d_1)\cos(r_2) - \cos(d_2)\cos(r_1) = 0$. Extend to the case where the circles meet in exactly one point.

13. Suppose three small circles meet pairwise in two points. Prove that the three great circles passing through each pair all pass through a pair of antipodal points.

14. Suppose that in a spherical triangle, we construct three arcs from each vertex to the opposite side in such a way as to bisect the area of the triangle. Prove that these three arcs meet at a point.

15. Suppose that in $\triangle^s ABC$, points D, E, and F are chosen on the sides opposite A, B, and C. Prove that the great circles perpendicular to these sides at D, E, and F pass through the same point if and only if

$$\cos(\overset{\frown}{AF})\cos(\overset{\frown}{BD})\cos(\overset{\frown}{CE}) = \cos(\overset{\frown}{FB})\cos(\overset{\frown}{DC})\cos(\overset{\frown}{EA}).$$

16. Suppose that we are given $\triangle^s ABC$ and a point P not lying on any of the great circles $\bigcirc AB$, $\bigcirc AC$, and $\bigcirc BC$. Suppose that $\bigcirc AP$, $\bigcirc BP$, and $\bigcirc CP$ meet $\bigcirc BC$, $\bigcirc AC$, and $\bigcirc AB$ in X, Y, and Z, respectively. Use the spherical law of sines repeatedly to prove that

$$\frac{\sin(\overset{\frown}{AZ})}{\sin(\overset{\frown}{ZB})}\frac{\sin(\overset{\frown}{BX})}{\sin(\overset{\frown}{XC})}\frac{\sin(\overset{\frown}{CY})}{\sin(\overset{\frown}{YA})} = 1. \tag{4.8}$$

holds. (This is Ceva's theorem on the sphere.)

17. Prove the converse of Exercise 16: that if X, Y, and Z are chosen so that (4.8) holds, then the great circles $\bigcirc AX$, $\bigcirc BY$, and $\bigcirc CZ$ all pass through the same pair of antipodal points.

18. Using the notation of Exercise 16, prove that

$$\frac{\sin(\prec CAX)}{\sin(\prec XAB)}\frac{\sin(\prec ABY)}{\sin(\prec YBC)}\frac{\sin(\prec BCZ)}{\sin(\prec ZCA)} = 1. \tag{4.9}$$

19. Prove the converse of Exercise 18: that if X, Y, and Z are chosen so that (4.9) holds, then the great circles $\bigcirc AX$, $\bigcirc BY$, and $\bigcirc CZ$ all pass through the same pair of antipodal points.

20. Show that the results of Exercise 11 and Exercise 18 are dual to each other via polar triangles. More specifically, the statement of Exercise 18 for $\triangle^s ABC$ and a point P where three cevians meet corresponds directly to the statement of Exercise 11 for $\triangle^s A'B'C'$ and the polar circle of P.

21. Prove that the three spherical rays which bisect the angles of a spherical triangle meet at a point in the interior of the triangle.

22. Suppose that $\triangle^s ABC$ has a right angle at B, that point D is between A and C, and that E is between A and C^a. Prove that $m \prec EBA = m \prec DBA$ if and only if $\sin(\overset{\frown}{CE})/\sin(\overset{\frown}{AE}) = \sin(\overset{\frown}{CD})/\sin(\overset{\frown}{AD})$.

23. Suppose that in $\triangle^s ABC$ $m\ \overset{\frown}{AB} < m\ \overset{\frown}{AC}$, and $m\ \overset{\frown}{AB} + m\ \overset{\frown}{AC} < \pi$. Let D be the midpoint of $\overset{\frown}{BC}$ and O a point of $\overset{\frown}{AD}$. Prove that $m \prec ABO > m \prec ACO$.

24. Suppose that in $\triangle^s ABC$ $m\ \overset{\frown}{AB} < m\ \overset{\frown}{AC}$, and $m\ \overset{\frown}{AB} + m\ \overset{\frown}{AC} < \pi$. Choose D and E on $\overset{\frown}{BC}$ so that $m\ \overset{\frown}{BD} = m\ \overset{\frown}{CE}$. Prove that $m \prec BAD > m \prec EAC$. (See §13, Exercise 39.)

Historical notes. We let X-Y denote Book X, Proposition Y of Menelaus' *Sphaerica.* Then Exercise 11 is III-1 (often referred to as Menelaus' Theorem or the Sector Theorem), Exercise 6 is III-6, Exercise 22 is III-10 and III-11, Exercise 21 is III-12, Exercise 9 is I-33, Exercise 23 is I-36, and Exercise 24 is part 1 of I-37. Exercise 10 is Book III, Proposition 6 of Thedosius' *Sphaerica,* a text from circa 100 BC. See [VB, p. 51].

16 Spherical law of cosines and analogue formula

We now have enough background to prove the spherical law of cosines. As with the law of cosines for planar triangles, it is a formula which relates the measures of the three sides of a triangle with one of the angles.

Theorem 16.1 (The spherical law of cosines) *In any spherical* $\triangle^s ABC$,

$$\cos(c) = \cos(a)\cos(b) + \sin(a)\sin(b)\cos(C). \tag{4.10}$$

We may rotate the letters to obtain similar formulas:

$$\cos(b) = \cos(a)\cos(c) + \sin(a)\sin(c)\cos(B) \tag{4.11}$$
$$\cos(a) = \cos(b)\cos(c) + \sin(b)\sin(c)\cos(A) \tag{4.12}$$

Proof. There exists a point D on $\overset{\hookrightarrow}{CA}$ such that $\bigcirc BD$ is perpendicular to $\bigcirc AC$. If $D = C$, $\prec C$ is a right angle, $\cos(C) = 0$, so (4.10) is a consequence of Theorem 15.3. If $D \neq C$, let $d = m\ \overset{\frown}{CD}$ and $h = m\ \overset{\frown}{BD}$. We claim that $\cos(c) = \cos(h)\cos(b-d)$. If $D = A$ then $b = d$ and $c = h$ so this is automatic. If $D \neq A$ then $m\ \overset{\frown}{AD} = \pm(b-d)$, depending on whether b or d is larger. Either

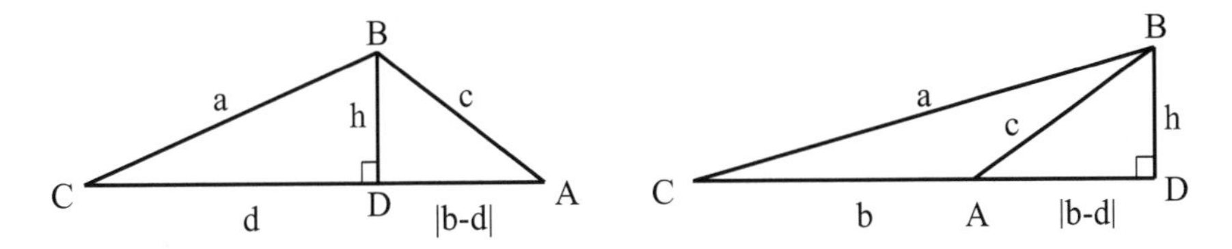

Figure 4.4: Theorem 16.1: $b < d$ (right) and $b > d$ (left).

way, $\triangle^s ADB$ has a right angle at D and $\cos(c) = \cos(h)\cos(b-d)$, making use of (1.6).

Similarly, $\triangle^s CDB$ has a right angle at D, so $\cos(a) = \cos(d)\cos(h)$ and $\cos(C) = \cos(h)\sin(\prec CBD) = \cos(h)\sin(d)/\sin(a)$ (by Theorem 15.3 and Prop. 15.2.) Then by (1.11),

$$\begin{aligned}
\cos(c) &= \cos(b)\cos(d)\cos(h) + \sin(b)\sin(d)\cos(h) \\
&= \cos(a)\cos(b) + \sin(a)\sin(b)(\sin(d)\cos(h)/\sin(a)) \\
&= \cos(a)\cos(b) + \sin(a)\sin(b)\cos(C).
\end{aligned}$$

◇

If $\prec C$ is a right angle, $\cos(C) = 0$, so the second term of (4.10) is zero, giving $\cos(c) = \cos(a)\cos(b)$. So the spherical law of cosines is a generalization of the spherical Pythagorean theorem. We may also see that Axiom 10.18 and Theorem 13.12 are special cases of the spherical law of cosines.

Example 1. Suppose that in $\triangle^s ABC$, $a = 40°$, $b = 65°$ and $C = 130°$. Determine the value of c.

Solution. We have $\cos(c) = \cos(40°)\cos(65°) + \sin(40°)\sin(65°)\cos(130°) \approx -0.05$. So $c \approx 92.9°$.

A simple application of the law of cosines may be found in the problem of determining the distance between two places on the surface of the earth — or on any planet. We have seen (Corollary 13.6; §13, Exercises 54a, 54b) that the shortest route between two points on the sphere follows a great circle arc, so let us see how to find the length of the great circle arc between two cities.

We assume that we are given the latitude and longitude in degrees for each of the cities A and B. We let C be the north pole. Given the latitude ℓ_A for A, the pole distance AC in degrees will be the complement $90° - \ell_A$. Similarly the pole distance in degrees for city B will be $90° - \ell_B$. (Note that under the convention that southern latitudes have negative latitude, these calculations still work.) Then one finds the difference in the longitudes of the cities; this will provide the angle measure of $\prec C$ in degrees. Note that under the convention that western longitudes are negative and eastern longitudes are positive, a simple difference d (larger minus smaller) in the longitudes will not always give $m \prec C$; if $d > 180°$, then the measure of $\prec C$ would be $360° - d$. But in the spherical law of cosines one takes the cosine of this angle and since $\cos(d) = \cos(360° - d)$ one does not need to worry about this distinction.

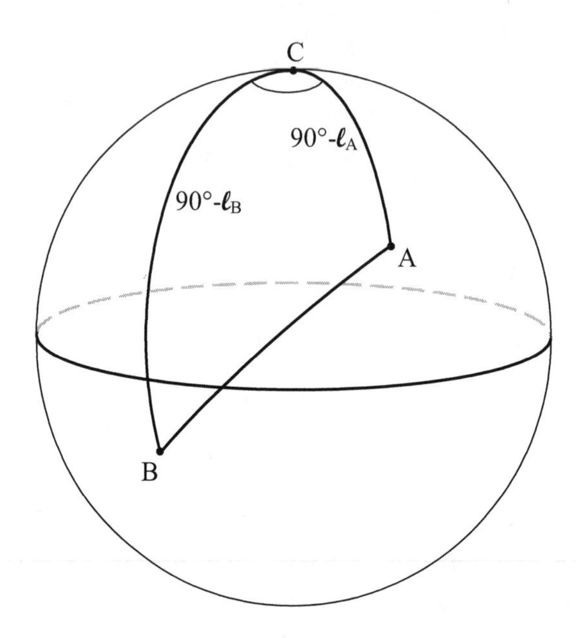

Figure 4.5: The distance between cities A and B.

Since $\cos(d) = \cos(-d)$ one also does not need to worry about the order of substraction.

Example 2. Find the length of the great circle arc from Los Angeles (latitude 34.05° N, 118.25° W) and Jakarta (latitude 6.2° S, 106.8° E).

Solution. Using the sign convention of §8, the latitude-longitude coordinates of Los Angeles are $(+34.05°, -118.25°)$, and Jakarta has coordinates $(-6.2°, +106.8°)$. Then for Los Angeles, the degree distance to the north pole is $90° - 34.05° = 55.95°$. For Jakarta, the degree distance to the north pole will be $90° + 6.2° = 96.2°$. The difference in longitudes is $106.8° - (-118.25°) = 225.05°$. Then if c is the spherical distance between the two cities,

$$\cos(c) = \cos(55.95°)\cos(96.2°) + \sin(55.95°)\sin(96.2°)\cos(225.05°)$$

We get $\cos(c) \approx -0.6424$, so if c (as the side of a spherical triangle) is understood to be a value between 0° and 180°, $c \approx 129.97°$. To find the distance in miles, we first convert c to radians: $c \approx 2.2684$. Then to find the distance between the cities in miles, we use the formula for the length of a circular arc (the length is the radius times the measure of the arc) to obtain the distance $(3,964)(2.2684) \approx 8,992$ miles.

Suppose that one is given the values of a, b, and C in a triangle. The spherical law of cosines permits the calculation of $\cos(c)$, from which one may obtain the value of c since $0 < c < \pi$. If one wishes to calculate the values of A and B, one may apply the law of cosines by permuting the values of A, B,

C, a, b, and c. This method results in the following formula[2].

Theorem 16.2 (Analogue formula) *In any spherical* $\triangle^s ABC$,

$$\sin(c)\cos(B) = \cos(b)\sin(a) - \sin(b)\cos(a)\cos(C). \tag{4.13}$$

By rotating the letters in (4.13), we also obtain:

$$\sin(a)\cos(B) = \cos(b)\sin(c) - \sin(b)\cos(c)\cos(A) \tag{4.14}$$
$$\sin(a)\cos(C) = \cos(c)\sin(b) - \sin(c)\cos(b)\cos(A) \tag{4.15}$$
$$\sin(b)\cos(C) = \cos(c)\sin(a) - \sin(c)\cos(a)\cos(B) \tag{4.16}$$
$$\sin(c)\cos(A) = \cos(a)\sin(b) - \sin(a)\cos(b)\cos(C) \tag{4.17}$$
$$\sin(b)\cos(A) = \cos(a)\sin(c) - \sin(a)\cos(c)\cos(B) \tag{4.18}$$

Proof. We begin with (4.11) and solve for $\sin(a)\sin(c)\cos(B)$:

$$\sin(a)\sin(c)\cos(B) = \cos(b) - \cos(a)\cos(c)$$

and substitute (4.10) to obtain

$$\begin{aligned}
\sin(a)\sin(c)\cos(B) &= \cos(b) - \cos(a)(\cos(a)\cos(b) + \sin(a)\sin(b)\cos(C)) \\
&= \cos(b) - \cos^2(a)\cos(b) - \cos(a)\sin(a)\sin(b)\cos(C) \\
&= \cos(b)(1 - \cos^2(a)) - \cos(a)\sin(a)\sin(b)\cos(C) \\
&= \cos(b)\sin^2(a) - \cos(a)\sin(a)\sin(b)\cos(C) \\
&= \sin(a)(\cos(b)\sin(a) - \cos(a)\sin(b)\cos(C)).
\end{aligned}$$

Since $\sin(a)$ is nonzero for all a between 0 and π, we may divide by it to obtain (4.13). ◇

The analogue formula may seem awkward in the sense that it involves five different elements of a triangle, so given three elements one cannot solve directly for a fourth. For example, if one is given a, b, and C in (4.13), one cannot solve for B without already knowing something about the value of c (which one would usually obtain from the law of cosines (4.12).) However, the analogue formula arises naturally with coordinate approaches to spherical geometry (see the derivation of (8.30) in Chapter 8) and thus is useful in changing coordinate systems in astronomy. If one wishes to solve for B directly without solving for c first, one may use a slight variation of the analogue formula called the four-parts formula (4.42).

Example 3. Suppose that in $\triangle^s ABC$, $a = 40°$, $b = 65°$ and $C = 130°$. Determine the values of A and B.

Solution. In Example 1 we found $c \approx 92.9°$. Using the analogue formula (4.13), $\sin(92.9°)\cos(B) = \cos(65°)\sin(40°) - \sin(65°)\cos(40°)\cos(130°)$,

[2] We follow [Gr1985] in using this name for the formula.

so $\cos(B) \approx 0.72$ and $B \approx 44.04°$. Similarly, $\sin(92.9°)\cos(A) = \cos(40°)\sin(65°) - \sin(40°)\cos(65°)\cos(130°)$ so $\cos(A) \approx 0.87$ and $A \approx 29.5°$.

Here is an application of the analogue formula closely related to Example 2 of this section. A *bearing* from a point on the surface of the earth is a direction along the surface of the earth based at that point. The directions of "north," "south," "east," and "west" (abbreviated N, S, E, and W) are the four most basic bearings at any point. We may prescribe other directions based on these four. In general, if a is a real number, D_1 is either N or S, and D_2 is either E or W, then $D_1 a D_2$ refers to the bearing found by facing in direction D_1 and turning a degrees toward the direction D_2. For example, the direction $N40°W$ (read: "north 40 degrees west") refers to the bearing found by facing north and turning 40° to the west (left). Similarly, $S25°E$ (read: "south 25 degrees east") refers to the bearing found by facing south and turning 25° to the east (left).

Suppose that one wished to travel from one point to another on the surface of the earth. We can use the analogue formula to determine the initial bearing for such a course of travel. Suppose that as above, the two places are A and B and the north pole is C. Then to find the bearing from A to B it suffices to determine the measure of angle A. If this number is smaller than 90°, then (in the notation above) the first letter of the bearing will be N and one will turn A degrees either east or west. If the number is larger than 90°, the first letter of the bearing will be an S and one will turn $180° - A°$ either east or west. Whether one turns east or west will depend on whether the other city's longitude difference is smaller in the easterly or westerly direction.

Example 4. For the cities in Example 2, determine the initial bearing from each city to the other.

Solution. Suppose A is Los Angeles, B is Jakarta, and C is the north pole. Because the difference of their longitude coordinates is $106.8° - (-118.25°) = 225.05°$ is greater than 180°, $\sphericalangle ACB$ contains in its interior the meridians with longitudes larger than 106.8° and smaller than −118.25°, so by Proposition 10.9 the shorter great circle arc from A to B passes west around the earth. We use the third formula in (4.14) to find

$$\begin{aligned} & \sin(129.97°)\cos(A) \\ = \; & \cos(96.2°)\sin(55.95°) - \sin(96.2°)\cos(55.95°)\cos(134.95°) \end{aligned}$$

so $A \approx 66.65°$. Thus the initial bearing from Los Angeles is $N66.65°W$. For the bearing at Jakarta, the argument above shows that the bearing will be east from Jakarta. We calculate $m\sphericalangle CBA$ from the fifth formula of (4.14) to obtain

$$\begin{aligned} & \sin(129.97°)\cos(B) \\ = \; & \cos(55.95°)\sin(96.2°) - \sin(55.95°)\cos(96.2°)\cos(134.95°) \end{aligned}$$

so $\cos(B) \approx 0.64$ and $B \approx 49.92°$. We conclude that the intial bearing from Jakarta is $N49.92°E$.

Example 4 is one of the types of ancient problems that first motivated the study of spherical trigonometry. The observant Muslim is to face the city of Mecca when praying. Determination of the appropriate bearing to Mecca from the position of the observer is the key calculation in making this manner of prayer possible.

Example 5. Suppose that in $\triangle^s ABC$, $a = 80°$, $b = 65°$, and $c = 110°$. Determine the value of C.

Solution. From the spherical law of cosines (4.10), $\cos(110°) = \cos(80°)\cos(65°) + \sin(80°)\sin(65°)\cos(C)$. We must solve for

$$\cos(C) = \frac{\cos(110°) - \cos(80°)\cos(65°)}{\sin(80°)\sin(65°)} \approx -0.47$$

so $C \approx 117.7°$.

Exercises §16

1. In each of the following, we are given the values of a, b, and C in $\triangle^s ABC$. Determine the values of A, B, and c.

   (a) $a = 70°$, $b = 40°$, $C = 35°$
   (b) $a = 105°$, $b = 45°$, $C = 40°$
   (c) $a = 130°$, $b = 140°$, $C = 35°$
   (d) $a = 115°$, $b = 30°$, $C = 125°$
   (e) $a = 70°$, $b = 40°$, $C = 155°$
   (f) $a = 170°$, $b = 140°$, $C = 155°$

2. In each of the following, we are given the values of a, b, and c in $\triangle^s ABC$. Determine the values of A, B, and C.

   (a) $a = 70°$, $b = 40°$, $c = 35°$
   (b) $a = 105°$, $b = 45°$, $c = 80°$
   (c) $a = 130°$, $b = 140°$, $c = 35°$
   (d) $a = 115°$, $b = 30°$, $c = 125°$
   (e) $a = 130°$, $b = 40°$, $c = 155°$
   (f) $a = 80°$, $b = 115°$, $c = 155°$

3. For each pair of cities given, determine (i) the distance between the cities, (ii) the intial bearing when travelling from each city to the other, (iii) the nearest approach to the poles, (iv) the distance from each of the cities to the points on the great circle between the two cities which are nearest the north and south poles, and (v) the latitude-longitude coordinates of the point on the great circle between the two cities which are nearest the north and south poles.

(a) Boston and Paris

(b) Moscow and Tokyo

(c) Mexico City and Bangalore

(d) Rio de Janeiro and Cairo

(e) Mexico City and Cape Town

4. Do any of the great circle routes in Exercise 3 cross each other?

5. Find the coordinates of the halfway point of the routes given in Exercise 3.

6. Which of the routes in Exercise 3 have a point where the route travels due east or west? Where are these points?

7. Suppose that a sphere is given spherical coordinates (ϕ, θ) with respect to a pole P. Given two points with coordinates (ϕ_1, θ_1) and (ϕ_2, θ_2), prove that the spherical distance d between them satisfies

$$\cos(d) = \cos(\phi_1)\cos(\phi_2) + \sin(\phi_1)\sin(\phi_2)\cos(\theta_1 - \theta_2) \tag{4.19}$$

8. Suppose that an airplane located at $10°S$, $50°E$ flies for 3 hours at 500 mph on a great circle course which is initially $N20°W$. What is its final position? What is the course it is following at arrival?

9. A ship sails 500 miles by great circle arc until it reaches $45°N$. At that moment its longitude is $40°W$ and its course is $N30°E$. What was its initial position and course?

   Exercises 10 and 11 show how to use the spherical law of cosines to prove a number of results discussed earlier with more elementary methods.

10. Take the formula (4.10) for the spherical law of cosines, solve for $\cos(C)$ and square both sides. Use this expression to prove that

$$\frac{\sin(C)}{\sin(c)} = \frac{\sqrt{1 - \cos^2(a) - \cos^2(b) - \cos^2(c) + 2\cos(a)\cos(b)\cos(c)}}{\sin(a)\sin(b)\sin(c)} \tag{4.20}$$

   and conclude an extension of the spherical law of sines.

11. Show how Theorem 15.3 is a special case of Theorems 15.4, 16.1, and 16.2.

12. Suppose that two small circles have radius ρ_1 and ρ_2 and their centers are spherical distance d apart. If the small circles meet, determine the angle between them. What if one of the circles is a great circle?

13. Show how Theorem 13.4 may be derived from the spherical law of cosines.

14. Suppose that in $\triangle^s ABC$, the value of a is no closer to $\frac{\pi}{2}$ than the value of b. Prove that $\sphericalangle A$ is acute, right, or obtuse if and only if the opposite side is acute, right, or obtuse, respectively.

15. Use trigonometry to prove Exercise 25 of §13.

16. Use trigonometry to prove Exercise 35 of §13.

17. Given spherical quadrilateral $\diamond ABCD$, and suppose that diagonals $\overset{\frown}{AC}$ and $\overset{\frown}{BD}$ meet at point E. Prove that

$$\cos(\sphericalangle AED) = (\cos(\overset{\frown}{AD})\cos(\overset{\frown}{BC}) - \cos(\overset{\frown}{AB})\cos(\overset{\frown}{CD}))\csc(\overset{\frown}{AC})\csc(\overset{\frown}{BD}).$$

18. Suppose that in $\triangle^s ABC$, point D is chosen between B and C. Let a, b, c, A, B, and C denote the usual measures of sides and angles. Let $d = m\ \overset{\frown}{AD}$, $a_1 = m\ \overset{\frown}{BD}$, $a_2 = m\ \overset{\frown}{CD}$, $A_1 = m \sphericalangle BAD$, and $A_2 = m \sphericalangle CAD$.

    (a) Prove that $\cos(b)\sin(a_1) + \cos(c)\sin(a_2) = \cos(d)\sin(a)$.

    (b) Prove that $\cot(b)\sin(A_1) + \cot(c)\sin(A_2) = \cot(d)\sin(A)$.

19. Suppose that the medians of a spherical triangle $\triangle^s ABC$ to sides $\overset{\frown}{AB}$ and $\overset{\frown}{AC}$ have the same measure. Prove that $b = c$ or $\sin^2(\frac{a}{2}) = \cos^2(\frac{b}{2}) + \cos^2(\frac{c}{2}) + \cos(\frac{b}{2})\cos(\frac{c}{2})$.

20. Suppose that in a spherical triangle the sum of the measures of two sides is less than π. Consider an arc from their common vertex to the opposite side. If this arc bisects the opposite side, prove that the arc has measure less than $\frac{\pi}{2}$. Furthermore, prove that the measure of this arc is less than half the sum of the measures of the two adjacent sides. (Compare Exercises 36 and 37 of §13.)

21. Suppose that in a spherical triangle the sum of the measures of two sides is less than π. Consider an arc from their common vertex to the opposite side. If this arc bisects the angle, prove that the arc has measure less than $\frac{\pi}{2}$. Furthermore, prove that the measure of this arc is less than half the sum of the measures of the two adjacent sides. (Compare Exercises 36 and 37 of §13, and Exercise 10 of §18.)

22. Suppose that in $\triangle^s ABC$, $m\ \overset{\frown}{AB} < m\ \overset{\frown}{AC}$, $m\ \overset{\frown}{AB} + m\ \overset{\frown}{AC} < \pi$, and M is the midpoint of $\overset{\frown}{BC}$. Suppose that D and E are chosen on $\overset{\frown}{BM}$ and $\overset{\frown}{MC}$, respectively, so that $\overset{\frown}{BD} \cong \overset{\frown}{EC}$. Prove that $m\ \overset{\frown}{AD} + m\ \overset{\frown}{AE} < m\ \overset{\frown}{AB} + m\ \overset{\frown}{AC}$ and $m\ \overset{\frown}{AD} < m\ \overset{\frown}{AE}$. (Compare with §13, Exercise 39.)

23. Suppose that a small circle has center P, and point Q is different from both P and the antipode of P. Show that the nearest point on the small circle to Q is the intersection of $\overleftrightarrow{PQ}$ with the circle, and the furthest point is the intersection of $\overleftrightarrow{PQ^a}$ with the circle. Furthermore, show that if points R and S are chosen on the circle then $m\ \widehat{QR} < m\ \widehat{QS}$ if and only if $m \prec RPQ < m \prec SPQ$.

24. Suppose that two small circles have spherical radii r_1 and r_2 and their spherical centers are at spherical distance d. Show that the circles meet in exactly two points if $|r_1 - r_2| < d < r_1 + r_2$.

25. Use the following procedure to prove the analogue formula (4.13) for the case when both angles B and C are acute.

    (1) Let A' be the pole of $\bigcirc BC$ on the same side of $\bigcirc BC$ as A.

    (2) The foot D of the shorter altitude from A to $\bigcirc BC$ is between B and C.

    (3) Let $x = m\ \widehat{CD}$. Apply the law of sines to $\triangle^s A'CA$ to obtain $\sin(b)\cos(C) = \cos(h)\sin(x)$.

    (4) Apply the law of sines to $\triangle^s A'BA$ to obtain

    $$\sin(c)\cos(B) = \cos(h)\sin(a - x).$$

    (5) Expand the expression in (4) and eliminate the x to obtain (4.13).

    If one assumes that the trigonometric functions and their inverses are continuous functions, one may use so-called "continuity" arguments to solve Exercises 26 to 29 by invoking the intermediate value theorem.

26. Prove that if a spherical arc has one endpoint in the interior of a small circle and one endpoint in the exterior of the same small circle, then some point on the arc lies on the small circle.

27. Prove that if a small circle intersects both the interior and exterior of a second small circle, then the small circles have a nonempty intersection.

28. Suppose that for every point D in the interior of $\triangle^s ABC$, $m \prec ADB > m \prec ACB$. Prove that $a \leq \frac{\pi}{2}$, $b \leq \frac{\pi}{2}$, $a + c \leq \pi$, and $b + c \leq \pi$. (This is a partial converse of §13, Exercise 14.)

29. Suppose that in $\triangle^s ABC$, $m \prec B + m \prec C < \pi$. Let D be a point in the interior of $\triangle^s ABC$. Prove that there exists a point E between B and C such that $m \prec DEC = m \prec ABC$.

Historical notes. We let X-Y denote Book X, Proposition Y of Menelaus' *Sphaerica*. Then Exercise 13 is I-6, Exercise 15 is I-24, Exercise 16 is I-26, Exercise 20 is part of I-32 and I-34, Exercise 21 is part of I-32 and I-34, Exercise 22 contains the second part of I-37, and Exercise 29 is II-1.

17 Right triangles

In plane trigonometry the student learns a whole series of formulas relating the measures of angles and sides in right triangles. There are a corresponding set of formulas for spherical triangles, three of which we encountered in §15. These are included in the complete list below.

Theorem 17.1 *Let $\triangle^s ABC$ be a spherical triangle with a right angle at C. Then the following ten equations hold.*

$$\begin{aligned}
\cos(c) &= \cos(a)\cos(b) && (4.21)\\
\cos(c) &= \cot(A)\cot(B) && (4.22)\\
\sin(a) &= \sin(A)\sin(c) && (4.23)\\
\cot(c) &= \cot(b)\cos(A) && (4.24)\\
\cot(A) &= \cot(a)\sin(b) && (4.25)\\
\cos(B) &= \sin(A)\cos(b) && (4.26)\\
\sin(b) &= \sin(B)\sin(c) && (4.27)\\
\cot(c) &= \cot(a)\cos(B) && (4.28)\\
\cot(B) &= \cot(b)\sin(a) && (4.29)\\
\cos(A) &= \sin(B)\cos(a) && (4.30)
\end{aligned}$$

These may be rewritten as follows where the expressions are defined:

$$\begin{aligned}
\cos(c) &= \cos(a)\cos(b) && (4.31)\\
\cos(c) &= \cot(A)\cot(B) && (4.32)\\
\sin(A) &= \sin(a)/\sin(c) && (4.33)\\
\cos(A) &= \tan(b)/\tan(c) && (4.34)\\
\tan(A) &= \tan(a)/\sin(b) && (4.35)\\
\sin(A) &= \cos(B)/\cos(b) && (4.36)\\
\sin(B) &= \sin(b)/\sin(c) && (4.37)\\
\cos(B) &= \tan(a)/\tan(c) && (4.38)\\
\tan(B) &= \tan(b)/\sin(a) && (4.39)\\
\sin(B) &= \cos(A)/\cos(a) && (4.40)
\end{aligned}$$

Proof. Equations (4.21) and (4.26) were proven in Theorem 15.3. Equations (4.23) and (4.33) were proven in Proposition 15.2. (But see also §16, Exercises 10 and 11.)

Note that (4.23) is never zero since the sine values are nonzero for angles and sides of a triangle. Thus we may divide (4.26) by (4.23) to obtain $\cos(B)/\sin(a) = \cos(b)/\sin(c)$. Multiplying both sides by $\cos(a)$ results in $\cot(a)\cos(B) = \cos(a)\cos(b)/\sin(c)$, which is $\cos(c)/\sin(c) = \cot(c)$ by (4.21). This is (4.28). Then (4.27), (4.30), and (4.24) are found when the roles of A, a

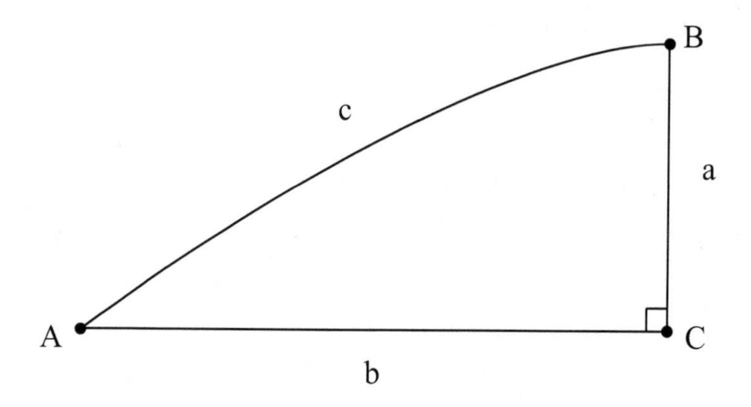

Figure 4.6: Theorem 17.1.

and B, b are reversed in (4.23), (4.26), and (4.28), respectively. Since then $\cos(A) = \sin(B)\cos(a) = \sin(b)\cos(a)/\sin(c)$ and $\sin(A) = \sin(a)/\sin(c)$, we obtain (4.25) by dividing these. Next, take (4.25) and switch the pairs A, a and B, b to get (4.29). Multiplying together (4.25) and (4.29), we get $\cot(A)\cot(B) = \cot(a)\sin(b)\cot(b)\sin(a) = \cos(a)\cos(b)$, which is $\cos(c)$ by (4.21); this gives (4.22).

This proves all of the formulas in the first group of ten. The second group of ten follows from the first group by appropriate solving and division where division by zero does not occur. $\diamondsuit$

Because there are so many right triangle formulas, one might wonder whether there is some convenient device for remembering them. In fact, John Napier invented such a device in the seventeenth century. We here give Napier's *rules of circular elements*, but leave to the reader the question of whether learning it makes remembering the rules easier.

Suppose a spherical right triangle $\triangle^s ABC$ has a right angle at C. (See Figure 4.7.) We again let A, B, and C denote the measures of the angles, and let a, b, and c denote the measures of the opposite sides. Place these quantities in order about a circle: $\frac{\pi}{2} - A, b, a, \frac{\pi}{2} - B, \frac{\pi}{2} - c$. These five quantities are often called "circular elements." Each element has two quantities which are adjacent around the circle, and each element has two quantities which are not adjacent. Note that the order of the elements is the same as that in which the associated letters appear around a spherical triangle, with C missing. Then Napier's rule states that (1) the sine of a given circular element on the circle equals the product of the tangents of the two adjacent circular elements and (2) the sine of any given circular element equals the product of the cosines of the two non-adjacent circular elements. The reader may check that, with suitable algebraic rearrangements, these two rules give rise to all the formulas of Theorem 17.1.

Napier introduced the rules of circular elements in his work *Mirifici Logarithmorum Canonis Descriptio* in 1614. This is the same work in which Napier

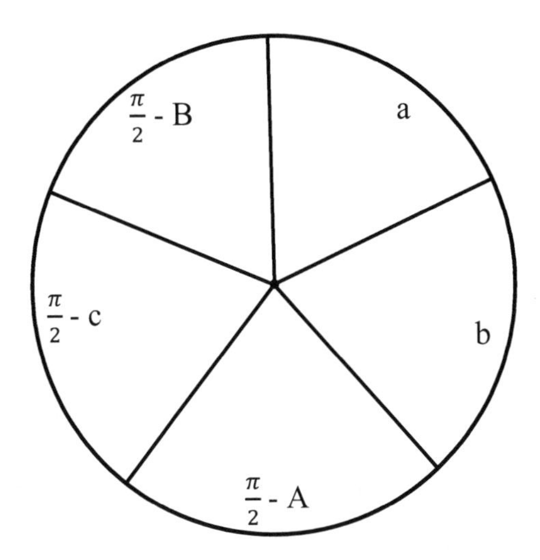

Figure 4.7: Napier's rules of circular elements.

introduced the logarithm. These rules have received mixed reviews historically from the mathematical community, partly on the belief that they do not make the formulas of Theorem 17.1 easier to remember, and partly because to some it was not clear that the rules of circular parts were anything but accidental. In 1843 prominent British mathematician Augustus De Morgan wrote that Napier's rules would "only create confusion instead of assisting the memory." Another mathematician, Robert Woodhouse, Lucasian Professor[3] at the University of Cambridge, wrote that "there is no separate and independent proof of these rules..." American William Chauvenet wrote in 1850 that the rules "though artificial, are very generally employed as aids to memory."[4] But Napier's rules are no mere accident; they have a surprising basis in the geometry of the sphere which is discussed in §19.

The equations in Theorem 17.1 are useful in solving right triangles. That is, given the measures of some of the sides and angles of a spherical right triangle, find the measures of the others. Given three of the five measures a, b, c, A, and B, precisely one of the equations in Theorem 17.1 involves those three. If two of those three are known, we can use the equations to solve for the third. We illustrate with some examples.

Example 1. Suppose that in $\triangle^s ABC$ with right angle at C, $a = 20°$ and $c = 40°$. Solve for b, A, and B.

Solution. By the spherical Pythagorean theorem,

$$\cos(40°) = \cos(20°)\cos(b),$$

so $\cos(b) \approx 0.815$ and $b \approx 35.39°$. We may solve for A using (4.33): $\sin(A) \approx$

[3]This is Newton's chair at Cambridge.

[4]As quoted in [Si2008].

0.53. By Proposition 11.12, A must be acute since a is acute. So $A \approx 32.14°$. We may solve for B by using (4.38) and find $\cos(B) \approx 0.36/0.84 \approx 0.43$ so $B \approx 64.29°$.

Example 2. Suppose that in $\triangle^s ABC$ with right angle at C, $a = 30°$ and $b = 40°$. Solve for c, A, and B.

Solution. We obtain $\cos(c) = \cos(30°)\cos(40°) \approx 0.66$ from the spherical Pythagorean theorem so $c \approx 48.4°$. From here, we could proceed as in Example 1, but in all our examples we will show how to find each part of the triangle directly from the given information. So to find A we use (4.35) and find $\tan(A) = \tan(30°)/\sin(40°) \approx 0.89$ so $A \approx 41.93°$. Similarly, $\tan(B) = \tan(40°)/\sin(30°) \approx 1.68$, so $B \approx 59.21°$.

Example 3. Suppose that in $\triangle^s ABC$ with right angle at C, $a = 30°$ and $B = 50°$. Solve for c, A, and b.

Solution. We may find c using (4.38): $\tan(c) = \tan(30°)/\cos(50°) \approx 0.9$ so $c \approx 41.93°$. We may find b from (4.39): $\tan(b) = \sin(30)\tan(50) \approx 0.59$ so $b \approx 30.79°$. To obtain A we use (4.40): $\cos(A) = \sin(50°)\cos(30°) \approx 0.66$, so $A \approx 48.4°$.

Example 4. Suppose that in $\triangle^s ABC$ with right angle at C, $c = 60°$ and $B = 130°$. Solve for a, A, and b.

Solution. To find a, we use (4.38): $\cos(130°) = \tan(a)/\tan(60°)$, so $\tan(a) = \cos(130°)\tan(60°) \approx -1.11$, and $a \approx 132.01°$. To find b, use (4.37): $\sin(130°) = \sin(b)/\sin(60°)$, so $\sin(b) = \sin(130°)\sin(60°) \approx 0.66$, and b is either $41.56°$ or the supplement $138.44°$. Since B is obtuse, by Proposition 11.12, b must be obtuse also so $b \approx 138.44°$.

Example 5. Suppose that in $\triangle^s ABC$ with right angle at C, $A = 60°$ and $B = 50°$. Solve for a, b, and c.

Solution. For c we use (4.32): $\cos(c) = \cot(50°)\cot(60°) \approx 0.48$, so $c \approx 61.02°$. For b we use (4.36): $\sin(60°) = \cos(50°)/\cos(b)$, so $\cos(b) = \cos(50°)/\sin(60°) \approx 0.74$, so $b \approx 42.07°$. Similarly, for a we have $\sin(50°) = \cos(60°)/\cos(a)$, so $\cos(a) = \cos(60°)/\sin(50°) \approx 0.65$ and $a = 49.25°$.

Example 6. Suppose that in $\triangle^s ABC$ with right angle at C, $b = 40°$ and $B = 50°$. Solve for a, b, and c.

Solution. It is useful to note at the start that the colunar triangle $\triangle^s AB^aC$ also has a right angle at C, an angle of measure $50°$ at B^a and an opposite side of measure $40°$. So we can expect at the start that there are two solutions to this problem (as exhibited in Figure 4.8.) For c we use (4.33): $\sin(50°) = \sin(40°)/\sin(c)$, so $\sin(c) = \sin(40°)/\sin(50°) \approx 0.84$. Then $c \approx$ either $57.04°$ or $122.95°$. For a we use (4.39): $\tan(50°) = \tan(40°)/\sin(a)$, so $\sin(a) = \tan(40°)/\tan(50°) \approx 0.7$. Then $a \approx$ either $44.75°$ or $135.24°$. For A we use (4.36): $\sin(A) = \cos(50°)/\cos(40°) \approx 0.84$, so $A \approx$ either $57.04°$ or $122.95°$. Now A is acute if and only if a is acute by Proposition 11.12. Furthermore a is acute if and only if c is acute (by Exercises 9b and 9c or §11, Exercises 8 and 9.) So a, A, and c are all acute or all obtuse and we obtain two possible solutions: $(a, A, c) \approx$ either $(44.75°, 57.04°, 57.04°)$ or $(135.24°, 122.95°, 122.95°)$.

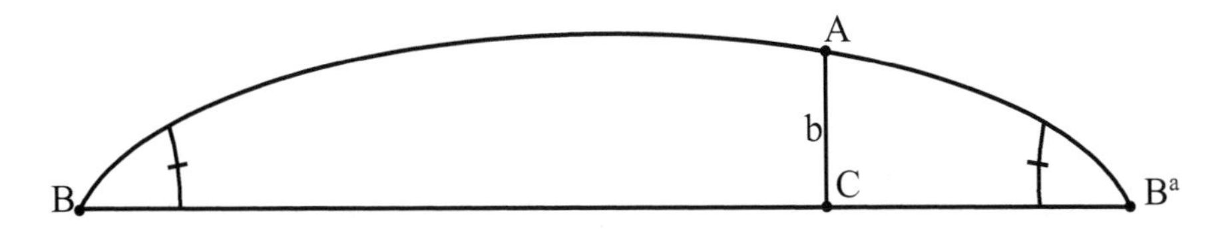

Figure 4.8: The ambiguous case for right spherical triangles: $\triangle^s ABC$ and $\triangle^s AB^aC$ are both solutions to the same problem where $B = m \prec ABC = m \prec AB^aC$ and $b = m \overset{\frown}{AC}$ are given.

We obtained multiple solutions in this example because the equation $\sin(x) = \alpha$ can have two solutions for $0 < x < 180°$. (It can also have zero or one solution.) Because of this ambiguity, we often refer to the problem where we are given a leg and an opposite angle as an "ambiguous case."

The first six examples illustrate cases where the parts of a triangle may be solved for. We further consider cases where the solution does not exist or there are infinitely many solutions.

Example 7. Suppose that in $\triangle^s ABC$ with right angle at C, $c = 40°$ and $a = 50°$. Solve for A, b, and B.

Solution. By (4.21), $\cos(b) = \cos(c)/\cos(a) \approx 0.76/0.64 \approx 1.19$. But the cosine cannot be larger than 1 so no such triangle exists.

Example 8. Suppose that in $\triangle^s ABC$ with right angle at C, $c = 90°$ and $A = 90°$. Solve for a, b, and B.

Solution. The triangle has two right angles at A and C so the opposite sides a and c are also right by Proposition 11.12. But then $B = b$ can be any value between $0°$ and $180°$ so there are infinitely many solutions.

Example 9. Suppose that in $\triangle^s ABC$ with right angle at C, $A = 130°$ and $B = 25°$. Solve for a, b, and c.

Solution. By §13, Exercise 1, we cannot have such a triangle unless $90° < A + B < 270°$ and $|A - B| < 90°$. But here $|A - B| = |130° - 25°| = 105°$ which is too large. So there is no such triangle.

Example 10. Suppose that in $\triangle^s ABC$ with right angle at C, $A = 130°$ and $a = 25°$. Solve for c, b, and B.

Solution. Since $A > 90°$ and $a < 90°$ by Proposition 11.12 no such triangle exists.

Example 11. Suppose that in $\triangle^s ABC$ with right angle at C, $A = 50°$ and $a = 60°$. Solve for c, b, and B.

Solution. By Exercise 12, no such triangle exists.

We organize the solution of a $\triangle^s ABC$ with right angle at C into six cases corresponding to Examples 1 through 6. When we write that the solution is "unique," we mean that it is unique only up to congruence; that is, any two solution triangles must be congruent to each other.

Case 1. Given c and a. If $a = 90°$ then $c = C = A = 90°$ and there are infinitely many solutions for B and b (with $B = b$). If $a \neq 90°$ and c lies between a and $180° - a$, then the triangle can be solved for uniquely. (See Exercise 14.)

Case 2. Given a and b. This can always be solved for any $0 < a, b < 180°$. (See Exercise 15.)

Case 3. Given a and B. This can always be solved for any $0 < a, B < 180°$. (See Exercise 16.)

Case 4. Given c and B. If $B = 90°$, we must have $B = C = b = c = 90°$ to obtain solutions, and then there are infinitely many solutions for A and a (with $A = a$). (If c was initially given to be different from $90°$, there are no solutions.) If $B \neq 90°$, there is a unique solution. (See Exercise 17.)

Case 5. Given A and B. There is a solution if and only if A and B satisfy the inequalities $|A - B| < 90°$ and $90° < A + B < 270°$. (See Figure 4.9 and Exercise 18.)

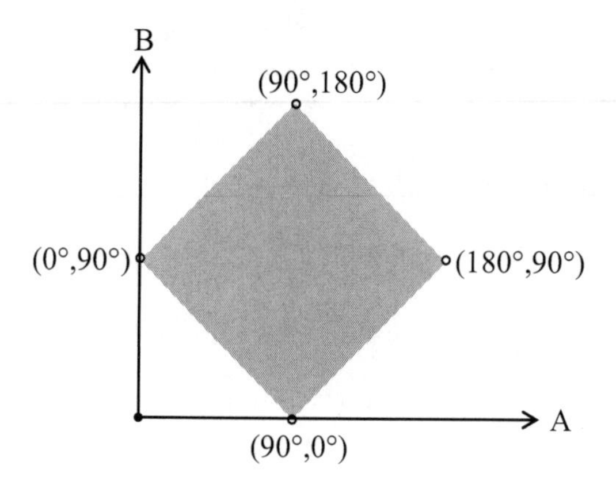

Figure 4.9: Case 5: Region of values for (A, B) for which the right triangle $\triangle^s ABC$ is solvable.

Case 6. Given A and a, the "ambiguous case." For there to be any solution, we have the following necessary conditions. If A is acute, we must have $a \leq A$, if A is right then $a = A = 90°$ and if A is obtuse, we must have $a \geq A$. (See Exercise 12.) If $a = A = 90°$ then $c = C = 90°$ and there are infinitely many solutions for B and b (with $B = b$). If $a = A \neq 90°$ then $b = c = B = C = 90°$. (See Exercise 13.) If $a \neq A$ then there are exactly two solutions of triangles which are colunar to each other. (See Figure 4.8 and Exercise 19.)

A useful (but not foolproof) check may be carried out as follows. In each case one has determined three unknown parts of the right triangle. These three must satisfy some equation among those in Theorem 17.1. For example, in Case 2, we solve for A, B, and c; these must satisfy (4.22). An error made in solution for one of A, B, and c would probably show up when checking (4.22) to be true. Similarly, in Case 4, we solve for a, A, and b. These must satisfy (4.25) and (4.35).

Exercises §17.

In Exercises 1 to 6, solve for the remaining parts of the $\triangle^s ABC$ which has a right angle at C. If this is not possible explain why it is not possible.

1. (a) $c = 70°$, $a = 25°$
   (b) $c = 125°$, $a = 25°$
   (c) $c = 70°$, $a = 120°$
   (d) $c = 90°$, $a = 40°$
   (e) $c = 40°$, $a = 90°$
   (f) $c = 40°$, $a = 60°$
   (g) $c = 40°$, $a = 150°$
2. (a) $a = 20°$, $b = 40°$
   (b) $a = 40°$, $b = 100°$
   (c) $a = 100°$, $b = 160°$
3. (a) $a = 20°$, $B = 40°$
   (b) $a = 90°$, $B = 30°$
   (c) $a = 20°$, $B = 90°$
   (d) $a = 40°$, $B = 110°$
   (e) $a = 120°$, $B = 130°$
4. (a) $c = 40°$, $B = 90°$
   (b) $c = 90°$, $B = 90°$
   (c) $c = 90°$, $B = 40°$
   (d) $c = 60°$, $B = 40°$
   (e) $c = 70°$, $B = 130°$
   (f) $c = 110°$, $B = 140°$
5. (a) $A = 80°$, $B = 20°$
   (b) $A = 80°$, $B = 70°$
   (c) $A = 47°$, $B = 46°$
   (d) $A = 135°$, $B = 50°$
   (e) $A = 135°$, $B = 133°$
   (f) $A = 45.1°$, $B = 45.1°$
   (g) $A = 65.2°$, $B = 154.9°$
   (h) $A = 90°$, $B = 40°$
   (i) $A = 100°$, $B = 70°$

(j) $A = 40°$, $B = 40°$

(k) $A = 140°$, $B = 40°$

(l) $A = 130°$, $B = 150°$

6. (a) $a = 70°$, $A = 100°$

   (b) $a = 80°$, $A = 70°$

   (c) $a = 145°$, $A = 135°$

   (d) $a = 80°$, $A = 80°$

   (e) $a = 140°$, $A = 140°$

   (f) $a = 40°$, $A = 60°$

   (g) $a = 150°$, $A = 110°$

   (h) $a = 10°$, $A = 30°$

   (i) $a = 140°$, $A = 100°$

   (j) $a = 70°$, $A = 100°$

7. Show that an equilateral spherical triangle with side a and angle A satisfies $2\sin\frac{A}{2}\cos\frac{a}{2} = 1$.

8. Verify that Napier's rules are equivalent to Theorem 17.1.

9. Show how to use Theorem 17.1 to prove the following, which were discussed earlier with more elementary methods.

   (a) If the hypotenuse of a spherical right triangle has measure $\frac{\pi}{2}$, then one of the angles opposite the other sides must be a right angle.

   (b) If the hypotenuse of a spherical right triangle is acute, then the other sides are not right and must be both acute or both obtuse.

   (c) If the hypotenuse of a spherical right triangle is obtuse, the other sides are not right, one must be acute, and the other obtuse.
   (Compare to §11, Exercises 6 through 9.)

   (d) Prove Proposition 11.12: that in $\triangle^s ABC$ with right angle at C, a and A are both right, both acute, or both obtuse.

   (e) Prove Theorem 12.10.

   (f) Prove Proposition 11.14.

10. Suppose that in a spherical right triangle with right angle at C, $a \neq 90°$. Show that $a < c < 180° - a$ or $180° - a < c < a$.

11. Suppose that in a spherical right triangle with right angle at C, B and b are complementary (that is, $B + b = 90°$.) Show that $A = c$.

12. In spherical right triangle with right angle at C, show that if A is acute then $a \leq A$ and if A is obtuse then $a \geq A$.

13. In spherical right triangle with right angle at C, show that if $a = A$ then either $a = A = 90°$ or all other sides and angles are right.

14. Prove the existence and uniqueness of solutions stated in Case 1.

15. Prove the existence and uniqueness of solutions stated in Case 2.

16. Prove the existence and uniqueness of solutions stated in Case 3.

17. Prove the existence and uniqueness of solutions stated in Case 4.

18. Prove the existence and uniqueness of solutions stated in Case 5.

19. Prove the conditions for two solutions stated in Case 6: assume that if A is acute, $a \leq A$ and if A is obtuse then $a \geq A$. Then if $a \neq A$ there are exactly two solutions which are colunar to each other.

20. What right triangle formulas can be derived from the analogue formula?

21. Suppose that in an isosceles triangle $\triangle^s ABC$ we have $a = b$ (so $A = B$). Show that $\cos(a) = \cot(B)\cot(\frac{C}{2})$ and $\cos(B)\cot(\frac{c}{2}) = \cot(a)$. Explain why these equations allow for the solution of C and c in terms of A, B, a and b in the event that A and B are not right angles. What happens if A and B are right angles?

22. Refer to Definition 10.16. Suppose that if E is a great circle and X is a point which is not a pole of E. Let Y be the foot of the shorter perpendicular from X to E and let Z be the foot of the longer perpendicular from X to E. Prove that for any W on E other than Y or Z, $m\,\overset{\frown}{XY} < m\,\overset{\frown}{XW} < m\,\overset{\frown}{XZ}$.

23. Let $\triangle^s ABC$ be a spherical triangle. Let the foot of the shorter perpendicular from A to $\bigcirc BC$ be D and let the foot of the longer perpendicular be E. Prove that $\sin(AD) = \sin(AE) = \sin(b)\sin(C) = \sin(c)\sin(B)$.

24. Prove that the great circles containing the three altitudes of a spherical triangle pass through a single point, provided those altitudes are well-defined.

25. Suppose that P is a pole of $\bigcirc AB$, that C is between P and B, B is between A and D, E is between P and D and C is between A and E. Prove that

$$\frac{\sin(\overset{\frown}{BD})}{\sin(\overset{\frown}{CE})} = \frac{\cos(\sphericalangle BAC)}{\cos(\overset{\frown}{BC})\cos(\overset{\frown}{DE})}. \tag{4.41}$$

Historical notes. We let X-Y denote Book X, Proposition Y of Menelaus' *Sphaerica*. Then Exercise 9e is I-12, Exercise 24 is III-13, and Exercise 25 is III-22.

18 The four-parts and half-angle formulas

In this section we derive several other formulas which express trigonometric relationships among the sides and angles of a spherical triangle. These will all be consequences of the spherical laws of sines and cosines and the analogue formula (which itself was derived from the law of cosines.) The first formula, which appears to be preferred by astronomers, is called the *four-parts formula*[5] because it relates the values of four of the six parts a, b, c, A, B, and C consecutively as one moves around the perimeter of a triangle. It is a mild variation of the analogue formula, merely reducing the number of variables from five to four.

Theorem 18.1 (The four-parts formula) *In any spherical* $\triangle^s ABC$,

$$\sin(C)\cot(B) = \cot(b)\sin(a) - \cos(a)\cos(C) \tag{4.42}$$

By rotating the letters in (4.42), we also obtain:

$$\sin(A)\cot(C) = \cot(c)\sin(b) - \cos(b)\cos(A) \tag{4.43}$$
$$\sin(B)\cot(C) = \cot(c)\sin(a) - \cos(a)\cos(B) \tag{4.44}$$
$$\sin(C)\cot(A) = \cot(a)\sin(b) - \cos(b)\cos(C) \tag{4.45}$$
$$\sin(B)\cot(A) = \cot(a)\sin(c) - \cos(c)\cos(B) \tag{4.46}$$
$$\sin(A)\cot(B) = \cot(b)\sin(c) - \cos(c)\cos(A) \tag{4.47}$$

Proof. We begin with the analogue formula (4.13)

$$\sin(c)\cos(B) = \cos(b)\sin(a) - \sin(b)\cos(a)\cos(C)$$

and divide by $\sin(b)$ to obtain

$$\frac{\sin(c)}{\sin(b)}\cos(B) = \cot(b)\sin(a) - \cos(a)\cos(C). \tag{4.48}$$

From the spherical law of sines (4.3), $\frac{\sin(c)}{\sin(b)} = \frac{\sin(C)}{\sin(B)}$, which we substitute into (4.48) to obtain

$$\frac{\sin(C)}{\sin(B)}\cos(B) = \cot(b)\sin(a) - \cos(a)\cos(C),$$

which reduces to (4.42) by the definition of the cotangent. ◇

For any spherical triangle with side lengths a, b, and c, the *perimeter* is the quantity $a+b+c$ and we let $s = (a+b+c)/2$ denote the *semiperimeter*.

[5] We follow [Gr1985] in using this name.

Theorem 18.2 (The half-angle formulas) *In any spherical* $\triangle^s ABC$:

$$\sin(\frac{C}{2}) = \sqrt{\frac{\sin(s-a)\sin(s-b)}{\sin(a)\sin(b)}} \quad (4.49)$$

$$\cos(\frac{C}{2}) = \sqrt{\frac{\sin(s)\sin(s-c)}{\sin(a)\sin(b)}} \quad (4.50)$$

$$\tan(\frac{C}{2}) = \sqrt{\frac{\sin(s-a)\sin(s-b)}{\sin(s)\sin(s-c)}} \quad (4.51)$$

Proof. We begin with the law of cosines (4.10) and solve for $\cos(C)$:

$$\cos(C) = \frac{\cos(c) - \cos(a)\cos(b)}{\sin(a)\sin(b)}. \quad (4.52)$$

From (1.14),

$$\begin{aligned} \sin^2(\frac{C}{2}) &= \frac{1-\cos(C)}{2} \\ &= \frac{\cos(a)\cos(b) + \sin(a)\sin(b) - \cos(c)}{2\sin(a)\sin(b)} \\ &= \frac{\cos(a-b) - \cos(c)}{2\sin(a)\sin(b)}. \end{aligned} \quad (4.53)$$

Using the sum-to-product formulas,

$$\cos(a-b) - \cos(c) = -2\sin(\frac{a-b+c}{2})\sin(\frac{a-b-c}{2}),$$

which is $2\sin(\frac{a-b+c}{2})\sin(\frac{b+c-a}{2})$, since the sine is odd. But this is $2\sin(s-b)\sin(s-a)$; substituting in (4.53), we find

$$\sin^2(\frac{C}{2}) = \frac{\sin(s-a)\sin(s-b)}{\sin(a)\sin(b)} \quad (4.54)$$

Since $\prec C$ is an angle of a triangle, $0 < C < \pi$, so $0 < \frac{C}{2} < \frac{\pi}{2}$, and for such a $\frac{C}{2}$, the sine is always positive, so we can solve (4.54) by taking a square root to obtain (4.49). The equation (4.50) is proven similarly and is left as an exercise (or see §19.) The equation (4.51) is found by dividing the first two. ♢

Similar formulas can be found for $\prec A$ and $\prec B$ by permuting the vertices. These formulas allow one to calculate an angle in terms of the sides. Of course, the reader might wonder why we would need such a formula, given that we can easily solve for $\cos(C)$ as in (4.52). The reason is partly historic: prior to the invention of electronic computers, a formula was valued if it was easy to use with logarithms. The mathematician would have at his disposal tables of values of trigonometric, inverse trigonometric, exponential, and logarithmic functions. Decimal values obtained from those tables would be easy to add and subtract, but difficult to multiply and divide where needed. With the half-angle formulas, one would take the logarithm of both sides and use the laws of logarithms to break the expression into sums of pieces, all of which would have the form of the composition of a logarithm and a trigonometric function. Each term could be calculated. Then they would all be added together, and one could get the value of $\frac{C}{2}$ by first exponentiating the right side and then taking an inverse trigonometric function. For example, with (4.49), if we take the natural logarithm ln of both sides, we obtain

$$\ln(\sin(\frac{C}{2})) = \frac{1}{2}(\ln(\sin(s-a)) + \ln(\sin(s-b)) - \ln(\sin(a)) - \ln(\sin(b))),$$

using the laws of logarithms. Given a, b and c, only addition and subtraction is required to obtain $s-a$ and $s-b$. If we have a table of the $\ln(\sin(\cdot))$ function, we may obtain the entire right side via addition, subtraction, and division by 2. We then have the value of $\ln(\sin(\frac{C}{2}))$. Using the (inverse of the) $\ln(\sin(\cdot))$ table again, we find the value of $\frac{C}{2}$, and then C.

The law of cosines formula does not work as well because it involves both multiplication and addition of trigonometric quantities which are, in most cases, decimal expressions. Taking the logarithm of the law of cosines does not make the formula more workable.

The half-angle formulas have some other uses that we explore in the exercises of this section and the next.

Exercises §18

1. Following the pattern used in the proof of (4.49), prove (4.50).

2. Let $\triangle^s ABC$ be any spherical triangle. Recalling the double angle formula for the sine (1.13), write $\sin(A) = 2\sin(\frac{A}{2})\cos(\frac{A}{2})$. Then use the half angle formulas from Theorem 18.2 to prove the extended law of sines

$$\frac{\sin(a)}{\sin(A)} = \frac{\sin(b)}{\sin(B)} = \frac{\sin(c)}{\sin(C)} = \frac{\sin(a)\sin(b)\sin(c)}{2\sqrt{\sin(s)\sin(s-a)\sin(s-b)\sin(s-c)}}. \tag{4.55}$$

3. Use sum-to-product formulas and double angle formulas from §3 to prove that

$$\begin{aligned} & 2\sqrt{\sin(s)\sin(s-a)\sin(s-b)\sin(s-c)} \qquad (4.56) \\ = \; & \sqrt{1-\cos^2(a)-\cos^2(b)-\cos^2(c)+2\cos(a)\cos(b)\cos(c)}. \end{aligned}$$

4. Let r be the radius of the inscribed circle of $\triangle^s ABC$. Prove that

$$\tan(r) = \sqrt{\frac{\sin(s-a)\sin(s-b)\sin(s-c)}{\sin(s)}} \tag{4.57}$$

where $s = \frac{1}{2}(a+b+c)$.

5. What formulas in Theorem 17.1 can be derived from the various versions of the four-parts formula of Theorem 18.1?

6. Use Theorem 18.2 and the sum-to-product formulas (1.15) to prove:

$$\frac{\sin\frac{1}{2}(A-B)}{\sin\frac{1}{2}(A+B)} = \frac{\tan\frac{1}{2}(a-b)}{\tan\frac{1}{2}c} \tag{4.58}$$

$$\frac{\cos\frac{1}{2}(A-B)}{\cos\frac{1}{2}(A+B)} = \frac{\tan\frac{1}{2}(a+b)}{\tan\frac{1}{2}c} \tag{4.59}$$

These are two of *Napier's analogies.*[6] See also §19, problem 11. Provided that A, B, a, and b are known, $A \neq B$ (so $a \neq b$) these equations allow one to solve for c. The case where $A = B$ is dealt with in §17, problem 21.

7. Use trigonometry to prove the SSAA congruence theorem (Theorem 12.6).

8. Suppose that in $\triangle^s ABC$, point D is chosen between B and C. Let a, b, c, A, B, C denote the usual measures of sides and angles. Let $d = m\,\overset{\frown}{AD}, a_1 = m\,\overset{\frown}{BD}$, $a_2 = m\,\overset{\frown}{CD}$, $A_1 = m\,\sphericalangle BAD$ and $A_2 = m\,\sphericalangle CAD$.

   (a) Prove that $\cot(b)\sin(A_1) + \cot(c)\sin(A_2) = \cot(d)\sin(A)$.

   (b) Prove that $-\cot(B)\sin(a_2) + \cot(C)\sin(a_1) = \cot(\sphericalangle ADB)\sin(a)$.

9. Suppose that a, b, and c are three real numbers between 0 and π such that $a+b > c$, $a+c > b$, $b+c > a$, and $a+b+c < 2\pi$. Use (4.51) and (4.10) to argue that there exists a spherical triangle with sides whose lengths are a, b, and c. You may assume that given a positive real number r there exists an angle $0 < \alpha < \frac{\pi}{2}$ such that $\tan(\alpha) = r$.

10. Suppose that in a spherical triangle the sum of the measures of two sides is less than π. Consider an arc from their common vertex to the opposite side. If this arc bisects the angle, prove that the arc has measure less than $\frac{\pi}{2}$. Furthermore, conclude that the measure of the arc is less than half of the sum of the measures of the adjacent sides. (Compare Exercises 36 and 37 of §13 and Exercise 21 of §16.)

[6]See §34 for a brief discussion of the archaic use of the word "analogy" here.

11. Use the four-parts formula to do §13, Exercise 44.

Historical notes. We let X-Y denote Book X, Proposition Y of Menelaus' *Sphaerica*. Exercise 11 is I-27, Exercise 10 is part of I-32 and I-34, and Exercise 7 is I-13 and I-17.

19 Dualization

We have seen already how the polar and colunar triangles can be used to create new theorems from existing theorems. By applying the statement of a theorem about spherical triangles to the polar or colunar triangle of a given triangle, we may discover a new fact about the given triangle.

Now that we have a number of formulas showing trigonometric relationships among the sides and angles of a spherical triangle, we will show how all of these formulas can be "dualized" via the polar and colunar triangles to obtain many other related formulas.

Let us begin with the law of cosines. Note that since (4.10) is true for all triangles, it holds for the polar triangle $\triangle^s A'B'C'$:

$$\cos(c') = \cos(a')\cos(b') + \sin(a')\sin(b')\cos(C').$$

But from Theorem 11.19, $c' = \pi - C$, $a' = \pi - A$, $b' = \pi - B$, $C' = \pi - c$, so

$$\cos(\pi - C) = \cos(\pi - A)\cos(\pi - B) + \sin(\pi - A)\sin(\pi - B)\cos(\pi - c).$$

Using standard properties of the cosine and sine (see (1.1)), (1.2)),

$$-\cos(C) = \cos(A)\cos(B) - \sin(A)\sin(B)\cos(c),$$

or

$$\cos(C) = -\cos(A)\cos(B) + \sin(A)\sin(B)\cos(c), \tag{4.60}$$

which expresses the cosine of an angle of $\triangle^s ABC$ in terms of the other two angles and their included side.

By making use of the polar triangle as done in section 16 we may attempt to dualize the law of sines also, but it produces the same formula.

In general, any formula in a spherical triangle involving the quantities a, b, c, A, B, and C may be replaced by the values $\pi - A$, $\pi - B$, $\pi - C$, $\pi - a$, $\pi - b$, and $\pi - c$, respectively, to obtain another formula.

We illustrate this idea with one other example. We can use the half-angle formulas in Theorem 18.2 to obtain what might be called the "half-side formulas." Let us begin with (4.49); it is true for the polar triangle of $\triangle^s ABC$, so

$$\sin(\frac{C'}{2}) = \sqrt{\frac{\sin(s' - a')\sin(s' - b')}{\sin(a')\sin(b')}} \tag{4.61}$$

where $s' = \frac{a'+b'+c'}{2}$. But using Theorem 11.19, $C' = \pi - c$, $a' = \pi - A$, $b' = \pi - B$, and

$$\begin{aligned} s' - a' &= \frac{b' + c' - a'}{2} \\ &= \frac{(\pi - B) + (\pi - C) - (\pi - A)}{2} \\ &= \frac{\pi - (A + B + C) + 2A}{2} \\ &= \frac{-2E + 2A}{2} \\ &= A - E, \end{aligned}$$

where E is the spherical excess as defined in Definition 14.5. Similarly we can show $s' - b' = B - E$. Then we substitute these into (4.61) to obtain

$$\sin(\frac{\pi - c}{2}) = \sqrt{\frac{\sin(A - E)\sin(B - E)}{\sin(\pi - A)\sin(\pi - B)}}$$

which, using the standard properties of sine ((1.1) and (1.3)) results in

$$\cos(\frac{c}{2}) = \sqrt{\frac{\sin(A - E)\sin(B - E)}{\sin(A)\sin(B)}} \tag{4.62}$$

Similar formulas for sine and tangent are left to the exercises. As before, we may permute the vertices to obtain formulas for $\cos(\frac{a}{2})$ and $\cos(\frac{b}{2})$, etc., as well.

A similar method applies with colunar triangles. A given triangle with parts a, b, c, A, B, C has a colunar triangle with corresponding parts a, $\pi - b$, $\pi - c$, A, $\pi - B$, $\pi - C$. Thus any relationship among the parts a, b, c, A, B, C is also true when these values are replaced by a, $\pi - b$, $\pi - c$, A, $\pi - B$, $\pi - C$, respectively.

Let us apply this to Theorem 18.2. There we proved (4.49) and left the similar argument to (4.50) to the reader. But we can derive (4.50) from (4.49) via a colunar triangle to $\triangle^s ABC$. We can replace the values a, b, c, A, B, C by a, $\pi - b$, $\pi - c$, A, $\pi - B$, $\pi - C$, respectively, in (4.49). Recalling that $s = (a + b + c)/2$, we obtain

$$\sin(\frac{\pi - C}{2}) = \sqrt{\frac{\sin\left(\frac{-a+(\pi-b)+(\pi-c)}{2}\right)\sin\left(\frac{a-(\pi-b)+(\pi-c)}{2}\right)}{\sin(a)\sin(\pi - b)}},$$

which simplifies to (4.50).

In §18, Exercise 6, we can see that (4.58) and (4.59) can be derived from each other via a colunar triangle. If we replace a, b, c, A, B, C by a, $\pi - b$, $\pi - c$, A, $\pi - B$, $\pi - C$, respectively, in (4.58), we obtain (4.59) (and vice versa).

In Exercise 11 the reader will see how to use the polar triangle dualization on these equations to produce two more of Napier's analogies from the first two.

We conclude this section with a discussion of an unusual construction for right triangles on the sphere used in the proof of the spherical Pythagorean Theorem that allows us to take a single theorem for a right triangle and create not just one, but four other theorems!

Let us begin with $\triangle^s ABC$ with right angle at C. (See Figure 4.10.) Assume that the measures of the sides are all acute, and A, B, C, a, b, and c represent the measures of the angles and sides of $\triangle^s ABC$ in the usual manner. Choose points D and E on $\overset{\hookrightarrow}{AB}$ and $\overset{\hookrightarrow}{AC}$, respectively, such that $m\ \overset{\frown}{AD} = m\ \overset{\frown}{AE} = \frac{\pi}{2}$. Let F be the point where $\overset{\hookrightarrow}{ED}$ meets $\overset{\hookrightarrow}{CB}$ and consider $\triangle^s BFD$. We may calculate the measures of its sides and angles. Since $m\ \overset{\frown}{AD} = m\ \overset{\frown}{AE} = \frac{\pi}{2}$, $\prec BDF$ and $\prec AEF$ are right, and $m\ \overset{\frown}{DE} = m \prec BAC$ by Definition 10.2. Since $\prec ACF$ and $\prec AEF$ are right, the opposite sides $\overset{\frown}{FC}$ and $\overset{\frown}{FE}$ are right (Theorem 11.11) and $m \prec BFE = m\ \overset{\frown}{CE}$ (Def. 10.2). Then $\triangle^s BDF$ has a right angle at D, $m \prec DBF = B$, $m \prec BFD = \frac{\pi}{2} - m\ \overset{\frown}{CE} = \frac{\pi}{2} - b$, $m\ \overset{\frown}{BF} = \frac{\pi}{2} - m\ \overset{\frown}{BC} = \frac{\pi}{2} - a$, $m\ \overset{\frown}{BD} = \frac{\pi}{2} - m\ \overset{\frown}{AB} = \frac{\pi}{2} - c$, and $m\ \overset{\frown}{DF} = \frac{\pi}{2} - m\ \overset{\frown}{DE} = \frac{\pi}{2} - A$. Because of the relationship between $\triangle^s ABC$ and $\triangle^s BDF$, they have sometimes been called *complemental* triangles.

Since $\triangle^s BDF$ is right, a statement relating its sides and angles produces a statement about the sides and angles of the original $\triangle^s ABC$. For example, in the proof of the spherical Pythagorean theorem we used the fact that

$$\sin(\prec BFD) = \sin(\overset{\frown}{BD}) / \sin(\overset{\frown}{BF})$$

in $\triangle^s BDF$, so $\sin(\frac{\pi}{2} - b) = \sin(\frac{\pi}{2} - c) / \sin(\frac{\pi}{2} - a)$ and $\cos(c) = \cos(a)\cos(b)$. But in $\triangle^s BDF$ we also have $\sin(B) = \sin(\frac{\pi}{2} - A) / \sin(\frac{\pi}{2} - a)$, or $\sin(B) = \cos(A)/\cos(a)$, which is (4.40). Each of the ten relationships among sides and angles in $\triangle^s BDF$ corresponds to such a relationship in $\triangle^s ABC$.

As we did with polar and colunar triangles, the natural question is: what happens if we apply the same process to right $\triangle^s BDF$ to obtain a third right triangle? Do we get $\triangle^s ABC$ back again? If we let vertex F play the role that A did in the original construction, then the answer is yes.

But what happens if we let B play the role that A did? A quick look shows that we could not possibly get $\triangle^s ABC$ back. Let us match the vertices of $\triangle^s ABC$ and $\triangle^s BFD$ by matching A, B, C with B, F, D, respectively, and create a table showing how the measures of corresponding sides and angles compare:

$\triangle^s ABC$	a	A	B	b	c
$\triangle^s BFD$	$\frac{\pi}{2} - A$	B	$\frac{\pi}{2} - b$	$\frac{\pi}{2} - c$	$\frac{\pi}{2} - a$

Note that in this ordering of sides and angles, the letters a, A, B, b, c in the line for $\triangle^s ABC$ appear in a similar ordering in the second line — merely shifted

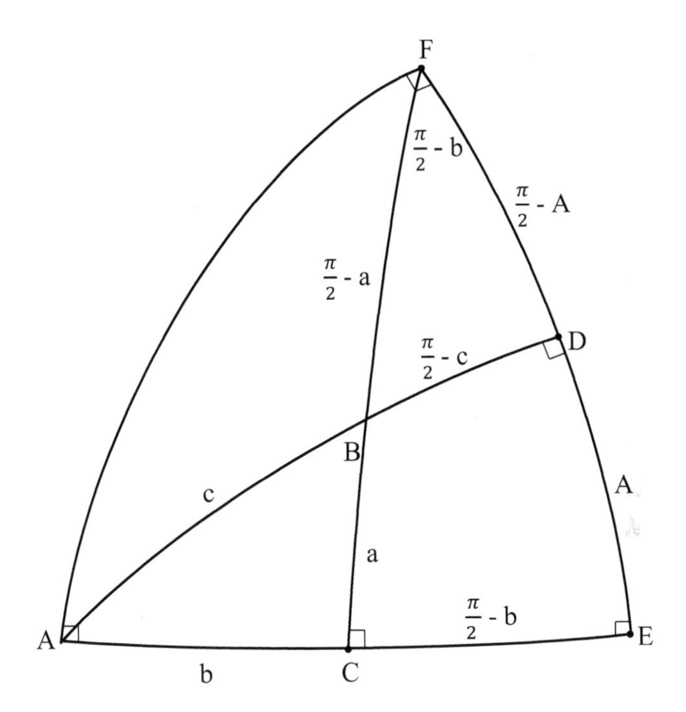

Figure 4.10: The "complemental" triangles $\triangle^s ABC$ and $\triangle^s BDF$.

just a column to the left (thinking of the c column as being just to the left of the a column.) Then if we apply the original transformation to $\triangle^s BFD$, it is easy to see what the measures of the corresponding sides and angles of the third triangle will be:

$\triangle^s ABC$	a	A	B	b	c
$\triangle^s BFD$	$\frac{\pi}{2} - A$	B	$\frac{\pi}{2} - b$	$\frac{\pi}{2} - c$	$\frac{\pi}{2} - a$
	$\frac{\pi}{2} - B$	$\frac{\pi}{2} - b$	c	a	A

The third triangle is different from $\triangle^s ABC$, as its sides and angles have different measures. But we still may apply a rule for right triangles to the third triangle to obtain a true relationship among the sides and angles in $\triangle^s ABC$.

Since the third triangle on the list is different from the first, we might wonder what would happen if we continue to apply the transformation successively to the triangles obtained. It is easy to compose an extended list showing the measures of corresponding sides and angles:

$\triangle^s ABC$	a	A	B	b	c
$\triangle^s BFD$	$\frac{\pi}{2} - A$	B	$\frac{\pi}{2} - b$	$\frac{\pi}{2} - c$	$\frac{\pi}{2} - a$
	$\frac{\pi}{2} - B$	$\frac{\pi}{2} - b$	c	a	A
	b	c	$\frac{\pi}{2} - a$	$\frac{\pi}{2} - A$	B
	$\frac{\pi}{2} - c$	$\frac{\pi}{2} - a$	A	$\frac{\pi}{2} - B$	$\frac{\pi}{2} - b$
	a	A	B	b	c

We may continue to apply any rule for sides and angles of a right triangle to the sides and angles on each row in the list to obtain true statements about the original right triangle, but the side/angle sequence returns to the side/angle sequence of the original triangle after five iterations of the transformation process. Thus there is a limit to how we can milk the triangles on the list for new theorems about the original right triangle.

The surprising fact is that not only is the last triangle on the list congruent to the original triangle, it is the same triangle!

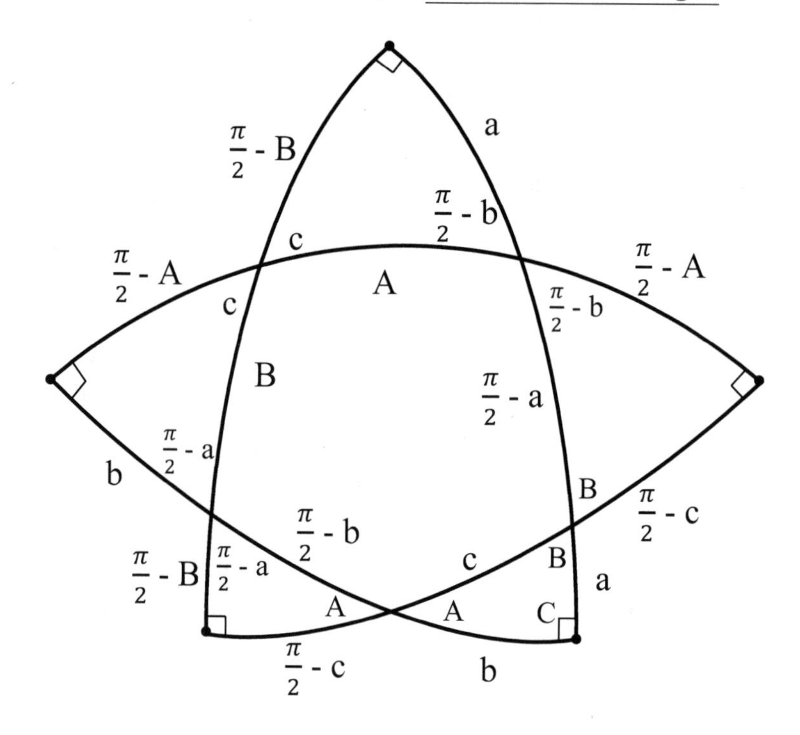

Figure 4.11: The *pentagramma mirificum.*

This fact is depicted in Figure 4.11: we have five spherical right triangles connected successively by a common vertex creating a curious five-pointed figure that John Napier introduced and which has become known as the *pentagramma mirificum.* If we take the spherical Pythagorean theorem and apply it to each of the five triangles, we obtain five of the identities in Theorem 17.1. If we apply the identity $\cos(c) = \cot(A)\cot(B)$ to all five of the triangles, we get the other five identities of Theorem 17.1. (Verification is left as an exercise.)

But where does Napier's diagram of circular elements come from? (See Figure 4.7.) Note that in this diagram, Napier's rules of circular elements applied centered at $\frac{\pi}{2} - c$ gives $\sin(\frac{\pi}{2} - c) = \cos(a)\cos(b) = \tan(\frac{\pi}{2} - A)\tan(\frac{\pi}{2} - B)$, or $\cos(c) = \cos(a)\cos(b) = \cot(A)\cot(B)$. In the diagram we can view the right angle as being located between a and b as they are in $\triangle^s ABC$. But why do the rules work when centered on the other parts?

The answer lies in another look at the table above. Let us take this table, delete the superfluous bottom row and reorder the columns so that the elements on the top row are ordered b, A, c, B, a, which is the order in which those elements appear around the triangles:

$\triangle^s ABC$	b	A	c	B	a
$\triangle^s BFD$	$\frac{\pi}{2} - c$	B	$\frac{\pi}{2} - a$	$\frac{\pi}{2} - b$	$\frac{\pi}{2} - A$
	a	$\frac{\pi}{2} - b$	A	c	$\frac{\pi}{2} - B$
	$\frac{\pi}{2} - A$	c	B	$\frac{\pi}{2} - a$	b
	$\frac{\pi}{2} - B$	$\frac{\pi}{2} - a$	$\frac{\pi}{2} - b$	A	$\frac{\pi}{2} - c$

Next, let us replace the values in the second, third, and fourth columns of elements by their complements:

$\triangle^s ABC$	b	$\frac{\pi}{2} - A$	$\frac{\pi}{2} - c$	$\frac{\pi}{2} - B$	a
$\triangle^s BFD$	$\frac{\pi}{2} - c$	$\frac{\pi}{2} - B$	a	b	$\frac{\pi}{2} - A$
	a	b	$\frac{\pi}{2} - A$	$\frac{\pi}{2} - c$	$\frac{\pi}{2} - B$
	$\frac{\pi}{2} - A$	$\frac{\pi}{2} - c$	$\frac{\pi}{2} - B$	a	b
	$\frac{\pi}{2} - B$	a	b	$\frac{\pi}{2} - A$	$\frac{\pi}{2} - c$

Note that every row contains the same five elements. These are the five elements of Napier's diagram!

By replacing the elements in the second, third, and fourth columns with their complements, the property $\cos(c) = \cot(A)\cot(B)$ (or $\sin(\frac{\pi}{2} - c) = \tan(\frac{\pi}{2} - A)\tan(\frac{\pi}{2} - B)$) for each of the five triangles in the pentagram guarantees that

$$\sin(3^{rd}\text{ column element}) = \tan(2^{nd}\text{ column element})\tan(4^{th}\text{ column element})$$

in every row. This is Napier's first rule (adjacent parts) applied around the circular diagram. Furthermore, the spherical Pythagorean theorem applied to each of the five triangles guarantees that

$$\sin(3^{rd}\text{ column element}) = \cos(1^{st}\text{ column element})\cos(5^{th}\text{ column element})$$

in every row. This is Napier's second rule (non-adjacent parts).

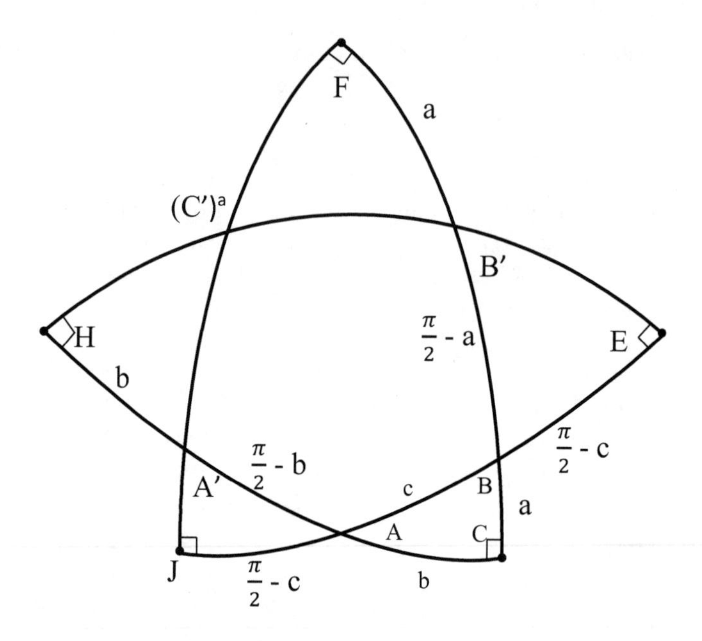

Figure 4.12: The *pentagramma mirificum.*

We conclude by discussing why the "miracle" of the pentagram occurs in the first place; that is, why does the sixth triangle on the list above turn out to be the same as the first? We here give a geometric explanation and in §32 another explanation.

Looking at Figure 4.12, suppose that the sides of $\triangle^s ABC$ are acute. Suppose that from the start we built the polar circles of A and B. Let E and H be the points where $\overleftrightarrow{AB}$ and $\overleftrightarrow{CA}$, respectively, meet the polar circle of A. Let F and J be the points where $\overleftrightarrow{CB}$ and $\overleftrightarrow{BA}$ meet the polar circle of B. The angles formed at E, F, H, and J are all right angles, and $\overset{\frown}{AE}$, $\overset{\frown}{AH}$, $\overset{\frown}{BF}$, and $\overset{\frown}{BJ}$ are all quarter circles. Then it may be laboriously checked (see Exercise 19) that the five successive right triangles of the pentagram constructed above are in fact $\triangle^s ABC$, $\triangle^s BB'E$, $\triangle^s B'(C')^a F$, $\triangle^s (C')^a A'H$, and $\triangle^s A'AJ$.

Exercises §19.

1. Find the dual via polar triangles of the analogue formula (4.13).

2. In each of the following, we are given the values of A, B and c in $\triangle^s ABC$. Determine the value of C. (Use the dual law of cosines (4.60).)

   (a) $A = 60°$, $B = 40°$, $c = 70°$

   (b) $A = 50°$, $B = 40°$, $c = 110°$

   (c) $A = 60°$, $B = 40°$, $c = 170°$

(d) $A = 60°$, $B = 140°$, $c = 70°$

(e) $A = 60°$, $B = 140°$, $c = 170°$

3. For each of the triangles in Exercise 2, determine the values of a and b. (Use the dual law of cosines (4.60) and the dual analogue formula in Exercise 1.)

4. Let E be half the spherical excess of $\triangle^s ABC$. Prove

$$\sin(\frac{c}{2}) = \sqrt{\frac{\sin(E)\sin(C-E)}{\sin(A)\sin(B)}} \tag{4.63}$$

5. Let E be half the spherical excess of $\triangle^s ABC$. Prove

$$\tan(\frac{c}{2}) = \sqrt{\frac{\sin(E)\sin(C-E)}{\sin(A-E)\sin(B-E)}} \tag{4.64}$$

6. Prove that a triangle is quadrantal if and only if its polar triangle is a right triangle.

7. Suppose that $\triangle^s ABC$ is quadrantal, where $\overset{\frown}{AB}$ is a right side. Using Exercise 6 and Theorem 17.1, prove the following identities: $\cos(C) = -\cos(A)\cos(B)$, $\sin(a) = \sin(A)/\sin(C)$, and $\tan(a) = \tan(A)/\sin(B)$.

8. Suppose that $\triangle^s ABC$ is quadrantal, where $\overset{\frown}{AB}$ is a right side. Using Exercise 6, and Theorem 17.1, prove the following identities: $\cos(C) = -\cot(a)\cot(b)$, $\cos(a) = -\tan(B)/\tan(C)$, and $\sin(a) = \cos(b)/\cos(B)$ (assuming no denominators are zero).

9. What formulas in Theorem 17.1 can be derived from (4.60)?

10. What formulas can be obtained from dualizing the various versions of the four-parts formula of Theorem 18.1?

11. Use §18 problem 6 to prove:

$$\frac{\sin\frac{1}{2}(a-b)}{\sin\frac{1}{2}(a+b)} = \frac{\tan\frac{1}{2}(A-B)}{\cot\frac{1}{2}C} \tag{4.65}$$

$$\frac{\cos\frac{1}{2}(a-b)}{\cos\frac{1}{2}(a+b)} = \frac{\tan\frac{1}{2}(A+B)}{\cot\frac{1}{2}C} \tag{4.66}$$

These are two more of *Napier's analogies.* Provided that A, B, a, and b are known, $A \neq B$ (so $a \neq b$) these equations allow one to solve for C. The case where $A = B$ is dealt with in §17, Exercise 21.

12. Suppose that the center of the circumcircle of $\triangle^s ABC$ is in the interior of the triangle. Then prove that its radius R satisfies

$$\tan(R) = \sqrt{\frac{\sin(E)}{\sin(A-E)\sin(B-E)\sin(C-E)}}, \tag{4.67}$$

where A, B, and C are the measures of the angles and $E = \frac{1}{2}(A+B+C-\pi)$. (See §13 Exercise 34 for the definition of circumcircle.)

13. Let E be half the spherical excess of $\triangle^s ABC$. Prove that $\sin(E) = \sin(\frac{a}{2})\sin(\frac{b}{2})\sin(C)/\cos(\frac{c}{2})$. Using §18, Exercise 2, conclude that

$$\sin(E) = \frac{\sqrt{\sin(s)\sin(s-a)\sin(s-b)\sin(s-c)}}{2\cos(\frac{a}{2})\cos(\frac{b}{2})\cos(\frac{c}{2})} \tag{4.68}$$

This is called *Cagnoli's formula.* It provides a way of calculating the area of a spherical triangle given the measures of the sides.

14. Suppose that the *pentagramma mirificum* is regular in the sense that the five sides of the star connecting the five right angles are all congruent. Show that (referring to Figure 4.11) $c = B = A = \frac{\pi}{2} - a = \frac{\pi}{2} - b$. Show that the interior pentagon of the pentagram is regular, and that the radius r of its inscribed circle satisfies $\sin(r) = 1/\sqrt[4]{5}$. Show that the five points of the star lie on a small circle with spherical radius R, where the center of the circle is the same as that of the pentagon. Show that the radius R of this circle satisfies $\sin(R) = \sqrt[4]{\frac{4}{5}}$. Also find expressions for the side of the pentagon and pentagram, and the radius of the circumscribed circle of the pentagon.

15. State and prove a criterion for the numbers A, B, and C which characterizes when there exists a spherical triangle with angle measures A, B, and C.

16. There exists a spherical triangle $\triangle^s ABC$ with side and angle (radian) values a, b, and C if and only if a, b, and C are all between 0 and π. Use this to prove a similar condition for having angle and side values A, B, and c.

17. Suppose that in $\triangle^s ABC$, the value of A is no closer to $\frac{\pi}{2}$ than the value of B. Prove that $\prec A$ is acute, right, or obtuse if and only if the opposite side is acute, right, or obtuse, respectively. (Hint. See §16, Exercise 14.)

18. Suppose that given $\triangle^s ABC$ and $\triangle^s DEF$, $\prec ABC \cong \prec DEF$ and $\prec ACB \cong \prec DFE$. Prove that $m \prec BAC$ is equal to, less than, or greater than $m \prec FDE$, respectively, if and only if $m\,\overset{\frown}{BC}$ is equal to, less than, or greater than $m\,\overset{\frown}{FE}$.

19. Check that as indicated in Figure 4.12 the five successive right triangles of the pentagram constructed above are in fact $\triangle^s ABC$, $\triangle^s BB'E$, $\triangle^s B'(C')^a F$, $\triangle^s (C')^a A'H$, and $\triangle^s A'AJ$.

Historical notes. We let X-Y denote Book X, Proposition Y of Menelaus' *Sphaerica*. Then Exercise 18 is the first part of I-19.

20 Solution of triangles

In this section we discuss how to solve the following problem: given some of the measures of the sides and angles of a spherical triangle, find the measures of the other sides and angles. Mostly this problem is solved like the corresponding problem in the plane. But there are some differences. The most important difference arises with the fact that the sum of the measures of the angles in a triangle depends on the triangle. Thus knowledge of the measures of two angles in a triangle does not immediately determine the third. Also, because the law of sines involves the sine of both sides and angles, it is not often helpful, because the solution of an equation of the form $\sin(x) = k$ for x usually results in two possible answers for x which are supplementary. Lastly, usage of the polar triangle dualization is often employed for spherical triangles.

Several separate issues arise that are worth considering. First, given certain elements of the triangle, does a triangle exist with those elements? Second, if such a triangle exists, is it unique? Third, given some elements, how would we find the other elements?

(1) (SAS) (Given two sides and the included angle.) This problem can be solved and is unique provided the two sides and angle given all have measures between 0 and π. (The proof is left to the reader.) As in the plane, it makes sense to use the spherical law of cosines to find the third side. Then one can apply the spherical law of cosines again to find the remaining angles. As noted in §16, this amounts to using the analogue formula after finding the third side. One could also use the four-parts formula if one wished to solve only for one of the unknown angles, avoiding the inconvenience of finding the unknown side.

(2) (ASA) (Given two angles and the included side.) By Theorem 11.19, this amounts to being given two sides and the included angle in the polar triangle. Thus we can solve for the other side and angles in the polar triangle, which promptly gives us the other angle and sides in the given triangle by Theorem 11.19 again. Again, the four parts formula is a shortcut if only some of the unknown parts are desired. Existence and uniqueness of a solution depends again on the unknown sides and angles having measures between 0 and π. (See Exercise 16.)

(3) (SSS) (Given three sides.) One can use the spherical law of cosines to obtain the other angles one at a time. Alternately, one could use the half-angle formulas. Here, existence and uniqueness of solutions are discussed in §18, Exercise 9.

(4) (AAA) (Given three angles.) By the dualization associated with the polar triangle, this is equivalent to (3). Existence and uniqueness of a solution are discussed in Exercise 17.

We next consider the case (5) where we are given two sides and an angle opposite one of the sides (SSA) and the case (6) two angles and a side opposite one of the angles. By means of polar triangles, the treatments of these cases are dual to each other. These cases, known as the ambiguous cases because they can lead to zero, one, two, or infinitely many solutions for the triangle, are more complex than the earlier cases.

(5) (SSA) (Given two sides and an angle opposite one of the sides.) We propose the following algorithm which will determine the solutions to the triangle, if there are any.

Algorithm for solving the ambiguous case SSA. Suppose that the known sides of a triangle are a and b and that the known angle is A.

(a) By use of the law of sines we may determine $\sin(B)$ from the equation

$$\frac{\sin(a)}{\sin(A)} = \frac{\sin(b)}{\sin(B)}. \tag{4.69}$$

Since all the letters have values between 0 and π, their sines are positive, so $\sin(B) > 0$.

(b) If we determine $\sin(B) > 1$, there is no solution triangle.

(c) If we determine $\sin(B) \leq 1$ then we have either one or two candidates for $\prec B$ between 0 and π. That is, if $\sin(B) = 1$, $\prec B$ is a right angle, and if $\sin(B) < 1$, we have two possibilities for $\prec B$, one acute and one obtuse. We keep only the remaining possibilites for a, b, A, and B where $a - b$ and $A - B$ have the same sign or are both zero.

(d) Suppose $a - b = A - B = 0$.

(d1) If A, B, a, and b are not all acute, all right, or all obtuse then there is no solution (by Proposition 11.14).

(d2) If A, B, a, and b are all right, then there are infinitely many solutions where $c = C$ is any value between 0 and π.

(d3) If A, B, a, and b are all acute or all obtuse then we can solve uniquely for c and C by means of the right triangle formulas in §17, Exercise 21.

(e) Suppose $A - B \neq 0$ and $a - b \neq 0$. By use of Napier's analogies (see (4.58), (4.65)), we can solve for uniquely determined values for c and C.

Theorem 20.1 *Given values for the measures of two sides and an angle opposite to one of those sides, the set of all spherical triangles which have such sides and angles are those which emerge from the above algorithm.*

Proof. As in the algorithm, we assume when we write $\triangle^s ABC$, that it has side lengths a, b, and c and angle measures denoted by A, B, and C. We suppose that the values of a, b, and A are the measures of the specified sides and angle.

We first check that the algorithm imposes necessary conditions on the values of A, B, and c. In part (b), it is necessary that $\sin(B)$ be less than or equal to 1 for there to be any solutions since the sine of any angle is less than or equal to 1. Similarly, making use of Theorem 11.10, Exercise 1 of §12, and Theorem 13.10 we conclude that it is necessary to discard any value of B where (in step (c)) we fail to have that $A - B$ and $a - b$ have the same sign or are both zero.

In part (d) we consider what happens when $A-B = a-b = 0$. In part (d1) we discard any candidates for B which would produce a triangle which would violate Proposition 11.14. In part (d2), we must have $c = C$ by Definition 10.2, and we must have $0 < c, C < \pi$ because all spherical triangles have sides with measure in that range. In part (d3), c and C must satisfy the values determined in §17, Exercise 21.

In part (e), where $a-b$ and $A-B$ are both nonzero, we find that the values of a, b, c, A, B, and C must satisfy Napier's analogies (see (4.58), (4.65)). Since $a-b$ and $A-B$ are both nonzero, the numerators and denominators are both nonzero, and we can solve for c and C.

Now suppose, conversely, that B, c, and C are values that arise from the algorithm. We claim that they in fact give rise to a real triangle with sides a, b, and c and opposite angles A, B, and C. It will be necessary separately to treat each case where B, c, and C are calculated.

The case that arises from part (d2) is easy: a triangle where $A = a = B = b = \frac{\pi}{2}$ and $c = C$ are arbitrary (between 0 and π) can be constructed: simply take any angle of measure C and choose points A and B on either side of the angle at distance $\frac{\pi}{2}$ from C. By Proposition 11.11, in $\triangle^s ABC$, $B = C = \frac{\pi}{2}$ and $c = C$ by Definition 10.2, as desired.

Construction of a triangle in the case (d3) where $A - B = a - b = 0$ is left as Exercise 18.

Suppose in part (e) of the algorithm, we obtain values for B, c, and C where $A - B$ (and $a - b$) are nonzero. We construct $\prec \tilde{C}$ with measure C, choose point $\tilde{A}$ on one side at spherical distance b from $\tilde{C}$, and choose point $\tilde{B}$ on the other side of the angle at spherical distance a from $\tilde{C}$. Let $\tilde{A}$ be the measure of $\prec\tilde{C}\tilde{A}\tilde{B}$, $\tilde{B}$ be the measure of $\prec\tilde{C}\tilde{B}\tilde{A}$, and $\tilde{c}$ be the measure of $\overset{\frown}{\tilde{A}\tilde{B}}$. Then we claim $\tilde{A} = A$, $\tilde{B} = B$ and $\tilde{c} = c$ as desired, which makes $\triangle^s\tilde{A}\tilde{B}\tilde{C}$ the triangle that we want.

We already know by definition of B, c, and C that they satisfy Napier's analogies (4.58) and (4.65). But now that we have constructed $\triangle^s\tilde{A}\tilde{B}\tilde{C}$, we have that the same equations hold for a, b, C, $\tilde{A}$, $\tilde{B}$, and $\tilde{c}$:

$$\frac{\sin\frac{1}{2}(\tilde{A}-\tilde{B})}{\sin\frac{1}{2}(\tilde{A}+\tilde{B})} = \frac{\tan\frac{1}{2}(a-b)}{\tan\frac{1}{2}\tilde{c}} \tag{4.70}$$

$$\frac{\sin\frac{1}{2}(a-b)}{\sin\frac{1}{2}(a+b)} = \frac{\tan\frac{1}{2}(\tilde{A}-\tilde{B})}{\cot\frac{1}{2}C} \tag{4.71}$$

Combining (4.65) with (4.71), we conclude that $\tan \frac{1}{2}(\tilde{A}-\tilde{B}) = \tan \frac{1}{2}(A-B)$. Since $\tilde{A} - \tilde{B}$ and $A - B$ both lie between $-\pi$ and π, taking half must give a value between $-\frac{\pi}{2}$ and $\frac{\pi}{2}$. There the tangent takes values only once, so we can conclude that $\frac{1}{2}(\tilde{A} - \tilde{B}) = \frac{1}{2}(A - B)$, so $\tilde{A} - \tilde{B} = A - B$. We proceed to show that $\tilde{A} + \tilde{B} = A + B$, which will establish that $A = \tilde{A}$ and $B = \tilde{B}$. To do this, we consider two cases: $\sin(A) = \sin(B)$ and $\sin(A) \neq \sin(B)$.

If $\sin(A) \neq \sin(B)$, then $\sin(A)/\sin(B) \neq 1$. (This means that A and B are neither the same nor supplementary.) By definition of B, $\sin(A)/\sin(B) = \sin(a)/\sin(b)$ and by the spherical law of sines $\sin(\tilde{A})/\sin(\tilde{B}) = \sin(a)/\sin(b)$, so $\sin(\tilde{A})/\sin(\tilde{B}) \neq 1$. Thus $\tilde{A}$ and $\tilde{B}$ are also neither the same nor supplementary. We then invoke §15, Exercise 2 to conclude that

$$\frac{\tan \frac{a+b}{2}}{\tan \frac{a-b}{2}} = \frac{\tan \frac{A+B}{2}}{\tan \frac{A-B}{2}}$$

and

$$\frac{\tan \frac{a+b}{2}}{\tan \frac{a-b}{2}} = \frac{\tan \frac{\tilde{A}+\tilde{B}}{2}}{\tan \frac{\tilde{A}-\tilde{B}}{2}}.$$

Thus

$$\frac{\tan \frac{A+B}{2}}{\tan \frac{A-B}{2}} = \frac{\tan \frac{\tilde{A}+\tilde{B}}{2}}{\tan \frac{\tilde{A}-\tilde{B}}{2}}.$$

Using the fact found above that $A - B = \tilde{A} - \tilde{B}$, we find $\tan \frac{A+B}{2} = \tan \frac{\tilde{A}+\tilde{B}}{2}$. Now A and B are assumed to be between 0 and π, and $\tilde{A}$ and $\tilde{B}$ are also as measures of angles in a spherical triangle. Thus $\frac{A+B}{2}$ and $\frac{\tilde{A}+\tilde{B}}{2}$ are between 0 and π (but not $\frac{\pi}{2}$) and conclude that $A + B = \tilde{A} + \tilde{B}$. Since we already have $A - B = \tilde{A} - \tilde{B}$, we conclude that $A = \tilde{A}$ and $B = \tilde{B}$. Then combining (4.58) and (4.70), we find that $\tan \frac{1}{2}c = \tan \frac{1}{2}\tilde{c}$, so $c = \tilde{c}$. Thus we have shown that the desired triangle exists.

If $\sin(A) = \sin(B)$ then $\sin(\tilde{A}) = \sin(\tilde{B})$; since $A \neq B$, we must have $A + B = \pi$. Since $0 \neq A - B = \tilde{A} - \tilde{B}$, we must also have $\tilde{A} + \tilde{B} = \pi$. Thus $A + B = \tilde{A} + \tilde{B}$. Since we already have $A - B = \tilde{A} - \tilde{B}$, we conclude that $A = \tilde{A}$ and $B = \tilde{B}$ and proceed as above to conclude that also $c = \tilde{c}$ and we have constructed the desired triangle. ◇

(6) (SAA) (Given two angles and a side opposite one of the angles.) Suppose that the measures of the two angles given are A and B and the opposite side is a. Then we consider the dual (SSA) problem: take the supplements $\pi - A, \pi - B$ and $\pi - a$ and attempt to find a triangle with two sides of measure $\pi - A$ and $\pi - B$ and angle with measure $\pi - a$ opposite the side of measure $\pi - A$. We use the above algorithm to solve this problem. Then the polar triangle of any solution to the dual (SSA) problem is a solution to the original (SAA) problem.

Example 1. Suppose that in $\triangle^s ABC$ we are given the measures of the angles: $A = 100°$, $B = 115°$ and $C = 70°$. Determine the value of c.

Solution. In the polar triangle, the measures of the sides are $a' = 80°$, $b' = 65°$ and $c' = 110°$. In Example 5 of §16 we found the measures of the angle $C' \approx 117.7°$. So $c \approx 180° - 117.7° \approx 62.3°$. If desired, the other sides may be found in a similar manner.

Example 2. Suppose that in $\triangle^s ABC$ we are given the measures $a = 40°$, $b = 80°$ and $A = 20°$. Determine the values of B, C and c.

Solution. Using the spherical law of sines,

$$\sin(B) = \sin(20°)\sin(80°)/\sin(40°) \approx 0.52.$$

So we could have $B \approx 31.6°$ or $B \approx 148.4°$. Since $a < b$, $A < B$, and both candidates for B satisfy this. In both cases we use Napier's analogies to find c and C. We first consider $B \approx 31.6°$. We have from (4.58)

$$\frac{\sin\frac{1}{2}(-11.6°)}{\sin\frac{1}{2}(51.6°)} = \frac{\tan\frac{1}{2}(-40°)}{\tan\frac{1}{2}c},$$

so $\tan\frac{1}{2}c \approx 1.57$ and $c \approx 114.9°$. We have from (4.65)

$$\frac{\sin\frac{1}{2}(-40°)}{\sin\frac{1}{2}(120°)} = \frac{\tan\frac{1}{2}(-11.6°)}{\cot\frac{1}{2}C},$$

so $\tan\frac{1}{2}C \approx 3.89$ and $C \approx 151.2°$.

We next consider $B \approx 148.4°$. We have from (4.58)

$$\frac{\sin\frac{1}{2}(-128.4°)}{\sin\frac{1}{2}(168.4°)} = \frac{\tan\frac{1}{2}(-40°)}{\tan\frac{1}{2}c},$$

so $\tan\frac{1}{2}c \approx 0.4$ and $c \approx 43.8°$. We have from (4.65)

$$\frac{\sin\frac{1}{2}(-40°)}{\sin\frac{1}{2}(120°)} = \frac{\tan\frac{1}{2}(-128.4°)}{\cot\frac{1}{2}C},$$

so $\tan\frac{1}{2}C \approx 0.19$ and $C \approx 21.6°$.

Exercises §20

Problems 1-5: Given each of the following sides and angles of $\triangle^s ABC$, find all possibilities for the other sides and angles. Where units are not stated explicitly, it should be assumed that they are radians.

1. (a) $a = 20°$, $b = 35°$, $C = 76°$
   (b) $b = 80°$, $c = 85°$, $A = 170°$
   (c) $c = 175°$, $a = 170°$, $B = 160°$
   (d) $a = \frac{\pi}{4}$, $b = \frac{\pi}{4}$, $C = \frac{\pi}{2}$
2. (a) $A = 60°$, $B = 40°$, $c = 70°$

(b) $A = 50°$, $B = 40°$, $c = 110°$

(c) $A = 60°$, $B = 40°$, $c = 170°$

(d) $A = 60°$, $B = 140°$, $c = 70°$

(e) $A = 60°$, $B = 140°$, $c = 170°$

3. (a) $a = 70°$, $b = 40°$, $c = 35°$

   (b) $a = 105°$, $b = 45°$, $c = 80°$

   (c) $a = 130°$, $b = 140°$, $c = 35°$

   (d) $a = 115°$, $b = 30°$, $c = 125°$

   (e) $a = 130°$, $b = 40°$, $c = 155°$

   (f) $a = 80°$, $b = 115°$, $c = 155°$

   (g) $a = 130°$, $b = 80°$, $c = 155°$

   (h) $a = 80°$, $b = 65°$, $c = 150°$

4. (a) $A = B = C = 120°$ (find exact values)

   (b) $A = B = C = 72°$ (find exact values)

   (c) $A = 100°$, $B = 110°$, $C = 140°$

   (d) $A = 60°$, $B = 70°$, $C = 100°$

5. (a) $a = \frac{\pi}{2}, b = \frac{\pi}{2}, A = \frac{\pi}{5}$

   (b) $a = 30°, b = 90°, A = 30°$

   (c) $a = 30°, b = 90°, A = 150°$

   (d) $a = 35°, b = 80°, A = 60°$

   (e) $a = 40°, b = 40°, A = 90°$

   (f) $a = 90°, b = 90°, A = 90°$

   (g) $a = 40°, b = 40°, A = 60°$

   (h) $a = 30°, b = 45°, A = 45°$

   (i) $a = 80°, b = 130°, A = 30°$

   (j) $a = 40°, b = 80°, A = 20°$

6. Which of the routes in §16, Exercise 3 crosses the zero meridian? Where are the crossing points?

7. A ship's initial position is $30°N, 130°W$. It takes a great circle course whose initial bearing is $S60°W$. If its final position is on longitude $140°W$, how far did it travel and what is its final position?

8. A ship's initial position is $30°N, 130°W$. It takes a great circle course whose initial bearing is $N60°W$. If its final bearing is $N80°W$, how far did it travel and what is the final position?

9. Which of the routes in §16, Exercise 3 crosses the equator? Where are the crossing points?

10. Which of the routes in problem 3 crosses the latitude of 45° north? Where are the crossing points?

11. A ship's initial position is $30°S, 135°W$. It takes a great circle course with initial bearing $S45°E$ and arrives at a position whose latitude is $45°S$. What is its final position?

12. A ship's initial position is $30°S, 135°W$. It takes a great circle course for 2000 miles and arrives at a position whose latitude is $45°S$. What is its final position? (Two answers are possible.)

13. A hijacked airliner is due north of a fighter jet. The airliner is travelling on a great circle course initially $N80°W$ at 500 mph. The fighter jet travels 1500 mph and needs to intercept the airliner within 15 minutes. How far south of the airliner could the fighter jet be for this to be possible?

14. Given a spherical quadrilateral $ABCD$ with $a = m\ \overset{\frown}{AB}$, $b = m\ \overset{\frown}{BC}$, $B = m \prec ABC$, $x = m \prec ADC$ and $y = m \prec BDC$ known quantities. Formulate a strategy to determine the angles and spherical distances among the four points. (Hint: you cannot apply the triangle solution methods to determine the angles and sides one by one. You will need a system of equations. It is possible to find a system of equations involving just two of the unknown sides and angles. Once this is solved, the other sides and angles can be found one by one.)

15. Given spherical quadrilateral $ABCD$ with $a = m\ \overset{\frown}{AB}$, $x = m \prec ADB, y = m \prec CDB$, $z = m \prec BCA$ and $w = m \prec DCA$ known quantities. Formulate a strategy to determine the angles and spherical distances among the four points.

16. Justify the assertion made above that an ASA problem has a unique solution when the given side and angles have measures between 0 and π.

17. Use the theorem stated as part of §18, Exercise 9 to derive conditions under which the solution of a triangle exists and is unique given all three angles.

18. Suppose that B, c, and C are values that arise out of part (d3) in the case where $A - B = a - b = 0$, and A, B, a, and b are all acute or all obtuse. Complete the proof of Theorem 20.1 by showing that there exists a triangle with side lengths a, b, and c and angle measures A, B, and C.

19. Suppose that in the algorithm for solving an SSA ambiguous case, B is a right angle and A is not. Then it is easier and more natural to solve for c and C using the right triangle formulas $\cos(c) = \cos(b)/\cos(a)$ and $\cos(C) = \cot(b)/\cot(a)$ (see (4.21) and (4.24), where the letters must be permuted to accommodate the right angle at B) instead of Napier's analogies in part (e). Prove that all candidates for solutions that arise from this can be realized as actual triangles.

Historical notes. Exercises 14 and 15 are spherical versions of the plane trignometry problems known as the Pothenot and Hansen problems. (See [Do1965], problem 40.) Here is an application of Exercise 14: determine the location of two accessible places based on the bearing from these points to inaccessible points whose positions are known. The Hansen problem has a similar application.

Chapter 5

Astronomy

The purpose of this chapter is to introduce the reader to an important application of spherical geometry and trigonometry in astronomy. We will study how to calculate the time of rise or set for an object in the sky such as the sun or a star, as well as the location of the object when it rises. We will also show how to determine the location of an object at other times of the day.

21 The celestial sphere

The stars and planets in the sky are at enormous distances from the earth. From the point of view of an observer at a location in space such as the earth, the stars and planets in the sky appear to lie on an imaginary sphere called the *celestial sphere* for that location. They are all so distant that they may be regarded as lying on a sphere of fixed large radius. We may then use the techniques of spherical geometry for the purposes of determining and monitoring the position of such objects as seen from that location. The exact positions of objects as seen by the observer will change as the position of the observer changes — an effect called *parallax.* Hence the position of objects on the imaginary celestial sphere changes as the observer moves around. But most objects in the heavens are so distant from the earth that their parallax resulting from the movement of the observer is insignificant and we will mostly disregard it.

Most of the daily, or *diurnal* motion of the stars is a result of the rotation of the earth, which rotates on its axis once each day. The axis of rotation of the earth is a line ℓ passing through the north and south poles of the earth. This line meets the celestial sphere in two points called the *north celestial pole* P and its antipode, the *south celestial pole* P^a. The earth rotates, and the observer on the earth perceives this rotation through the movement of stars. This movement is perceived as a daily rotation of the stars about the axis through the north and south celestial poles. As the stars move, the celestial poles are fixed and the stars move in a manner so as to stay a fixed distance from these poles. As the earth rotates, the stars trace the course of a small or great circle with poles equal to the north and south celestial poles. Here we assume for the time being that the stars (other than the sun) have negligible motion in the sky other than that resulting from the earth's rotation, and the earth's axis of rotation stays the same. For this reason we will refer to the stars as "fixed." For most of the purposes of this chapter, this assumption will be adequate. (We will be mostly considering movement of objects over periods of less than a year; for astronomers looking to study the movement of stars or the earth's axis over a period of centuries, other motions would be significant.)

The motion of the other objects in the sky (the sun, moon, planets, and other objects in our solar system) is generally not quite so simple. These objects generally have a motion relative to the fixed stars that is significant on a daily basis, and must be accounted for if we wish to determine their locations in the sky at a particular time.

There is a reasonably bright star called *Polaris*, or simply the *north star* near the north celestial pole. There is no similar star near the south celestial pole.

The equator of the earth is a great circle on the surface of the earth; this great circle is the intersection of the surface of the earth with a plane p. This plane is perpendicular to ℓ since the equator has its poles at the north and south poles of the earth. The plane p intersects the celestial sphere in a great circle called the *celestial equator*. Because $p \perp \ell$, the poles of the celestial equator are the north and south celestial poles.

Let O be the center of the earth. Then O may also be regarded as the center of the celestial sphere. Given any point A on the surface of the earth, the ray $\overrightarrow{OA}$ intersects the celestial sphere in a point we shall call the *projection* $\mathcal{P}(A)$ of A to the celestial sphere. We will say that A projects to the point $\mathcal{P}(A)$. Then the north and south celestial poles are projections of the geographic north and south poles of the earth, and every point on the earth's equator projects to a point on the celestial equator. If an observer is located at point A, then $\mathcal{P}(A)$ is called the *zenith* Z of the observer. The antipode of the zenith of the observer is called the *nadir* of the observer. The projection $\mathcal{P}$ also has an important distance-preserving property: if A and B are points on the earth, then the spherical distance between $\mathcal{P}(A)$ and $\mathcal{P}(B)$ on the celestial sphere is the same as the spherical distance between A and B on the earth. The reason for this is that the spherical distance between A and B is $m\angle AOB$, and the spherical distance between $\mathcal{P}(A)$ and $\mathcal{P}(B)$ is the measure of $\angle\mathcal{P}(A)O\mathcal{P}(B)$. Since $\mathcal{P}(A)$ is on $\overrightarrow{OA}$ and $\mathcal{P}(B)$ is on $\overrightarrow{OB}$, the angles $\angle AOB$ and $\angle\mathcal{P}(A)O\mathcal{P}(B)$ are the same angle, so have the same measure. Thus the two distances are the same. An important application of this is the following.

Proposition 21.1 *If an observer has latitude ϕ then the spherical distance from the north celestial pole to the observer's zenith is $90° - \phi$.*

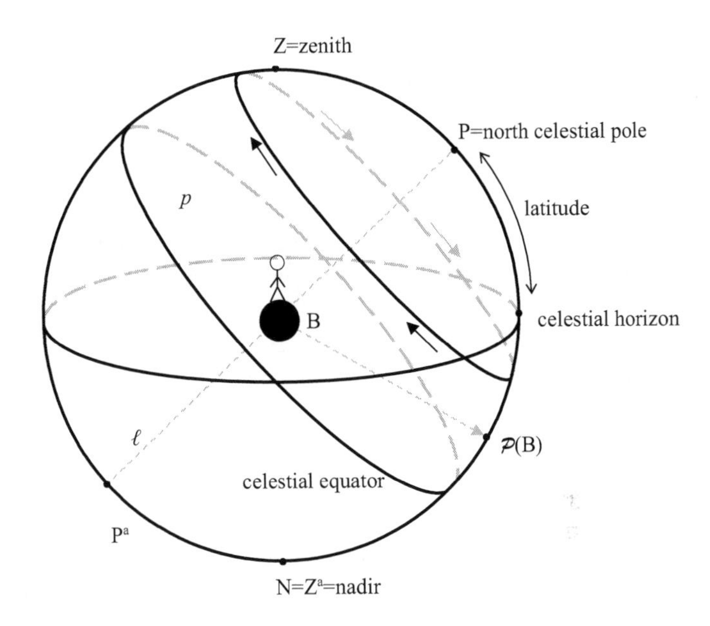

Figure 5.1: The celestial sphere.

Proof. Suppose the observer is located at point A on the earth. Note that A projects to Z and the earth's north pole projects to P. The distance from A to the earth's north pole is $90° - \phi$; this then is the spherical distance $m\,\overset{\frown}{PZ}$. ♢

At a location on the celestial sphere, one may also speak of directions "east," "west," "north," and "south" from that point as being the direction inherited from the projection of the associated directions from the earth. As an example, suppose that an object is moving on a path on the celestial sphere. Then there is a corresponding path on the surface of the earth whose points project via $\mathcal{P}$ to the points on the path being followed on the celestial sphere. If the corresponding path on the surface of the earth travels in a westerly direction, then we will say that the object on the celestial sphere is moving west.

In order to aid in the study of the positions of stars, planets, and other objects in the sky, astronomers have developed various coordinate systems for the celestial sphere. These are all coordinate systems of the type used in the latitude-longitude coordinate system used for the earth, described in §8. In each case, we have to choose a great circle from which the "latitude" variable will be measured and choose a great semicircle perpendicular to the above great circle from which the "longitude" variable will be measured. We also must specify in which directions the latitude and longitude variables are positive.

Before discussing these coordinate systems, knowledge of a few astronomical facts will be helpful. Over the course of a year, the earth revolves in a planar elliptical orbit (which is almost circular) once about the sun. From the point of view of an observer on the earth, the sun appears to move in a great circle in the sky known as the *ecliptic.* Another way of describing the ecliptic is that it is the intersection of the plane of the earth's orbit with the celestial sphere. Because the ecliptic is a great circle, it meets the celestial equator in two antipodal points known as the vernal (spring) and autumnal *equinoxes.* The vernal equinox is denoted with the zodiac symbol for Aries Υ. The measure of the acute angle (known as the *obliquity of the ecliptic*) between the ecliptic and the celestial equator, denoted by ε, is approximately 23.5°. Although the diurnal motion of the sun is east to west, if we remove the rotation of the earth, the sun moves west to east on the celestial sphere against the background of the fixed stars. This is a consequence of the direction of the earth's revolution about the sun.

Here are several standard coordinate systems for the celestial sphere. Astronomical references use various units of measurement. The unit of degrees (°) seems to be the most common, along with its subdivisions into minutes (′) and seconds (″), where a minute is $\frac{1}{60}$ of a degree, and a second is $\frac{1}{60}$ of a minute. However, one other important set of units is common: hours (h), minutes (m), and seconds (s), where 24 hours correspond to 360 degrees, a minute is $\frac{1}{60}$ of an hour, and a second is $\frac{1}{60}$ of a minute. Unfortunately, this means that we have two different types of minutes and seconds, but the reader may find the following conversion helpful: $1h = 15°$, $1m = 15'$ and $1s = 15''$.

1. Equatorial coordinates. (See Figure 5.2.) Let P be the north celestial pole and let Υ be the vernal equinox. For equatorial coordinates, the pole is P, and the zero meridian is the great semicircle $P\Upsilon P^a$. The "latitude" of a point X is called the *declination* of the star, denoted by δ. By Exercise 11 of §10, the declination is then the signed spherical distance from X to the celestial equator; this distance is measured as positive when north of the celestial equator and negative when south of the celestial equator. The spherical distance $m\,\overset{\frown}{PX}$ (called the *polar distance*) is $90° - \delta$. Note that this formula works for both positive and negative δ.

The "longitude" for equatorial coordinates is called the *right ascension.* The right ascension increases in the direction "east" of Υ and decreases in the direction "west."

The generally easterly motion of the sun against the stars ensures that its right ascension is always increasing (though not at a constant rate, as we shall see).

Astronomers commonly use units of degrees/minutes/seconds for declination and hours/minutes/seconds for right ascension. Standard values for right ascension range from 0 to 24 hours (though sometimes it is more convenient to use the range 0° to 360°). However, as with calculations of position on an arc of a circle (where values run from 0 to 2π or 0° to 360° and then cycle back to 0), frequently we encounter an hour value less than 0 and larger than

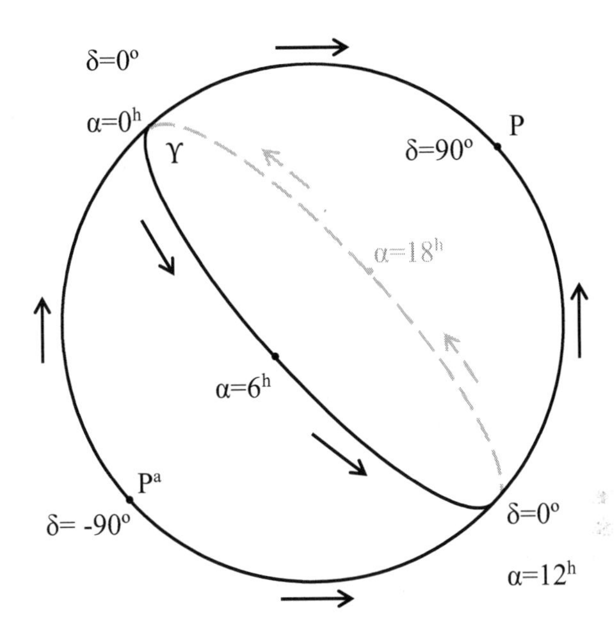

Figure 5.2: Equatorial coordinates.

24. This value is equivalent to a value from 0 to 24, found simply by adding or subtracting an appropriate multiple of 24 hours. Occasionally we have a number larger than 24 or less than 0 arise in a calculation intended to produce a time of the current day. We then must subtract or add a multiple of 24 to obtain the appropriate time of the current day. When we do this we will say we are *reducing modulo 24*. We reduce a number x modulo 24 by dividing by 24 and taking the remainder. We mean something similar when speaking of reducing a number modulo 360 degrees.

The equatorial coordinate system is an important system for cataloguing positions of stars. The right ascension and declination of a star are here considered to be fixed — independent of both time and the position of the observer. The astronomer can consult a catalogue for a star's equatorial coordinates and then point a telescope at that position to see the star. However, the way one sets up a telescope depends on the observer's position on the earth. For this, one sometimes uses altitude-azimuth coordinates.

2. Altitude-azimuth coordinates. (See Figure 5.3.) For any observer on the earth with an unobstructed view of the horizon, the horizon projects out to a great circle on the celestial sphere which we call the *(celestial) horizon*. The celestial horizon has the zenith Z and nadir as its poles. We assume Z is different from P and P^a (i.e., that the observer is not at either the north pole or the south pole). Because the planets and stars are at a very large distance relative to the size of the earth, the observer can then see a full hemisphere of the celestial sphere — all stars within a spherical distance of 90° of the zenith.

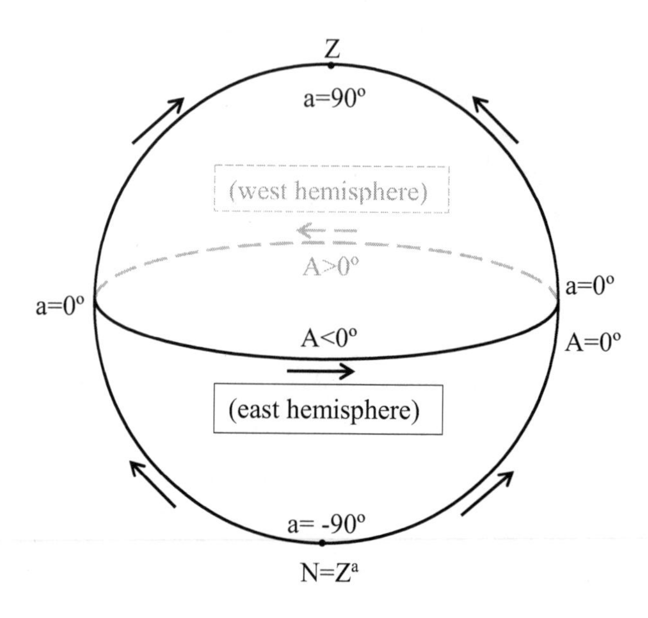

Figure 5.3: Altitude-azimuth coordinates.

In altitude-azimuth coordinates, the pole is Z, and the zero meridian is the great semicircle ZPZ^a. Thus the "latitude" of a point X is (by Exercise 11 of §10) the signed spherical distance from the horizon (positive when visible to the observer; negative when not visible to the observer) and is called the *altitude* a. The altitude ranges from $-90°$ to $90°$. The so-called *zenith distance* z is the spherical distance $m\,\widehat{ZX}$; then we have $a = 90° - z$. The "longitude" of the point X in this coordinate system is called the *azimuth* A. Officially we take $-180° < A \leq 180°$. Then A is positive if X is west of the zero meridian and negative if X is east of the zero meridian.

This definition of azimuth may be termed "azimuth from the north"; the reader should be aware that some references measure azimuth from the south.

Frequently the azimuth is described as being a particular angle value "east of north," "west of north," "east of south," or "west of south." These all simply indicate an angle through which one must turn from a given direction. For example, if the azimuth is 40 degrees east of south, that means the azimuth direction is found by turning 40 degrees to the east from the southerly direction; then $A = -140°$.

The altitude and azimuth of a star are dependent on the location of the observer. These coordinates are those that are used to identify an object in the sky at a location, and point a telescope at that object. The altitude and azimuth of an object also change with time. Degrees are the typical units for both altitude and azimuth.

The following fact will be helpful.

Proposition 21.2 *The altitude of the north celestial pole is equal to the observer's latitude.*

Proof. Let ϕ be the latitude of the observer. As noted in Proposition 21.1, the distance from the observer's zenith Z to the north pole is $90^\circ - \phi$. Since the altitude of a point is the complement of the distance from the point to the zenith (by Exercise 11 of §10), the altitude of the north celestial pole is $90^\circ - (90^\circ - \phi)$, which is ϕ, as desired. Note that this holds both for observers in the northern and southern hemispheres. ◇

3. Declination-hour angle coordinates. (See Figure 5.4.) We assume again that Z is different from both P and P^a, so that the observer is at neither the north pole nor the south pole. For declination-hour angle coordinates the pole is the north celestial pole (as was the case with equatorial coordinates) and the zero meridian is PZP^a. As with equatorial coordinates, the "latitude" coordinate is the declination δ. The "longitude" coordinate, known as the *hour angle* h, is different. The zero meridian PZP^a is called the *observer's meridian.* Typically we assume that $-12^h < h \leq 12^h$ (but as before we obtain an equivalent hour angle by adding or subtracting a multiple of 24^h). Then if X is the point being observed, its hour angle is positive if X is west of the observer's meridian and negative if X is east of the observer's meridian. Because the earth rotates west to east about its axis, the hour angle of a fixed star increases at a constant rate. These coordinates are specially designed for a telescope which has been set up with its axis of rotation parallel to the earth's axis of rotation. In order to keep the telescope pointed at the star, the telescope must merely rotate at a constant speed, as the hour angle increases at a constant rate.

The declination of an object does not depend on either time or location of the observer, but the hour angle depends on both time and location. The units for declination are usually degrees/minutes/seconds and hour angle is usually hours/minutes/seconds.

Since equatorial coordinates and declination-hour angle coordinates have the same pole, there is an important relationship between the hour angle and right ascension. Suppose that two points X and Y have hour angles h_X and h_Y and right ascensions α_X and α_Y. Hour angle and right ascension both measure angles measured with pole P, with the caveat that hour angle increases to the west and right ascension increases to the east. Thus the difference in hour angle between X and Y is the same as the difference in right ascension, except for sign:

$$h_X - h_Y = \alpha_Y - \alpha_X. \tag{5.1}$$

4. Ecliptic coordinates. (See Figure 5.5.) As defined above, the ecliptic is the great circle on the celestial sphere traveled by the sun during the course of a calendar year. The pole of the ecliptic which sits in the north celestial sphere is the *ecliptic north pole.* We label it M. The ecliptic splits the celestial sphere into two hemispheres; the *ecliptic northern hemisphere* is the

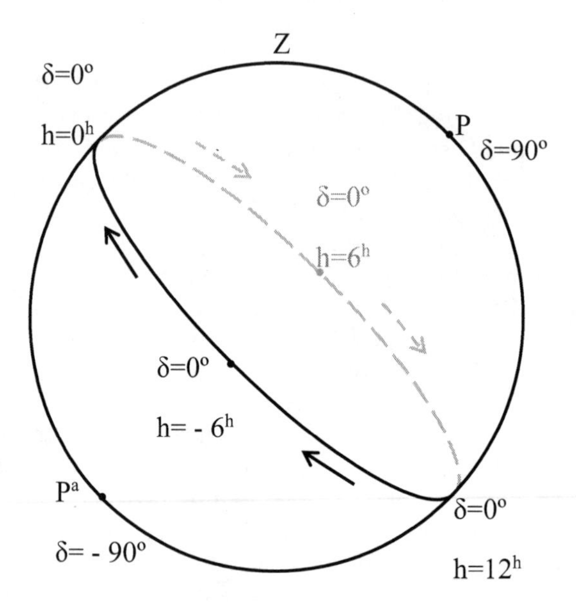

Figure 5.4: Declination-hour angle coordinates.

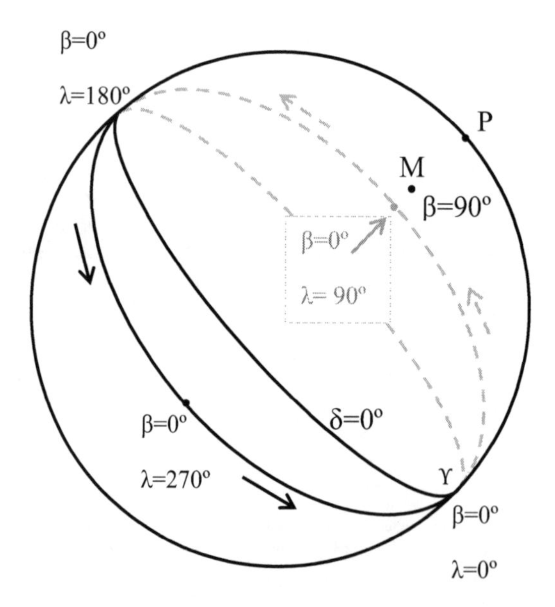

Figure 5.5: Ecliptic coordinates.

hemisphere containing M; the other hemisphere is the *ecliptic southern hemisphere.* Ecliptic coordinates are defined as follows. The pole of the coordinate system is the ecliptic north pole and the zero meridian is the great semicircle $M\Upsilon M^a$. Then the *ecliptic latitude* of a point X is its signed spherical distance from the ecliptic — measured as positive if X is in the ecliptic northern hemisphere and negative if X is in the ecliptic southern hemisphere. Note that the ecliptic latitude is $90° - m\,\widehat{MX}$. The value of the *ecliptic longitude* increases east of $M\Upsilon M^a$ and decreases west. This is designed so that the sun's ecliptic longitude is always increasing as it moves west to east against the fixed stars.

Ecliptic coordinates, like equatorial coordinates, do not depend on the position of the observer. Nor are they affected by the diurnal rotation of the earth. The ecliptic coordinates of fixed stars are assumed to be constant in this book.[1]

Once chosen this way, most planets and the moon generally have increasing ecliptic longitude as they move against the background of the fixed stars. This is a consequence of the fact that the planets revolve in the same direction about the sun, and the moon revolves in the same direction about the earth. This motion is called *direct motion.* Occasionally, due to relative position and velocity of the earth and another planet, the ecliptic longitude of that planet may briefly decrease; this is called *retrograde motion.* The principal use we shall have with ecliptic coordinates is in keeping track of the position of the sun during the course of a calendar year. The units for ecliptic latitude and longitude are usually degrees, minutes, and seconds.

We conclude this section with a few observations concerning the effect of parallax on celestial position. As noted above, the position of objects in the sky does change depending on the position of the observer. Because earth-based observers are on the surface of the earth and not its center, a parallax arises due to the daily rotation of the earth on its axis. That is, the apparent position of objects in the solar system changes slightly due to the rotation of the earth. This effect, negligible for the stars, becomes quite significant in the case of the moon since the moon is so close to the earth. In this case, accurate observations require a distinction between a *geocentric* celestial sphere (on which one may represent the positions of objects from the point of view of an observer at the center of the earth), and *topocentric* celestial sphere (on which one represents the positions of objects in the sky from a particular location on the earth). The position of the moon could be calculated from the point of view of the center of the earth and then this position could be corrected to represent the moon's position on the earth's surface. Similarly, one may speak of a *heliocentric* celestial sphere on which one may represent the positions of objects from the point of view of an observer at the sun. The positions of the stars are mostly the same on these celestial spheres because they are so far away when compared to the distance between the sun and the earth,

[1] The so-called precession of the equinoxes modifies both ecliptic and equatorial coordinates of the fixed stars slowly over the course of thousands of years.

but the positions of solar system objects are much different because they are much closer. In this book we shall largely be concerned with the appearance of objects from an earth-based observer, but an astronomer studying stars and galaxies might find the heliocentric celestial sphere better, owing to the inconvenience of the parallax caused by the earth's orbit.

Exercises §21.

1. Plot the following in equatorial coordinates (δ, h).
   (a) $(30°, 6^h)$ (b) $(-45°, 10^h)$ (c) $(60°, -8^h)$ (d) $(-50°, -4^h)$

2. Plot the following in altitude-azimuth coordinates (a, A).
   (a) $(30°, 40°)$ (b) $(-45°, 140°)$ (c) $(60°, -150°)$ (d) $(-50°, -45°)$

3. Plot the following in declination-hour angle coordinates (δ, h).
   (a) $(30°, 6^h)$ (b) $(-45°, 10^h)$ (c) $(60°, -8^h)$ (d) $(-50°, -4^h)$

4. Plot the following in ecliptic coordinates (β, λ).
   (a) $(30°, 40°)$ (b) $(-45°, 140°)$ (c) $(60°, -150°)$ (d) $(-50°, -45°)$

5. The moon revolves about the earth on a plane inclined approximately 5° with the ecliptic. When the moon is full in the winter, how large/small can its declination be? For latitude $40°N$, what is the maximum/minimum altitude of the full moon then? What about summer? Is the situation any different at a southern location? What about the equator? Use this information to justify the statement that the full moon travels very high in the sky in the winter, but low in the sky in the summer. What are the latitudes nearest the equator which might not see a full moon rise in summer (owing to the moon being too close to the opposite pole)?

6. Suppose we assume that the sun rotates around the poles 360° in 24 hours, so 15° per hour. We create a rudimentary sundial to keep track of time by placing a rod into the ground which points directly at the (visible) pole star. When the sun crosses the observer's meridian (i.e., at "apparent solar noon"; see §23) the rod creates a shadow on the ground pointing at the point on the horizon directly below the pole. Assume the rod is at latitude ϕ. Show that at x o'clock in the afternoon (i.e., when the sun has rotated $15x$ degrees) the rod creates a shadow whose angle s with the original shadow satisfies $\tan(s) = \sin(\phi)\tan(15x°)$. Such a sundial is said to be a *horizontal sundial.*

7. We now create a *vertical south sundial*, where the face of the dial is vertical and faces north-south. The rod again is parallel to the axis of the earth. At noon, the shadow it casts is vertical. Under the assumptions and notation in Exercise 6, show that at x o'clock in the afternoon the shadow forms an angle s with the vertical, where $\tan(s) = \cos(\phi)\tan(15x°)$.

8. Suppose we create another vertical sundial but turn the face of the dial by a given angle from the north-south position in Exercise 7. Show how to calculate the angle between the hour marks on the dial.

9. Assume that the sun does not change its position during a day relative to the fixed stars. A rod is placed vertically in the ground and the tip of its shadow traces a curve on the ground. Assuming that the surface of the ground is planar, show that the curve traced must be a conic section (i.e., an ellipse, hyperbola, parabola, or line.) What circumstances result in each case?

10. Interpret the result of §13, Exercise 39 in the following context. Assume that during the springtime the sun's position is observed on two separate intervals of time where its ecliptic longitude increases by the same amount. What conclusion can we make about the relative change of the declination of the sun on these two occasions? What conclusion can we make about the relative change of its right ascension?

22 Changing coordinates

An astronomer frequently must change coordinates among the coordinate systems given in §21. For example, suppose the astronomer is given the equatorial coordinates of an object. To spot the object in the sky, the altitude-azimuth coordinates would be useful. This might be accomplished first by using the time to determine the hour angle, and then switch from declination-hour angle to altitude-azimuth coordinates. With the sun, knowing the date and time, one can determine its ecliptic longitude. Then one can switch from ecliptic to equatorial coordinates to find its place against the fixed stars. In order to determine information related to rise and set of the sun, one would need information about the altitude. We now proceed to illustrate with several examples how one would switch coordinates. We always let ϕ be the latitude of the observer.

1. Change of coordinates between declination-hour angle and altitude-azimuth coordinates. (See Figure 5.6.) Let Z be the zenith point of the observer, P the north celestial pole, and X the location of an object in the sky. Suppose that X has altitude-azimuth coordinates (a, A) and declination-hour angle coordinates (δ, h). We assume these three points form a spherical triangle (which they do most of the time) and call $\triangle^s XPZ$ the *astronomical triangle.* Then by definition, $h = \pm m \prec ZPX$ and $A = \pm m \prec PZX$. By Proposition 21.1, $m \overset{\frown}{PZ} = 90° - \phi$. If the altitude of X is a then $m \overset{\frown}{ZX} = 90° - a$ (see Exercise 11 of §10). Similarly, if the declination of X is δ then $m \overset{\frown}{PX} = 90° - \delta$.

If we are given (a, A) then in $\triangle^s XPZ$ then we have $m \overset{\frown}{ZP}$, $m \overset{\frown}{XZ}$ and $m \prec PZX$. Then if we determine the measures of the other sides and angles in the triangle, we obtain $m \overset{\frown}{PX} = 90^\circ - \delta$ and $h = \pm m \prec ZPX$, so we get (δ, h). On the other hand, if we are given (δ, h), then in $\triangle^s XPZ$ we have $m \overset{\frown}{PZ}, m \overset{\frown}{PX}$, and $m \prec ZPX$. If we determine the measures of the other sides and angle in the triangle, we obtain $m \overset{\frown}{ZX} = 90^\circ - a$ and $A = \pm m \prec PZX$, so we get (a, A). Thus the process of switching coordinates may usually be reduced to solving for the measures of the unknown sides and angles of a triangle, where we are given two sides and an included angle. If the azimuth and hour angle have values in the prescribed ranges given above, we can easily determine the appropriate sign for each.

Let us see in general how this would be done. Suppose that in $\triangle^s ABC$ we are given b, c, and A and we need to determine a and B. We may use the law of cosines (4.12) to obtain a; then we can use the values of a, b, and c in (4.11) to obtain $\cos(B)$ and then B. But this is exactly what was done in the proof of the analogue formula to obtain (4.13). So we might as well simply use (4.13) directly. If we do this, we obtain (4.12) and (4.13). If we do this on $\triangle^s XPZ$, assuming (a, A) known, we obtain

$$\cos(90^\circ - \delta) = \cos(90^\circ - a)\cos(90^\circ - \phi) + \sin(90^\circ - a)\sin(90^\circ - \phi)\cos(A)$$

and

$$\begin{aligned} & \sin(90^\circ - \delta)\cos(h) \\ = \; & \cos(90^\circ - a)\sin(90^\circ - \phi) - \sin(90^\circ - a)\cos(90^\circ - \phi)\cos(A) \end{aligned}$$

so

$$\sin(\delta) = \sin(a)\sin(\phi) + \cos(a)\cos(\phi)\cos(A) \quad (5.2)$$

$$\cos(\delta)\cos(h) = \sin(a)\cos(\phi) - \cos(a)\sin(\phi)\cos(A) \quad (5.3)$$

Since $-90^\circ \leq \delta \leq 90^\circ$, (5.2) determines δ uniquely. If $\delta = \pm 90^\circ$ then the value of h is irrelevant. If $\delta \neq \pm 90^\circ$ then $\cos(\delta) \neq 0$ and we have the value of $\cos(h)$ from (5.3). Since $-12^h < h \leq 12^h$, there could be two possibilities for h — a positive or negative value.

Note that when X is east of the observer's meridian, both A and h are negative. When X is west of the observer's meridian, both A and h are positive. So A and h have the same sign if X is not on the observer's meridian. Thus when X is not on $\bigcirc PZ$, signs may be ignored while solving the triangles, but then h is given the same sign as A.

We illustrate with an example.

Example 1. An observer is located at latitude 10° south. On a given night a star has altitude 41° and azimuth 23° north of east. Determine the declination and hour angle of the star.

Solution. The latitude of 10° south means that $\phi = -10^\circ$. We have $a = 41^\circ$ and $A = -67^\circ$. Using (5.2) and (5.3), we get

$$\begin{aligned} \sin(\delta) &= \sin(41^\circ)\sin(-10^\circ) + \cos(41^\circ)\cos(-10^\circ)\cos(-67^\circ) \\ \cos(\delta)\cos(h) &= \sin(41^\circ)\cos(-10^\circ) - \cos(41^\circ)\sin(-10^\circ)\cos(-67^\circ), \end{aligned}$$

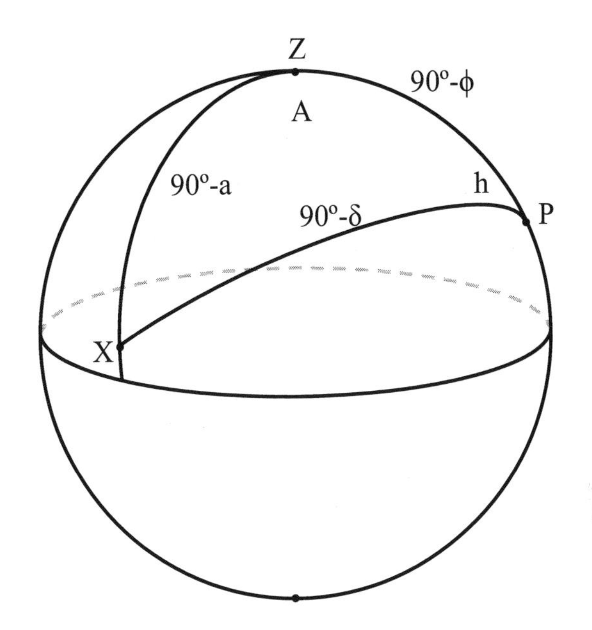

Figure 5.6: Changing coordinates: between declination-hour angle and altitude-azimuth coordinates.

so $\sin(\delta) \approx 0.1765$ and $\delta \approx 10.1651°$ from the first equation. The second equation gives $\cos(10.165°)\cos(h) \approx 0.6973$, $\cos(h) \approx 0.7084$ and so $h \approx \pm 44.89° \approx \pm 2.9929^h$. Since h has the same sign as A, $h \approx -2.9929^h$.

The ambiguity involved in determining the value of h from $\cos(h)$ is sometimes resolved in another way. For A and h both positive we use the spherical law of sines on the astronomical $\triangle^s XPZ$ and obtain

$$\frac{\sin(90° - \delta)}{\sin(A)} = \frac{\sin(90° - a)}{\sin(h)}. \tag{5.4}$$

If A and h are both negative we obtain

$$\frac{\sin(90° - \delta)}{\sin(-A)} = \frac{\sin(90° - a)}{\sin(-h)}. \tag{5.5}$$

But since the sine is an odd function (see (1.5)) we may divide out the negative from (5.5) to obtain (5.4). Clearing the fractions and using (1.3) we obtain $\cos(\delta)\sin(h) = \cos(a)\sin(A)$. Together with (5.2) and (5.3) we obtain a system of equations which allows us completely to change variables from (a, A) to (δ, h):

$$\begin{aligned}
\sin(\delta) &= \sin(a)\sin(\phi) + \cos(a)\cos(\phi)\cos(A) &(5.6)\\
\cos(\delta)\cos(h) &= \sin(a)\cos(\phi) - \cos(a)\sin(\phi)\cos(A) &(5.7)\\
\cos(\delta)\sin(h) &= \cos(a)\sin(A) &(5.8)
\end{aligned}$$

In (5.6), given (a, A), we can calculate δ as noted above. When $\cos(\delta) \neq 0$ we may divide by $\cos(\delta)$ in (5.7) and (5.8) to obtain $(\cos(h), \sin(h))$, which permits us to determine h (up to a multiple of 24^h.)

The above arguments only work if X, P, and Z determine a triangle. Of course, it is possible that they do not (i.e., that they lie on the same great circle). In these cases the formulas still work, and we leave verification of this to the exercises. The use of the law of sines may seem like an awkward way to determine the appropriate sign for h, but as we shall see later (§31), it is the natural way to do so when using spatial coordinates to do spherical trigonometry.

An entirely analogous sort of argument permits the derivation of formulas to change from equatorial to altitude-azimuth coordinates:

$$\begin{aligned} \sin(a) &= \sin(\delta)\sin(\phi) + \cos(\delta)\cos(\phi)\cos(h) \\ \cos(a)\cos(A) &= \sin(\delta)\cos(\phi) - \cos(\delta)\sin(\phi)\cos(h) \\ \cos(a)\sin(A) &= \cos(\delta)\sin(h) \end{aligned} \tag{5.9}$$

Example 2. An observer is located at latitude 10° south. The star Antares has a declination of approximately $-26°19'$. Suppose that on a given night its hour angle is -1^h30^m. Determine its altitude and azimuth at that time.

Solution. We have $\phi = -10°$, $\delta \approx -26.317°$ and $h \approx -22.5°$. Using (5.9), we find that

$$\begin{aligned} \sin(a) &\approx \sin(-26.317°)\sin(-10°) + \cos(-26.317°)\cos(-10°)\cos(-22.5°) \\ &\approx 0.8801, \end{aligned}$$

so $a \approx 63.1925°$. Then $\cos(a) \approx 0.451$; substituting this into the second two equations we obtain

$$\begin{aligned} &(\cos(A), \sin(A)) \\ = &\left(\frac{\sin(\delta)\cos(\phi) - \cos(\delta)\sin(\phi)\cos(h)}{\cos(a)}, \frac{\cos(\delta)\sin(h)}{\cos(a)}\right) \\ \approx &\left(\frac{\sin(-26.317°)\cos(-10°) - \cos(-26.317°)\sin(-10°)\cos(-22.5°)}{0.451},\right. \\ &\left.\frac{\cos(-26.317°)\sin(-22.5°)}{0.451}\right) \\ \approx &(-0.6491, -0.7606) \end{aligned}$$

from which A can be determined to be an angle in the third quadrant, $A \approx -130.5°$.

2. Change of coordinates between equatorial and ecliptic coordinates. (See Figure 5.7.) Let P be the north celestial pole, M the north ecliptic pole, ♈ the vernal equinox and X the location of an object in the sky. Because the acute angle between the celestial equator and the ecliptic is $\varepsilon \approx 23.5°$, P

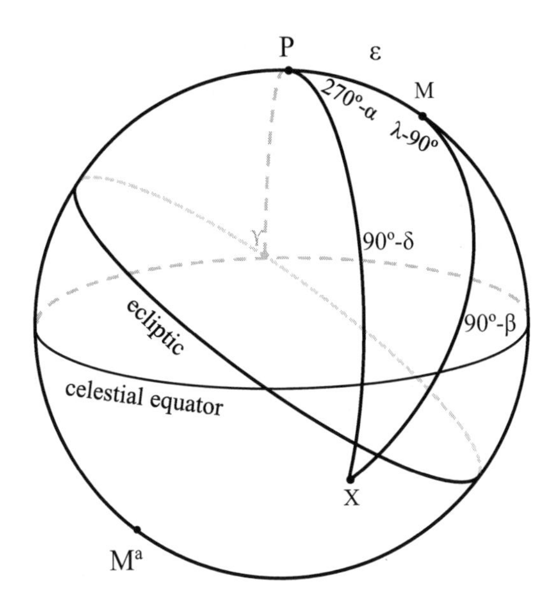

Figure 5.7: Changing coordinates between equatorial and ecliptic coordinates.

and M are a distance of ε apart on the sphere. Thus P and M are not antipodal and the great circle $\bigcirc PM$ is well-defined. Note that since Υ is on the intersection of the ecliptic and the celestial equator, it is a spherical distance of $90°$ from both P and M. By Proposition 9.26, Υ is a pole of $\bigcirc PM$. Suppose that X has equatorial coordinates (δ, α) and ecliptic coordinates (β, λ). We again assume these three points form a spherical triangle (which they do unless X is on $\bigcirc PM$). Suppose X is on the opposite side of $\bigcirc PM$ from Υ. Then by definition, α is between 6^h and 18^h ($90°$ and $270°$). Since $m \prec \Upsilon PM^a = 90°$, $m \prec XPM^a = \alpha - 90°$, so $m \prec XPM = 180° - m \prec XPM^a = 270° - \alpha$. Similarly, since $m \prec \Upsilon MP + m \prec XMP = \lambda$ and $m \prec \Upsilon MP = 90°$, $m \prec XMP = \lambda - 90°$.

The declination of X is δ so $PX = 90° - \delta$. Similarly, if the ecliptic latitude of X is β then $MX = 90° - \beta$.

We may now relate the values of (δ, α) to (β, λ) by applying the spherical law of cosines, spherical law of sines, and analogue formula to spherical $\triangle^s MPX$:

$$\cos(90° - \delta) = \cos(90° - \beta)\cos(\varepsilon) + \sin(90° - \beta)\sin(\varepsilon)\cos(\lambda - 90°),$$

$$\sin(90° - \delta)\cos(270° - \alpha) = \cos(90° - \beta)\sin(\varepsilon) - \sin(90° - \beta)\cos(\varepsilon)\cos(\lambda - 90°),$$

$$\sin(90° - \delta)\sin(270° - \alpha) = \sin(\lambda - 90°)\sin(90° - \beta).$$

So

$$\sin(\delta) = \sin(\beta)\cos(\varepsilon) + \cos(\beta)\sin(\varepsilon)\sin(\lambda) \quad (5.10)$$

$$\cos(\delta)\sin(\alpha) = -\sin(\beta)\sin(\varepsilon)+\cos(\beta)\cos(\varepsilon)\sin(\lambda) \tag{5.11}$$
$$\cos(\delta)\cos(\alpha) = \cos(\lambda)\cos(\beta). \tag{5.12}$$

Switching coordinates in the other direction can be done similarly:

$$\cos(90^\circ-\beta)=\cos(90^\circ-\delta)\cos(\varepsilon)+\sin(90^\circ-\delta)\sin(\varepsilon)\cos(270^\circ-\alpha),$$

$$\sin(90^\circ-\beta)\cos(\lambda-90^\circ)=\cos(90^\circ-\delta)\sin(\varepsilon)-\sin(90^\circ-\delta)\cos(\varepsilon)\cos(270^\circ-\alpha),$$

$$\sin(90^\circ-\beta)\sin(\lambda-90^\circ)=\sin(270^\circ-\alpha)\sin(90^\circ-\delta).$$

So

$$\sin(\beta) = \sin(\delta)\cos(\varepsilon)-\cos(\delta)\sin(\varepsilon)\sin(\alpha) \tag{5.13}$$
$$\cos(\beta)\sin(\lambda) = \sin(\delta)\sin(\varepsilon)+\cos(\delta)\cos(\varepsilon)\sin(\alpha) \tag{5.14}$$
$$\cos(\beta)\cos(\lambda) = \cos(\alpha)\cos(\delta). \tag{5.15}$$

It now remains to verify that these equations work in the case where X is on the same side of $\bigcirc PM$ as Υ. This may be seen by noting that the point on the other side of great circle $\bigcirc MP$ corresponding to X has the same declination and ecliptic latitude but right ascension $180^\circ - \alpha$ and ecliptic longitude $180^\circ - \lambda$ (modulo 360°). But it can be easily checked that (5.13), (5.14), and (5.15) still hold if λ and α are replaced by their supplements.

3. Change of coordinates between equatorial and declination-hour angle coordinates. This change of coordinates is easier because these two coordinate systems have the same pole, hence the same first coordinate (the declination). However, unlike the above changes of coordinates, the coordinate change requires us to consider the time. Let Υ be the vernal equinox, and suppose its hour angle is denoted by h_Υ. Then h_Υ is known as the *local sidereal time* for the observer. Let X be the position of an object in the sky with equatorial coordinates (δ_X, α_X) and declination-hour angle coordinates (δ_X, h_X). Because the hour angle of any object increases constantly with time at the same rate for every object, $h_\Upsilon - h_X$ is the same regardless of time, modulo multiples of 24 hours. Furthermore, $h_\Upsilon - h_X$ does not depend on the location of the observer because if an observer moves from one point on the earth to another, the hour angle (modulo multiples of 24 hours) of every object changes by the same amount (the amount being the change in longitude, measured in hours). Specifically, if an observer increases longitude by d°, this is a movement of d° to the east, which causes a motion of every object d° to the west, and an increase of $\frac{1}{15}d$ hours in the hour angle. But by (5.1) $h_\Upsilon - h_X = \alpha_X - \alpha_\Upsilon = \alpha_X$ since Υ has zero right ascension, so we have

$$h_\Upsilon = h_X + \alpha_X \tag{5.16}$$

modulo 24^h. Formula (5.16) allows us to switch between the hour angle h_X and right ascension α_X of an object via the local sidereal time. Here we regard the vernal equinox as the hand of a clock whose changing hour angle as seen by an observer measures the passage of time. This coordinate conversion via a measurement of time provides a reason for measuring right ascension in hours. To make use of this equation, we must know the position of the vernal equinox, which depends on the time and position of the observer. While the sidereal time is a good way to measure time, due to the constant rate of rotation of the earth, it is different from the usual way of measuring time via the position of the sun. Let us define the *sidereal day* as the amount of time required for the hour angle of the vernal equinox to change by 360 degrees (24 "sidereal" hours). Then the sidereal day is different from the civil "day" as measured by the movement of the sun once around the celestial sphere. This is due to the fact that the sun is moving relative to the fixed stars but the vernal equinox is not. Sidereal time can be calculated from civil time on a clock; for this we refer the reader to [Du1981].

The local sidereal time at the prime meridian (longitude zero) on the earth is called *Greenwich sidereal time.* This may be used as a universal time independent of location. Since the hour angle of an object at longitude $\lambda°$ is $\lambda/15$ hours greater than the hour angle of the object on the prime meridian, we conclude that local sidereal time may be found by adding $\lambda/15$ hours to Greenwich sidereal time.

There are numerous other variable changing scenarios which can arise. For example, suppose we are given the altitude of a star and its hour angle, and the latitude of the observer. How do we find the declination of the star? This does not fit into any of the above coordinate changes, and in fact it involves solving a triangle which falls into an ambiguous case. In situations like this and others, solution will simply require the ingenuity of the student.

Exercises §22

1. Given the following locations in declination-hour angle coordinates, convert to altitude-azimuth coordinates. Assume the latitude of the observer is $40° N$ in (1a) to (1d) and $20° S$ in (1e) to (1h).

   (a) $\delta = 30°,\ h = -4^h$
   (b) $\delta = 30°, h = 4^h$
   (c) $\delta = -60°, h = 2^h$
   (d) $\delta = -60°, h = -2^h$
   (e) $\delta = 70°, h = -10^h$
   (f) $\delta = -5°, h = 1^h$
   (g) $\delta = -80°, h = 8^h$
   (h) $\delta = 10°, h = -11^h$

2. Given the following locations in altitude-azimuth coordinates, convert to declination-hour angle coordinates. Assume the latitude of the observer is $40° S$.

(a) $a = 30°, A = -40°$
(b) $a = 30°, A = 40°$
(c) $a = -60°, A = 20°$
(d) $a = -60°, A = -20°$

3. Given the following locations in equatorial coordinates, convert to declination-hour angle coordinates, assuming the local sidereal time is 5^h.

   (a) $\delta = 30°, \alpha = 2^h$
   (b) $\delta = 30°, \alpha = 8^h$
   (c) $\delta = -60°, \alpha = 16^h$
   (d) $\delta = -60°, \alpha = 20^h$

4. Given the following locations in ecliptic coordinates, convert to equatorial coordinates.

   (a) $\beta = 30°, \lambda = 20°$
   (b) $\beta = 30°, \lambda = 100°$
   (c) $\beta = -60°, \lambda = 240°$
   (d) $\beta = -60°, \lambda = 330°$

5. Suppose the latitude of an observer is $40° N$, the declination of the sun is $10°$ and the altitude of the sun is $45°$. What are the hour angle and azimuth of the sun?

6. Suppose that the latitude of an observer is $40° S$, the hour angle of a star is 2^h and its azimuth is $120°$. Determine its declination and altitude.

7. Suppose that a star has altitude $30°$, declination $20°$, and its azimuth is $-100°$. Determine its hour angle and the latitude of the observer.

8. Suppose that a star has azimuth $-80°$, hour angle $-150°$ and altitude $-80°$. Find its declination and the latitude of the observer.

9. Given that a star has altitude $25°$, declination $-5°$ and hour angle $50°$, find its azimuth and the latitude of the observer.

10. Given that a star has azimuth $130°$, hour angle $80°$ and declination $-50°$, find its altitude and the latitude of the observer.

11. Given the right ascension and declination for each of two stars, how would one find the angular separation between them?

12. Suppose that an observer measures the altitudes of two stars whose positions (declination and right ascension) are known. Determine the latitude of the observer and the hour angles of the stars. (This is known as the Gauss two-altitude problem.)

13. Suppose that we are given the declination of a given star, its altitude at two different times, and the time interval between the observations. Explain how to determine the latitude of the observer. (This is known as Douwes' problem.)

14. Suppose that an observer knows the positions (declinations, right ascensions) of three stars and measures the time that elapses between the moments at which they attain a given altitude. Explain how to determine the latitude of the observer, the common altitude attained, and the time at which each attains that altitude. (This is known as the Gauss three-altitude problem.)

15. Suppose that we are given the declinations of two stars which rise at the same time. If the time between which each star reaches the observer's meridian (i.e., "culminates") is known, show how to determine the latitude of the observer.

16. Suppose we are given the declination δ of a star. Suppose also that at a particular moment its tangent trajectory is perpendicular to the horizon. That is, the star is located at a point on its small circle path about the pole where the small circle is tangent to a great circle passing through the zenith. From this information, show how to determine the hour angle, azimuth, and altitude.

23 Rise and set of objects in the sky

An object in the celestial sphere is said to *rise* if its altitude in the sky changes from negative to positive. Similarly, an object in the celestial sphere is said to *set* if its altitude in the sky changes from positive to negative. For all the fixed stars, rise and set is the result of the constant increase in hour angle due to the rotation of the earth. (See Figure 5.1, where the small circle with arrows shows this motion of a star.) Thus stars rise on the eastern portion of the horizon and set in the west. Objects nearer to the earth such as the sun, moon, and planets have a motion relative to the fixed stars. But the effect of this motion on the hour angle of the object is generally so small relative to that of the rotation of the earth that their rising and setting is similar to that of the stars: they rise in the east and set in the west.

If an observer is not on the celestial equator, one of the celestial poles is fixed above the horizon and the other is fixed below the horizon as the earth rotates. Thus these poles neither rise nor set. The same is true for a point which is near one of the poles. In order for a point X to rise or set, it cannot be too close to a pole that the observer can see (lest it be above the horizon all the time) and cannot be too close to a pole that the observer cannot see (lest it be below the horizon all the time.) In order to rise and set, a point must be further away from each pole than the pole is from the horizon. That is, in order to rise and set, we must have $|\delta| < 90° - |\phi|$, where δ is the declination of the point and ϕ is the latitude of the observer. A point which neither rises nor sets is called *circumpolar.*

This section is the first in a series of sections where we will consider the question of how to determine the time and place (azimuth) of rise and set for an object such as the planets, the sun, and other stars. It is important for the reader to understand that there are many factors that affect what one would call the time and place that an object such as the sun rises and sets. Obviously the rate of rotation of the earth (which we assume to be constant) is the most important of these. But the rise and set time is also affected by the apparent diameter of the object in the sky, the extent of atmospheric refraction,[2] the diurnal motion of the object (if any) relative to the fixed stars, the position of the earth in its orbit, and so on. The way in which one defines precisely the measurement of time is another factor. The decision as to what exactly is the final answer for the time of rise or set is a subjective one and depends on how many of these factors one wishes to include as significant. Here we consider each variable contributing to rise and set and reach a final answer which is usable for most observational purposes.

Proposition 23.1 *Let ϕ be the latitude of an observer and let δ be the declination of a point in the sky at the moment when it rises or sets, respectively. Assume that refraction in the atmosphere has no effect on the position of the point from the observer's perspective. Let H be the hour angle of the point at the moment the point rises or sets, and let A be the azimuth of the object at the same moment. Then we have*

$$\begin{aligned} \cos(H) &= -\tan(\phi)\tan(\delta) \\ \cos(A) &= \sin(\delta)/\cos(\phi) \end{aligned} \qquad (5.17)$$

Proof. (See Figure 5.8, left column.) Let X be the point in the sky of interest. (Note that X cannot be simply any point in the sky; it must be a point which can rise or set.) The assumption that there is no refraction means that X's apparent position (from the point of view of the observer) is not distorted by the presence of atmosphere. We first discuss the situation of the set of X at a northern hemisphere location. On the celestial sphere, assume that P is the north pole, Q is the point where X meets the horizon when it sets, and N is the point on the horizon due north of the observer. (Then N is the point where the great circle $\bigcirc ZP$ meets the horizon north of the observer.) Then $\triangle^s PQN$ is a right triangle with a right angle at N because Z is a pole of the horizon and any great circle through the pole of another great circle must be perpendicular to the latter by Proposition 10.12. Next, $m\,\overset{\frown}{PN} = \phi$ by Proposition 21.2. Since X has declination δ when it sets, its distance to P is $90° - \delta$, so $m\,\overset{\frown}{PQ} = 90° - \delta$. Next, the azimuth of X when

[2]Atmospheric refraction is discussed in detail later in §23.

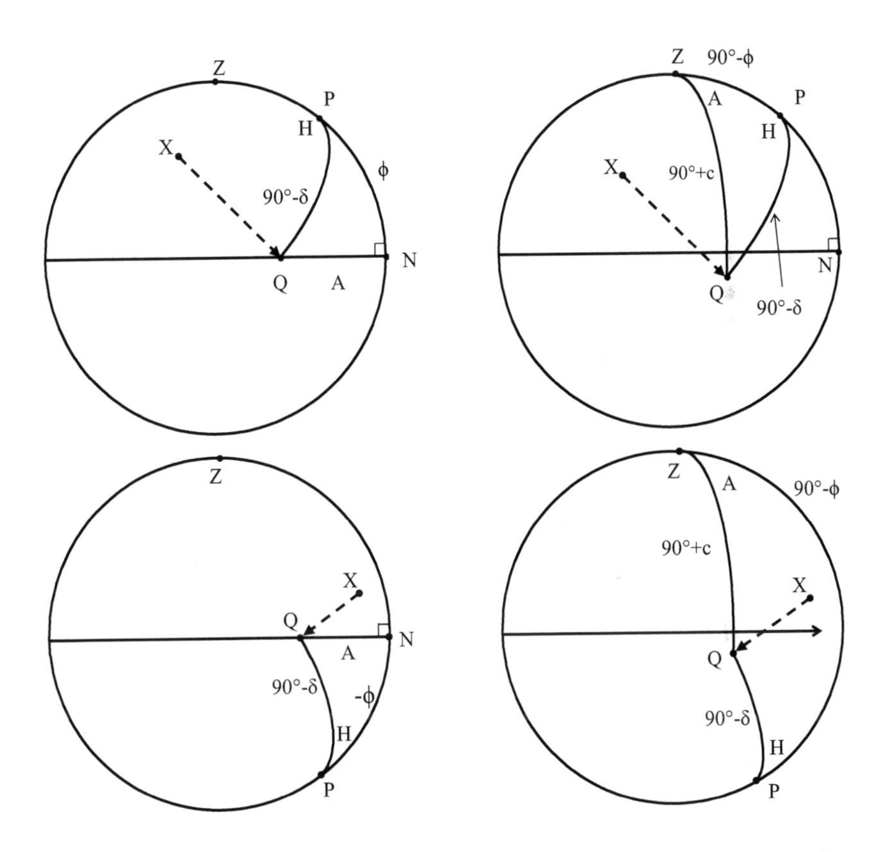

Figure 5.8: Figures for a setting object in Proposition 23.1 (left column) and Proposition 23.2 (right column), where the celestial horizon is viewed edge-on. The top row is for a northern latitude and the bottom row is for a southern latitude.

it sets is $A = m\,\overset{\frown}{QN}$ by Definition 10.2 (being positive since the azimuth is positive for an object west of the ZPZ^a meridian). Lastly, $\prec ZPQ$ is the hour angle H of X when it is about to set, and $m \prec QPN$ is its supplement, so $m \prec QPN = 180° - H$.

Now since $\triangle^s QPN$ has a right angle at N, we may apply Theorem 17.1 — in particular, (4.21) and (4.34) — to obtain, respectively,

$$\cos(90° - \delta) = \cos(\phi)\cos(A)$$

and

$$\cos(180° - H) = \tan(\phi)/\tan(90° - \delta).$$

Using elementary properties of the trigonometric functions,

$$\sin(\delta) = \cos(\phi)\cos(A)$$

$$-\cos(H) = \tan(\phi)/\cot(\delta),$$

which proves (5.17) in the event that X sets. The proof for the case where X rises is the same except that A and H are negative — but since $\cos(-x) = \cos(x)$ for all X, the same formulas hold. The argument is similar for a southern hemisphere location and is left to the reader. ♢

Example 1. Determine the hour angle and azimuth of the star Rigel when it rises and sets at Santiago, Chile, assuming no refraction in the atmosphere.

Solution. Rigel has declination $S8°12'5.91''$, i.e., $\approx S8.2°$, so $\delta \approx -8.2°$ and Santiago has latitude $S33°27'$, or $S33.45°$, so $\phi \approx -33.45°$. By (5.17), the hour angle and azimuth satisfy

$$\cos(H) \approx -\tan(-33.45°)\tan(-8.2°) \approx -0.095,$$

and

$$\cos(A) \approx \sin(-8.2°)/\cos(-33.45°) \approx -0.17,$$

so $H \approx -95.46° \approx -6^h21^m51.3^s$ at rise, $H \approx 95.46° \approx 6^h21^m51.3^s$ at set, $A \approx -99.84°$ at rise and $A \approx 99.84°$ at set.

It is easy to use Proposition 23.1 for a fixed star since its declination and right ascension change only slowly with time and are assumed constant in this book. However, objects such as the sun, moon, planets, and other objects in the solar system are close enough that their declination changes much more quickly. In this book our approach will be first to find an approximation to the declination of the desired object at its time of rise or set and use that in formulas such as (5.17).

Before doing this, we briefly discuss measurement of time. Following [Gr1985], we define *apparent solar noon* to be the moment that the center of the sun reaches the observer's meridian. We measure time forward and backward from this on a given day. Note that if a full day (360° of rotation) occurs over a period of 24 hours, we would use $\frac{24}{360} = \frac{1}{15}$ as a multiplication factor to convert a degree measure rotation angle to hours of time. Let $\odot$ be a symbol to denote the sun and let $h_\odot$ denote the hour angle of the sun (where the angle units are hours). Assuming approximately 24 hours between two consecutive times when the sun crosses the observer's meridian, we define the *local apparent solar time* T_A to be

$$T_A = 12^h + h_\odot \tag{5.18}$$

Note that the hour angle is negative when the sun is east of the observer's meridian and positive when west of that meridian, so the time is smaller than 12^h in the morning and larger than 12^h in the afternoon. In the next example we find local apparent solar time for the rising and setting of the sun.

Example 2. Let X be the center of the sun. Determine the local apparent solar time of rise and set for X at Johnstown, Pennsylvania, on the summer and winter solstices, assuming the sun does not move significantly relative to the fixed stars on these days and that there is no atmospheric refraction of the position of the sun. Determine the azimuth of X at this time.

Solution. We assume that the sun has declination $\delta \approx 23.5°$ for the entire day on the summer solstice and $\delta \approx -23.5°$ for the entire day on the winter solstice. The latitude of Johnstown is $40°16'42''$ north, or $40 + 16/60 + 42/3600 \approx 40.278°$ north, so take $\phi \approx 40.278°$. On the summer solstice $\delta \approx 23.5°$, so for sunset, $\cos(A) \approx \sin(23.5°)/\cos(40.278°)$, so $A \approx 58.489°$. Thus the sun's center sets $31.511°$ north of west. Also, $\cos(H) \approx -\tan(23.5°)\tan(40.278°)$, so $H \approx 111.621° \approx 7^h 26^m 27.6^s$. So the sun's center sets at about $19:26$ local apparent solar time, or $7:26$ p.m. The angles for rise are the negatives of those just found, so the sun's center rises at $4:34$ local apparent solar time, at an azimuth $31.511°$ north of east.

For the winter solstice, ϕ is the same but $\delta \approx -23.5°$. Then $\cos(A) \approx \sin(-23.5°)/\cos(40.278°)$, so $A \approx 121.511°$ (note that this is the supplement of the value of A for the summer solstice, since the cosine value for the winter is the negative of that for the summer). Thus the sun's center sets $31.511°$ south of west. Also, $\cos(H) \approx -\tan(-23.5°)\tan(40.278°)$, so $H \approx 68.379° \approx 4^h 33^m 30.96^s$. So the sun's center sets at about $16:34$ local apparent solar time, or about $4:34$ p.m. Similarly, the sun's center rises at around $7:26$ local apparent solar time.

The above examples have drawbacks: in fact, refraction in the atmosphere does distort the position of objects in the sky, especially at the horizon. The common understanding of sunset is the moment when the entire disk of the sun diappears, not the center of the sun. The sun's position does change a bit during the day. Lastly, we generally do not measure time in local apparent solar time; we would usually prefer to convert to standard time for the local time zone. We shall make corrections for all of these presently, but mastering of the above examples will be helpful as we do so. In any case, the above examples do give sunrise and sunset to within a few minutes on the given dates.

We first address the matter of refraction of light in the atmosphere. As the light from an object enters the atmosphere of the earth, the beam of light is bent from its path in a downward direction, toward the surface of the earth. The result is that an observer perceives an object's position as having a higher altitude than it would have were the atmosphere not present. This altitude change is higher for objects closer to the horizon. For objects near the horizon, it is standard to assume an altitude increase of $34'$ of arc, although this number depends on conditions in the atmosphere. The result is that an object is said to "set" or "rise" when it is $34'$ of arc *below* the horizon, as it is still visible to the observer until it is $34'$ below the horizon!

The sun (and the moon) are different from the rest of the objects in the sky in that they have a diameter that is perceptible to the observer, usually taken to be $32'$ of arc on the celestial sphere for both the sun and the moon, although this number varies somewhat during the orbits of the earth and the moon since the distance from the earth to the sun and moon varies. Thus for the sun, it is standard to speak of "sunset" as occurring when the top of the sun is $34'$ below the horizon (this being the moment the last sliver of the refracted sun disappears from the view of the observer). "Sunrise" occurs when the first sliver of the refracted sun comes into the view of the observer. Thus the center of the sun is about $50'$ below the horizon at sunrise and sunset. (The $50'$ is found by adding the $34'$ of refraction to the $16'$ spherical radius of the sun.) The same is the case at moonrise and moonset.

Proposition 23.2 *Let ϕ be the latitude of an observer. Assume that any point is refracted by an angle c at the horizon. At the moment the refracted image of a point in the sky rises or sets, let δ, H, and A be the declination, hour angle, and azimuth, respectively, of the (unrefracted) point. Then we have*

$$\cos(H) = -\frac{\sin(c) + \sin(\phi)\sin(\delta)}{\cos(\phi)\cos(\delta)} \quad (5.19)$$

$$\cos(A) = \frac{\sin(\delta) + \sin(c)\sin(\phi)}{\cos(c)\cos(\phi)}. \quad (5.20)$$

Proof. (See Figure 5.8.) We follow the approach used in Proposition 23.1, with some modifications. We assume that X is the point of interest, P is the north pole, Z is the zenith, and Q is the point where X reaches a spherical distance c below the horizon; this is the moment the refracted object disappears. By Proposition 21.1, $\overset{\frown}{ZP}$ has measure $90° - \phi$. As in Proposition 23.1, $\overset{\frown}{PQ}$ has measure $90° - \delta$. The hour angle of Q is $m \prec ZPQ$. Since Q is angle c below the horizon, the arc $\overset{\frown}{ZQ}$ has measure $90° + c$. Now by definition $m \prec PZQ$ is the azimuth $A > 0$ of X when it sets so $m \prec PZQ = A$. In triangle $\Delta^s PZQ$, all the sides are known, and in order to find H and A we must find two of the angles of the triangle. This may be easily done in two ways: either with the spherical law of cosines (4.10) or with a half-angle formula in Theorem 18.2. We use the former here:

$$\cos(90° + c) = \cos(90° - \phi)\cos(90° - \delta) + \sin(90° - \phi)\sin(90° - \delta)\cos(H),$$

so

$$-\sin(c) = \sin(\phi)\sin(\delta) + \cos(\phi)\cos(\delta)\cos(H)$$

and we solve for $\cos(H)$ to get (5.19). Using the spherical law of cosines again on $\Delta^s PZQ$ again,

$$\cos(90^\circ - \delta) = \cos(90^\circ + c)\cos(90^\circ - \phi) + \sin(90^\circ + c)\sin(90^\circ - \phi)\cos(A),$$

so

$$\sin(\delta) = -\sin(c)\sin(\phi) + \cos(c)\cos(\phi)\cos(A)$$

and solving for $\cos(A)$ we get (5.20). ♢

Example 3. Determine the hour angle and azimuth of the star Rigel when it rises and sets in Santiago, Chile, taking into account refraction in the atmosphere.

Solution. We apply (5.19) and (5.20) with $\phi \approx -33.45^\circ$, $\delta \approx -8.2016^\circ$ and that c is $34'$ of arc, or 0.567°. We obtain

$$\begin{aligned}\cos(H) &\approx -\frac{\sin(0.567^\circ) + \sin(-33.45^\circ)\sin(-8.2016^\circ)}{\cos(-33.45^\circ)\cos(-8.2016^\circ)} \approx -.1072 \\ \cos(A) &\approx \frac{\sin(-8.2016^\circ) + \sin(0.567^\circ)\sin(-33.45^\circ)}{\cos(0.567^\circ)\cos(-33.45^\circ)} \approx -0.1775,\end{aligned}$$

so $A \approx -100.225^\circ$ at rise, $A \approx 100.225^\circ$ at set, $H \approx -96.154^\circ \approx -6^h24^m36.96^s$ at rise and $H \approx 96.154^\circ \approx 6^h24^m36.96^s$ at set.

Proposition 23.3 *Let ϕ be the latitude of an observer looking at a circular object in the sky with angular radius R. Assume that any point is refracted by an angle c at the horizon. At the moment the (refracted) object rises or sets, let $\delta, H,$ and A be the declination, hour angle, and azimuth, respectively, of its (unrefracted) center. Then we have*

$$\cos(H) = -\frac{\sin(c+R) + \sin(\phi)\sin(\delta)}{\cos(\phi)\cos(\delta)} \tag{5.21}$$

$$\cos(A) = \frac{\sin(\delta) + \sin(c+R)\sin(\phi)}{\cos(c+R)\cos(\phi)}. \tag{5.22}$$

Proof. The proof is essentially the same as Proposition 23.2. The only difference is that the arc $\overset{\frown}{ZQ}$ now has measure $90^\circ + c + R$, since the center of the object has to be a distance of $c + R$ below the horizon for the object to set. Thus we must merely replace the c in (5.19) and (5.20) with $c + R$ to obtain (5.21) and (5.22). ♢

Example 4. Determine the local apparent solar time of sunrise and sunset at Johnstown, Pennsylvania, on the summer and winter solstice, assuming the sun does not move significantly relative to the fixed stars in a given day. Determine the azimuth of the sun at this time.

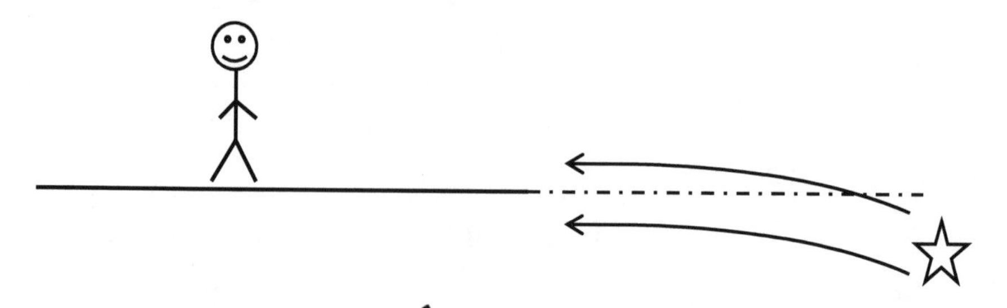

Figure 5.9: Refraction of the sun at sunrise or sunset.

Solution: We again assume that the sun has declination $\delta \approx \pm 23.5^\circ$ for the entire day, and that $\phi \approx 40.278^\circ$. The value of $c + R$ is taken to be $50'$ as noted above.

Using (5.21), we obtain

$$\cos(H) \approx -\frac{\sin(50') + \sin(40.278^\circ)\sin(23.5^\circ)}{\cos(40.278^\circ)\cos(23.5^\circ)} \approx -0.389 \tag{5.23}$$

for sunset so the hour angle H at sunset is approximately $112.908^\circ = 7^h 31^m 37.8^s$. So the sun sets at about $7:32$ p.m. local apparent solar time, about 6 minutes later than that of Problem 1. Similarly, the sunrise comes out to $4:28$ a.m. For the azimuth of sunrise and sunset, using (5.22),

$$\cos(A) = \frac{\sin(23.5^\circ) + \sin(50')\sin(40.278^\circ)}{\cos(50')\cos(40.278^\circ)} = 0.535, \tag{5.24}$$

so $A = \pm 57.653^\circ$. Thus the sun sets 32.347° north of west — and similarly rises 32.347° north of east — on the summer solstice. For the winter solstice in Johnstown, we change δ to -23.5° and obtain $\cos(H) = 0.348$, so $H = \pm 69.655^\circ = 4^h 38^m 37.2^s$. So we obtain sunset on the winter solstice at around $4:39$ p.m. and sunrise at $7:21$ a.m. in local apparent solar time. For the azimuth, we again change δ to -23.5° and obtain $\cos(A) = -0.510$, so $A = \pm 120.690^\circ$. Thus the sun rises 30.69° south of east and sets 30.69° south of west.

We next consider the rise and set for the sun on a general date. For this we have to consider how to find the equatorial coordinates of the sun from its ecliptic longitude. We prove a simple lemma regarding coordinates for the position of the sun. This lemma is a special case of a coordinate change (between ecliptic and equatorial coordinates) which we saw in §22.

Lemma 23.4 *Let ε be the obliquity of the ecliptic with the celestial equator. Suppose that an object on the ecliptic has ecliptic longitude λ. Then its declination δ and right ascension α satisfy*

$$\sin(\delta) = \sin(\lambda)\sin(\varepsilon) \tag{5.25}$$

$$\tan(\alpha) = \tan(\lambda)\cos(\varepsilon). \tag{5.26}$$

provided $\tan(\lambda)$ *is defined.*

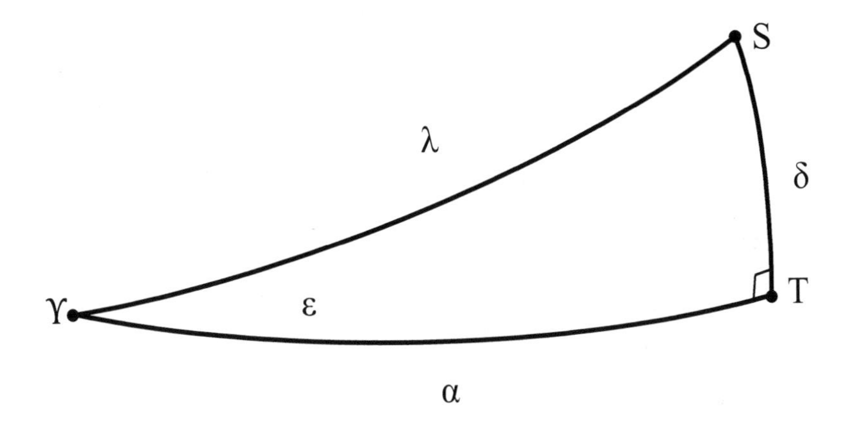

Figure 5.10: Declination δ and right ascension α at ecliptic longitude λ.

Proof. Suppose the object is at point S, and right ascension and ecliptic longitude are chosen so that $-12^h < \alpha \leq 12^h$ and $-180° < \lambda \leq 180°$. Since the ecliptic latitude is 0, α and λ have the same sign (or are both zero.) Let T be the point on the celestial equator with the same right ascension as S. Suppose $0 < \lambda < 180°$. Then $\triangle^s \Upsilon TS$ is a spherical right triangle with right angle at T. We have $m \sphericalangle S\Upsilon T = \varepsilon$, $m \overset{\frown}{ST} = \delta$, $m \overset{\frown}{T\Upsilon} = \alpha$ and $m \overset{\frown}{S\Upsilon} = \lambda$. By (4.23) we obtain (5.25) and by (4.34) we obtain

$$\cos(\varepsilon) = \frac{\tan(\alpha)}{\tan(\lambda)} \tag{5.27}$$

which implies (5.26) if $\tan(\lambda)$ is nonzero and defined. In the event that $-180° < \lambda < 0°$, so are δ and α, so we obtain similar equations with α, δ, and λ replaced by their negatives. But since for all x, $\sin(-x) = -\sin(x)$ and $\tan(-x) = -\tan(x)$, the signs in the fractions cancel and (5.25) and (5.26) hold for $\lambda \neq 0°, 90°, 180°$. If $\lambda = 0°$ or $\lambda = 180°$, we also have $\delta = 0°$ and $\alpha = 0^h$ or $\alpha = 12^h$ so (5.25) and (5.26) merely state that $0 = 0$. $\diamondsuit$

We first calculate time of sunrise and sunset under the simplifying assumption that the earth has a circular orbit around the sun, and so the sun moves along the ecliptic at a constant rate.

Example 5. Assuming the earth has a circular orbit, determine the local apparent solar time of rise and set for the sun at Johnstown, Pennsylvania, on May 6. Find the azimuth of the sun at these times.

Solution: The most recent vernal equinox occurred on March 20. If the earth's orbit is assumed to be circular the sun's ecliptic longitude $\lambda_\odot$ changes by approximately the same amount $(\frac{360}{365.25})°$ every day, so on May 6, $\lambda_\odot \approx (360)(47)/365.25 \approx 46.32°$. Letting $\varepsilon = 23.5$ we next calculate δ and α for rise from (5.25) and (5.26) to get

$$\sin(\delta) = \sin(\lambda_\odot)\sin(\varepsilon) \approx 0.29, \tag{5.28}$$

and

$$\tan(\alpha) = \tan(\lambda_\odot)\cos(\varepsilon) \approx 0.96 \tag{5.29}$$

so $\delta \approx 16.76°$ and $\alpha \approx 43.84° \approx 2^h 56^m$. Next, we calculate the hour angle $h_\odot$ at rise (from Proposition 23.3) to be

$$\cos(h_\odot) \approx -\frac{\sin(50') + \sin(40.278°)\sin(16.76°)}{\cos(40.278°)\cos(16.76°)} \approx -.275, \tag{5.30}$$

so, $h_\odot \approx \pm 105.97° \approx \pm 7^h 4^m$.

Thus the local apparent solar time T_A for sunrise is 4 : 56 a.m. and sunset is 7 : 04 p.m.

The azimuth is found from (5.22); we obtain

$$\cos(A) \approx \frac{\sin(16.76) + \sin(50')\sin(40.278)}{\cos(50')\cos(40.278)} \approx 0.39. \tag{5.31}$$

For azimuth at sunrise, we choose the negative value of A and obtain $A \approx -67°$; the rise is 23° north of east. The set is then 23° north of west.

We now prove a theorem that will help calculate apparent solar time in terms of the position of an object other than the sun. Our main use for it will be to calculate rise and set times for that object.

Theorem 23.5 *Suppose that at a particular moment the center of an object has declination and right ascension (δ, α). Let H be the hour angle of the center of the object, and $\alpha_\odot$ the right ascension of the sun. Then the apparent solar time T_A satisfies*

$$T_A = 12^h + H + (\alpha - \alpha_\odot), \tag{5.32}$$

reduced modulo 24^h.

Proof. We begin with (5.18) and write $T_A = 12^h + h_\odot = 12^h + H + (h_\odot - H)$. By (5.1) the difference $h_\odot - H$ in the hour angles of the sun and the center of the object is the same as the difference in their right ascensions in the reverse order (reduced modulo 24). Thus $T_A = 12^h + H + (h_\odot - H) = 12^h + H + (\alpha - \alpha_\odot)$ (mod 24^h), as desired. ◇

Example 6. Determine in local apparent solar time when the star Rigel rises and sets in Johnstown, Pennsylvania, on May 6, taking into account refraction in the atmosphere. Assume the earth has a circular orbit.

Solution: We apply (5.19) with $\phi \approx 40.278°$, $\delta \approx -8.2016°$ and that c is $34'$ of arc, or 0.567°. We obtain

$$\cos(H) \approx -\frac{\sin(0.567°) + \sin(40.278°)\sin(-8.2016°)}{\cos(40.278°)\cos(-8.2016°)} \approx 0.109 \tag{5.33}$$

so the hour angle of Rigel is $H \approx \pm 83.74^\circ \approx 5^h 35^m$. The right ascension of Rigel is $\alpha \approx 5^h 14^m$ (using a catalogue.) We have the right ascension of the sun on this date from Example 5: $\tan(\alpha_\odot) \approx 0.96$ so $\alpha_\odot \approx 43.8^\circ \approx 2^h 55^m$. Now using (5.32),

$$T_A \approx 12^h - 5^h 35^m + (5^h 14^m - 2^h 55^m) = 8^h 44^m$$

when Rigel rises, and

$$T_A \approx 12^h + 5^h 35^m + (5^h 14^m - 2^h 55^m) = 19^h 54^m$$

when Rigel sets. So the apparent solar time of rise and set for Rigel are approximately $8:44$ a.m. and $7:54$ p.m.

For a period of time before sunrise and after sunset, it is light outside due to light from the sun hitting the upper atmosphere; this is a period of *twilight*. The intensity of this light only changes gradually, making a definition of beginning and end of twilight a matter of preference. The brightness of twlight depends on how far below the horizon the sun is. There are three standard levels of twlight:

(1) When the (center of the) sun is 6° below the horizon, this marks the beginning (or end) of *civil twilight*. Around this time, outside work with sunlight becomes (or ceases to be) possible.

(2) When the (center of the) sun is 12° below the horizon, this marks the beginning (or end) of *nautical twilight*. At this time, both stars and the horizon are visible, and hence this time is a good one to make observations of objects in the sky to determine their positions.

(3) When the (center of the) sun is 18° below the horizon, this marks the beginning (or end) of *astronomical twilight*. At this time, the horizon begins (ceases) to be visible and astronomical work may stop (begin).

Determining the moments at which the various kinds of twilight begin or end can be done with (5.21) and is left to the exercises.

In the next two sections we will learn to correct these rise and set times to account for the sun's motion against the fixed stars, and to correct to local standard time.

Exercises §23.

For all exercises in this section where the rise and set of an object is sought, the reader should assume that the earth's orbit around the sun is circular as done in the examples.

1. Consider the following cities and their (latitude, longitude) coordinates: (a) New York City $(40^\circ 40' N, 73^\circ 56' W)$ (b) Shanghai $(31^\circ 12' N, 121^\circ 30' E)$ (c) Sydney $(33^\circ 52' S, 151^\circ 13' E)$ (d) Buenos Aires $(34^\circ 36' S, 58^\circ 23' W)$ (e) Oslo $(59^\circ 57' N, 10^\circ 45' E)$ (f) Los Angeles $(34^\circ 3' N, 118^\circ 15' W)$ (g) Chicago $(41^\circ 50' N, 87^\circ 38' W)$.

   Also consider the following stars and their right ascension, declination coordinates:

(i) Rigel ($5^h14^m32^s, -8°12'6''$)
(ii) Altair ($19^h50^m47^s, 8°52'6''$)
(iii) Vega ($18^h36^m56^s, 38°47'1''$)
(iv) Sirius ($6^h45^m9^s, -16°42'58''$)
(v) Canopus ($6^h23^m57^s, -52°41'44''$)
(vi) Arcturus ($14^h15^m40^s, 19°10'56''$)
(vii) Deneb ($20^h41^m26^s, 45°16'49''$)
(vii) Antares ($16^h29^m24^s, -26°25'55''$)
(viii) Achernar ($1^h37^m48^s, -57°14'12''$)
(ix) Betelgeuse ($5^h55^m10^s, 7°24'25''$)
(x) Aldebaran ($4^h35^m55^s, 16°30'33''$)
(xi) Spica ($13^h25^m12^s, -11°9'41''$)
(xii) Procyon ($7^h39^m18^s, 5°13'29''$)

For each (city, star) pair, determine whether the star is circumpolar when observed from the city. For those pairs where the star is not circumpolar determine the hour angle and azimuth when the star rises and sets at the city assuming (A) refraction does not occur and (B) refraction does occur. (Follow Example 1 for (A) and Example 3 for (B).)

2. For each city in Exercise 1, determine the local apparent solar time of rise and set for the center of the sun on the summer and winter solstices, assuming that the (center of the) sun has declination ± 23.5 on these dates. Also determine the azimuth of the sun at these moments. Assume there is no refraction in the atmosphere. (Follow Example 2.)

3. For each city in Exercise 1, determine the local apparent solar time of rise and set of the sun on the summer and winter solstices, assuming that the (center of the) sun has declination ± 23.5 on these dates. Also determine the azimuth of the sun at these moments. Assume there is refraction in the atmosphere. (Follow Example 4.)

4. For each city given in Exercise 1, determine for the date May 6 the local apparent solar time of rise and set for the sun assuming a circular orbit for the earth. (Follow Example 5.)

5. Repeat Exercise 4 for a date of your choice.

6. For each star given in Exercise 1, determine for the date May 6 the local apparent solar time of rise and set for the star, assuming a circular orbit for the earth. (Follow Example 6.)

7. Repeat Exercise 6 for a date of your choice.

8. For each city in Exercise 1, determine the local apparent solar time of the beginning of civil twilight on the morning and its end in the evening on the summer and winter solstices. What locations never have an end to civil twilight on the summer solstice?

9. Use the following reasoning to answer the following questions: on what day of the year is twilight (of whichever type) the shortest, and what is the length of that twilight? We let δ be the declination of the sun, ϕ the observer's latitude, h the distance under the horizon that the sun must be for twilight to end (respectively, begin), and t the time between sunrise (respectively, sunset) and the beginning (respectively, end) of twilight. Assume the celestial sphere is oriented so that the sun is fixed at the top (point X), the north pole P on the right, and the zenith rotates with the earth between the horizontal (point A) and the end (or beginning) of twilight (point B). Let x be the spherical distance $m\,\widehat{AB}$ and let $\alpha = m \prec AXB$,

   (a) Using $\triangle^s PAB$, find a relationship among x, t, and ϕ.
   (b) Using $\triangle^s AXB$, find a relationship among x, h, and α.
   (c) Use these relationships to find a relationship among t, h, α, and ϕ.
   (d) For t to be smallest, what conclusion can we make? What is the value of t in terms of h and ϕ?
   (e) Using $\triangle^s XAP$, find a relationship among δ, ϕ, and $m \prec XAP$.
   (f) Using $\triangle^s APB$, find a relationship among $m \prec XAP$, h, and ϕ.
   (g) Find a relationship among δ, h, and ϕ. What is the value of δ in terms of h and ϕ?
   (h) For a location in the northern (respectively, southern) hemisphere, explain why the shortest twilight occurs when the declination of the sun is negative (respectively, positive).

10. Suppose a star has declination δ and an observer has latitude ϕ. Let x be the acute angle that the (small circle) path of a star forms with the horizon as it rises or sets. Prove that $\cos(x) = \sin(\phi)/\cos(\delta)$.

11. The moon revolves about the earth on a plane which is inclined 5° with the ecliptic. Determine the maximum and minimum azimuth for a full moon rise in the winter. Do the same for summer.

24 The measurement of time

The proper measurement of time requires one to find a unit of time. One way of doing this is to find an event that occurs with precise regularity; the time interval between two successive occurrences of the event then becomes a unit of time.

The daily rotation of the earth serves as one such possibility. The earth rotates on its axis at a rate that is essentially constant for most purposes.[3] An

---

[3] Modern atomic clocks show that the rate of the earth's rotation does vary slightly with time. The reader interested in further exploration of this topic will find a discussion of this in [Gr1985].

observer may verify this by measuring the length of time between two crossings of a star across the observer's meridian — and make sure this length of time stays the same from day to day. The time the earth takes to make one full 360° rotation is called a *sidereal day.* The sidereal day makes a very good unit of time, and it is sometimes used for astronomical purposes.

However, human society has long made use of the daily passage of the sun through the sky as a measurement of time. One might attempt to make use of two successive crossings of sun across the observer's meridian to make a unit of time (of a day.) However, it is here that we encounter a problem. If we were to measure the time between two crossings of the sun across the observer's meridian — an *apparent solar day* — we would soon discover that the amount of time between such crossings varies from day to day. The sun can cross the observer's meridian as much as 15 minutes before or after civil noon on a given day. Thus an apparent solar day does not serve as a useful exact unit of time.

Nevertheless, the common usage by civil society of the daily movement of the sun as a unit of a "day" leads us to make some use of the sun for time-keeping purposes. We recall that the earth orbits the sun in an ellipse which is nearly circular. The point at which the earth is nearest the sun is called *perihelion* and the point where it is furthest from the sun is called *aphelion.*

The central reason for the irregularity in the length of the apparent solar day is that as the sun moves relative to the fixed stars, its right ascension does not increase at a constant rate. This means that the sun's hour angle also does not increase at a constant rate, so the time between successive crossings of the observer's meridian is not the same every day. The sun's movement in the sky has two features responsible for the changing length of the apparent solar day.

The first feature is that the rate of the sun's motion along the ecliptic is not constant. The reason for this is that the earth's orbit around the sun is elliptical, and the earth moves more quickly around the sun when it is closer to the sun. Thus the sun moves more quickly along the ecliptic when the earth is at perihelion than when it is at aphelion.

The second feature is that the sun does not follow the celestial equator. The result is that as the sun moves further away from the celestial equator, the meridians are closer and hence the sun crosses them more quickly than it would were it moving around the celestial equator. (Compare Figure 5.11 with Figure 5.12.) This affects significantly the time between the sun's crossings of the observer's meridian for the observer.

In order to use the sun's motion as a way to measure time, we introduce two fictitious bodies that will have the desired regularity and yet behave like the sun. First, the *dynamical mean sun* is a moving point which travels on the ecliptic as the sun does, but at a *constant angular velocity.* This is the path the sun would take in the sky if only the earth had a circular instead of a slightly elliptical orbit. The dynamical mean sun is defined to coincide with the sun at perihelion (hence it does also at aphelion, since the sun travels

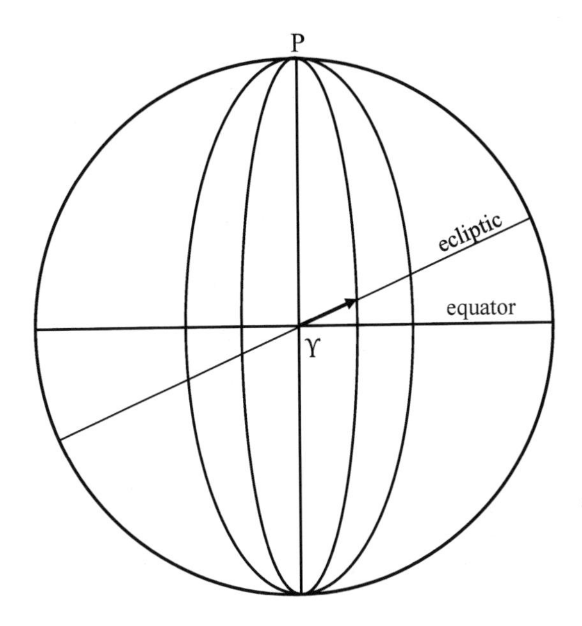

Figure 5.11: At ♈, the sun crosses equally meridians obliquely, and meridians are further apart than they are near the pole P.

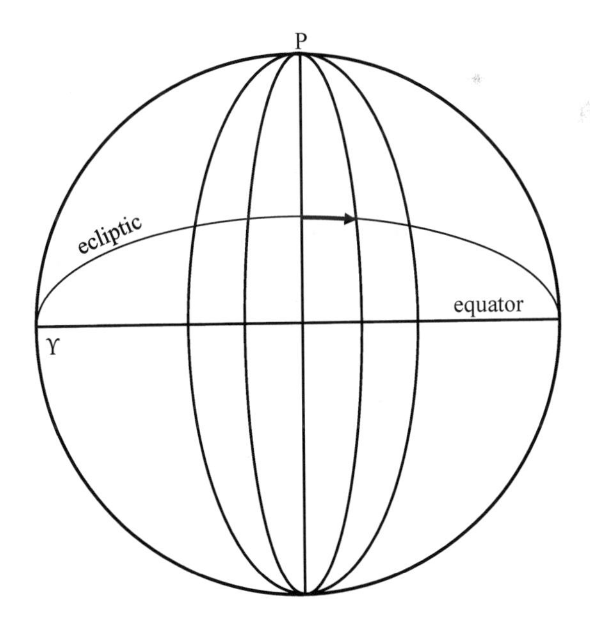

Figure 5.12: At the summer solstice, the sun crosses meridians at approximately a right angle, and meridians are closer together than they are on the celestial equator. The arrow is slightly longer in Figure 5.11.

from perihelion to aphelion in a half year).

The second fictitious body introduced is called the *fictitious mean sun* (FMS) It is defined (a) to coincide with the dynamical mean sun at the vernal equinox and yet (b) travel around the *celestial equator* at a constant angular velocity.

It is the fictitious mean sun that we will use as a means to measure time. It moves around the celestial equator against the fixed stars to the east at a constant rate of about 1 degree per day. Combined with the earth's constant rate of rotation, the fictitious mean sun has an hour angle which increases at a constant rate.

We define the *mean solar day* to be the length of time between two crossings of the fictitious mean sun across the observer's meridian. It is important that, by its definition, the mean solar day has the same length when measured from any point on earth.

It is the mean solar day that is the length of civil time that is commonly understood to be a "day." It is divided up into what is commonly understood to be 24 "hours." We let N denote the number of mean solar days in a year; then N is approximately equal to 365.2422.[4] The sidereal day mentioned earlier is slightly shorter, comprising approximately 23 hours and 56 minutes of "solar" time.

Let h_{FMS} denote the hour angle of the fictitious mean sun (measured in hours). We define the *local mean solar time* (MST) for an observer to be

$$T_M = 12^h + h_{FMS}. \tag{5.34}$$

Then the local mean solar time is $12^h = 12:00$ when the fictitious mean sun is on the observer's meridian, is larger than 12^h when FMS is west of the observer's meridian and is less than 12^h when FMS is east of the observer's meridian. The local mean solar time for any location can be used as a reliable measure of time; however, given its definition based on a particular location, it is natural to pick the local mean solar time of one special location as a standardized measure of time. We define *universal time* (UT) or *Greenwich mean time* (GMT) to be the local mean solar time of the prime meridian on earth. Lastly, we can define standard time for the various time zones around the earth. Time zones are roughly defined relative to meridians which are at multiples of 15° east and west. In the USA, Eastern Standard Time (EST) is the local mean solar time of the 75° west meridian, and Central, Mountain, and Pacific Standard times (CST, MST, PST) are the local mean solar times of the 90°, 105° and 120° west meridians. For areas which observe Daylight Saving Time, it is obtained by adding an hour to the standard time of the local time zone.

[4] This is the number of days in the so-called "tropical year" which is the amount of time the sun takes to go around the ecliptic once from vernal equinox to vernal equinox. Because of the precession of the equinoxes, this is slightly shorter (about 20 minutes) than the amount of time it takes the earth to complete one orbit of the sun. The calendar year is oriented around the tropical year in order to keep the position of seasons stable.

The *equation of time* is the difference between local apparent solar time (5.18) and mean solar time (5.34), which can be calculated as either a difference in hour angle or difference in right ascension between the fictitious mean sun and the sun (see (5.1)):

$$T_Q = T_A - T_M = h_\odot - h_{FMS} = \alpha_{FMS} - \alpha_\odot, \tag{5.35}$$

where α_X, h_X are the right ascension and hour angle of an object X. (See Figure 5.13.) We will also write

$$T_M = T_A - T_Q \tag{5.36}$$

for conversion from apparent solar time to mean solar time.

We will need one other time conversion: from local mean solar time T_M to the local time zone standard time T_Z. For this we assume that the observer is at longitude ψ and the local time zone is based at the meridian with longitude ψ_z. (We assume these longitudes follow the sign convention of west longitudes being negative and east longitudes being positive.) At longitude ψ_z, the hour angle of the fictitious mean sun is $\psi_z - \psi$ degrees greater than the hour angle of the fictitious mean sun at longitude ψ (modulo 360°). (Note that as longitude increases, the hour angle of an observed object also increases at the same rate.) Converting to hours by taking $\frac{1}{15}$, we obtain

$$T_Z = T_M + \frac{1}{15}(\psi_z - \psi) \tag{5.37}$$

The equation of time is somewhat difficult to calculate exactly because of the number of different factors that affect it. These factors include the declination of the sun, the elliptical orbit of the earth, uneven number of days in the year, and changing dates for the equinoxes and perihelion. This first two of these are the most important. See Figure 5.13 for a rough graph of the equation of time as it varies through the year. What is important to remember is what it means about the clock relative to the sun. The clock is given by the hour angle of the fictitious mean sun. When the hour angle of the sun is larger (smaller) than that of the fictitious mean sun, we say that the "sun is ahead of (behind) the clock." For us, $T_Q > 0$ when the sun is ahead of the clock, so when the sun is on the observer's meridian, the clock says it is still morning. In Figure 5.13, note particularly the periods of time near the solstices (June and December): there the sun is far from the celestial equator and hence crossing meridians of constant right ascension quickly relative to the time when it is close to the equinoxes. So the sun's hour angle decreases relative to the fictitious mean sun in these periods and the equation of time decreases. This decrease is more dramatic near the winter solstice because at that time the sun is also near perihelion (early January) when the sun moves most quickly on the ecliptic.

Many household globes and maps use a graphic called the *analemma* to represent the equation of time at various times during the year. (See Figure

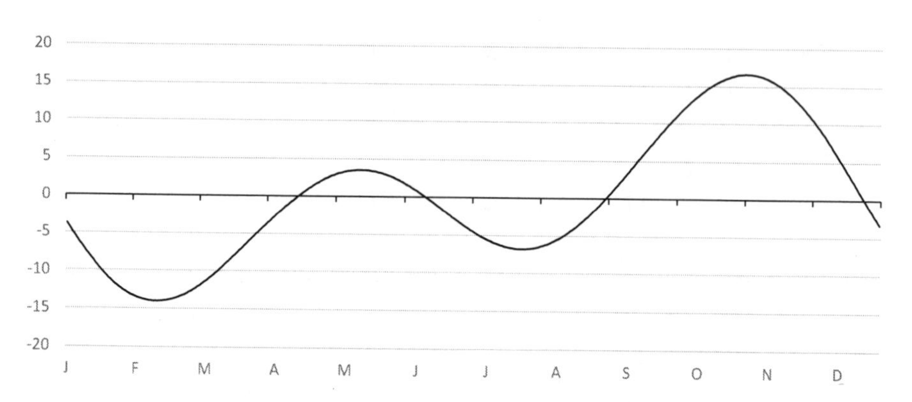

Figure 5.13: The equation of time gives the date (month) on the horizontal axis and the quantity $h_\odot - h_{FMS}$ in minutes on the vertical axis.

5.14.) An analemma is a plot of the position of the sun in the sky at the same mean solar time (typically noon) each day throughout the course of the year. The analemma may be thought of as a curve $t \mapsto (T_Q(t), \delta(t))$ which plots the declination of the sun along with the equation of time. The declination $\delta(t)$ of the sun is typically represented as a latitude on the globe or map and the equation of time $T_Q(t)$ is typically represented as a longitude east or west of a meridian which represents the observer's meridian. When the sun is to the left of the meridian, it is behind the clock and when the sun is to the right of the meridian it is ahead of the clock. The result is a figure "8" which is labelled with the value of t (typically the month).

We now discuss in detail how to calculate the equation of time. In order to do this, we make use of the Kepler equations for the elliptical orbit of a planet around a star. We do not derive the equations here, as this would take us afield from the subject of spherical geometry, but we refer the reader to [Gr1985] for a derivation of these formulas. Here we merely summarize them and show how to use them to determine the position of the sun.

Kepler's first law of planetary motion states that the planets all have elliptical orbits around the sun, where the sun is at one focus of the ellipse. The ellipse has a center and has eccentricity e. The line through the center and the focus intersects the elliptical orbit in two points, one of which is the closest point on the planet's orbit to the sun, and the other of which is the furthest point on the orbit from the sun. With the earth, the nearest point is called *perihelion* and the furthest point is called *aphelion*. The segment between the nearest and furthest points is called the major axis. The earth is at perihelion in early January each year, generally between January 2 and 5. The earth reaches aphelion sometime between July 4 and 6. Because of the symmetry of the ellipse and the nature of the earth's orbit speed, the amount of time from perihelion to aphelion is the same as the time from aphelion to perihelion. Let P be the point of perihelion (where the earth is at time

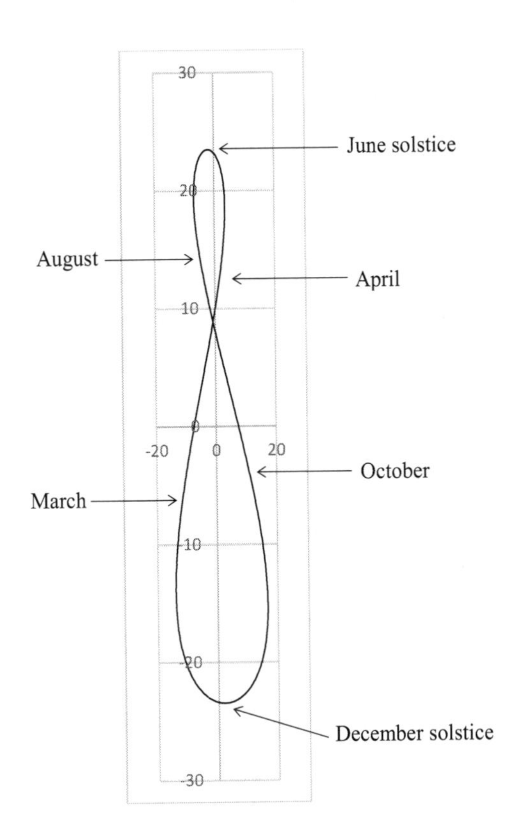

Figure 5.14: The analemma: the observer's view of the noonday sun's changing position in the sky relative to the celestial equator and the observer's meridian during the year. The vertical axis is the declination in degrees and the horizontal axis is the equation of time in minutes.

0,) let ⊙ be the sun, and ⊕ be the earth at some later time d. We let ν be the measure of the plane angle $\angle P \odot \oplus$. Then ν depends on d and is called the *true anomaly*. Since this is the angle through which the earth has moved since perihelion, it may be viewed as a feature of the earth's orbit; however, from the point of view of an observer at the earth, it is the spherical distance through which the sun has moved on the ecliptic from time 0 to time d. Note that if N is the number of mean solar days in a year, then the sun travels on average a spherical distance $M = \frac{360}{N}d$ in d days. Then the quantity M is known as the *mean anomaly*; it is the spherical distance traveled by both the fictitious mean sun and the dynamical mean sun in d days. If the earth had a circular orbit, then the mean anomaly and the true anomaly would be the same, and it would be easy to track the position of the sun in the sky (then the sun would travel the path of the dynamical mean sun.) However, the mean and true anomalies are slightly different — different enough to make a noticeable difference in the time of sunrise and sunset.

The Kepler equations determine a relationship between M and ν. However, the relationship is indirect and may be stated as follows. Introduce a third variable E, called the *eccentric anomaly*. This variable is also an increasing function of d, and the triple of variables (ν, M, E) satisfies the following pair of equations (where angles are measured in radians):

$$M = E - e\sin(E) \tag{5.38}$$

$$\nu = 2\tan^{-1}\left(\sqrt{\frac{1+e}{1-e}}\tan\left(\frac{E}{2}\right)\right) \tag{5.39}$$

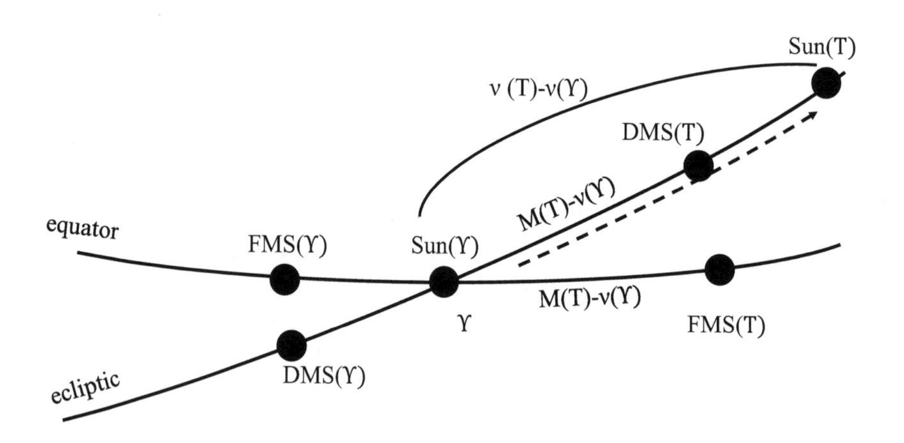

Figure 5.15: Positions of the sun, the dynamical mean sun (DMS), and ficitious mean sun (FMS) when the sun is at ♈ and at time T.

We will assume that we are given the dates of perihelion and the vernal equinox and propose to obtain the equation of time on another date we call

T. (See Figure 5.15.) The mean anomaly $M(T)$ on date T is a quantity that is then generally easy to obtain. Then one may use (5.38) and (5.39) to calculate $E(T)$ and then $\nu(T)$ on date T. The variable $\nu(T)$ is the key variable needed to determine the equatorial coordinates of the sun on T. We then calculate $\nu(\Upsilon)$ as the true anomaly on the date of the vernal equinox. Then we have that the ecliptic longitude $\lambda_{\odot}(T)$ of the sun on date T is given by $\lambda_{\odot}(T) = \nu(T) - \nu(\Upsilon)$, and from this we may calculate the sun's declination $\delta_{\odot}(T)$ and right ascension $\alpha_{\odot}(T)$ on T as desired. We will also need the right ascension $\alpha_{FMS}(T)$ of the fictitious mean sun on date T so that we have the equation of time $\alpha_{FMS}(T) - \alpha_{\odot}(T)$. But $\alpha_{FMS}(T)$ is the ecliptic longitude of the dynamical mean sun on T; this is $M(T) - \nu(\Upsilon)$.

Example 1. Determine the equation of time for the date May 6, assuming that the date of the vernal equinox is March 20 and perihelion is January 3.

Solution. Date T (May 6) is about 123 days after perihelion, so the mean anomaly $M(T) \approx (360)(123)/365.25 \approx 121.23°$, or 2.116 radians. Now the eccentricity of the earth's orbit is e which is approximately .01671123. We use a computer to solve $E - e\sin(E) = 2.116$ and get $E(T) \approx 2.134$. We take this value of $E(T)$ and solve to get $\nu(T) \approx 2.148$ radians.

The most recent vernal equinox occurred on March 20. So Υ is 76 days after perihelion, so the mean anomaly when the sun is at Υ is

$$M(\Upsilon) \approx \frac{(360)(76)}{365.25} \approx 74.91°,$$

or 1.307 radians. Proceeding as above, we obtain the eccentric anomaly at Υ to be $E(\Upsilon) \approx 1.323$ and then find the true anomaly $\nu(\Upsilon) \approx 1.339$ radians.

Letting $\varepsilon = 23.5$ and $\lambda_{\odot}(T) = \nu(T) - \nu(\Upsilon) \approx 0.8087$ radians, or 46.34° we next calculate the declination and right ascension $\delta_{\odot}(T)$ and $\alpha_{\odot}(T)$ for the sun on T from (5.25) and (5.26) to get

$$\sin(\delta_{\odot}(T)) = \sin(\lambda_{\odot}(T))\sin(\varepsilon) \approx 0.288, \tag{5.40}$$

and

$$\tan(\alpha_{\odot}(T)) = \tan(\lambda_{\odot}(T))\cos(\varepsilon) \approx 0.961 \tag{5.41}$$

so $\delta_{\odot}(T) \approx 16.77°$ and $\alpha_{\odot}(T) \approx 43.86° \approx 2^h55^m26^s$. We calculate the right ascension of the fictitious mean sun $\alpha_{FMS}(T) = M(T) - \nu(\Upsilon) \approx 2.116 - 1.339 \approx 0.777$ in radians, 44.52°, or $2^h58^m5^s$. So the equation of time is $T_Q = \alpha_{FMS}(T) - \alpha_{\odot}(T) \approx 2^h58^m5^s - 2^h55^m26^s = 2^m39^s$.

Exercises §24.

1. Following Example 1, determine the equation of time for a date of your choice, assuming that the date of the vernal equinox is March 20 and perihelion is on January 3. Compare your answer with that in Figure 5.13.

2. Assume that the dynamical mean sun and the fictitious mean sun coincide at the vernal equinox. Determine the maximum difference between their right ascensions. Show that this maximum occurs when the sum of these right ascensions is 90°. (Calculus is the easiest way to do this problem, but it is also possible without such methods, using a substitution. Let α be the right ascension of the fictitious mean sun and d the right ascension of the dynamical mean sun. Then $d - \alpha = d - \arctan((\cos(\varepsilon))(\tan(d)))$. Let $x = \tan(d)$; must maximize $\arctan(x) - \arctan((\cos(\varepsilon)x))$, which is the same as

$$\arctan(\frac{x(1-\cos(\varepsilon))}{1+\cos(\varepsilon)x^2}).$$

Since the arctangent is increasing, we must maximize the argument — or minimize the reciprocal of the argument. The latter can be broken up into a sum of 2 fractions which is bounded below by the arithmetic mean/geometric mean inequality. The case where equality occurs is the desired value of x. Menelaus also has an intriguing non-calculus approach to this problem in Book III, Proposition 23 of *Sphaerica*.)

25 Rise and set times in standard time

In this section we show how to use Theorem 23.5 and time conversions (5.36) and (5.37) to calculate rise and set times in local mean solar time and standard time for objects in the sky, and apply it to the sun and the stars. This will take into account refraction in the atmosphere and the elliptical orbit of the earth.

Example 1. Determine the local mean solar time and local standard time of sunrise and sunset at Johnstown, Pennsylvania, on May 6. (Take the date of perihelion to be January 3 and vernal equinox to be March 20.)

Solution. We recall that in Example 1 of §24 we obtained the right ascensions of the sun and fictitious mean sun on May 6: $\alpha_\odot(T) \approx 43.86° \approx 2^h55^m$ and $\alpha_{FMS}(T) \approx 0.777$ in radians, 44.52°, or $2^h58^m5^s$. We also found the sun's declination $\delta_\odot(T) \approx 16.77°$. Next, we calculate the hour angle $H_\odot$ of the sun at rise and set (from Proposition 23.3) to be

$$\cos(H_\odot) \approx -\frac{\sin(50') + \sin(40.278°)\sin(16.77°)}{\cos(40.278°)\cos(16.77°)} \approx -.275, \tag{5.42}$$

so $H_\odot \approx \pm 105.98° \approx \pm 7^h3^m54^s$. So the local apparent solar times of sunrise and sunset are $12 \pm H_\odot$, or $4^h56^m6^s$ and $19^h3^m54^s$.

To calculate the local mean solar time for rise, we use the equation of time conversion (5.36), using the equation of time $T_Q \approx 2^m39^s$ found in Example 1 of §24. We find $T_M = T_A - T_Q$ which is $4^h53^m27^s$ for rise and $19^h1^m15^s$

for set. So local mean solar time of sunrise is about $4:53$ a.m. and sunset is about $7:01$ p.m.

The longitude of Johnstown is $\psi = -78.92$ and the time zone is Eastern, so $\psi_z = -75$. We now have all the variables needed to calculate (5.37) for rise: $T_Z = 4^h53^m27^s + \frac{1}{15}(-75 - (-78.92)) = 4^h53^m27^s + 0^h15^m40^s = 5^h9^m7^s$, or $5:09$ a.m. EST. (In daylight saving time, this is $6:09$ a.m.) For set, we obtain $T_Z = 19^h1^m15^s + 0^h15^m40^s = 19^h16^m55^s$, or about $7:17$ p.m. EST. (In daylight saving time, this is $8:17$ p.m.)

We next find the local standard time of rise and set for a star. Again the main tool is Theorem 23.5 combined with the time conversions (5.36) and (5.37).

Example 2. Determine the local mean solar time and local standard time of rise and set for Rigel in Johnstown, Pennsylvania, on May 6.

Solution: In Example 6 of §23 we found the hour angle of Rigel to be $H \approx 5^h35^m$. The right ascension of Rigel is $\alpha \approx 5^h14^m$. In Example 1 of §24 we found $\alpha_\odot \approx 2^h55^m$. Then invoking Theorem 23.5 we find $T_A \approx 12 + 5^h35^m + (5^h14^m - 2^h55^m) \approx 19^h54^m$ when Rigel sets and $T_A \approx 12^h - 5^h35^m + (5^h14^m - 2^h55^m) \approx 8^h44^m$ when Rigel rises. Using the equation of time (5.36) conversion and the fact that $T_Q \approx 2^m39^s$ found in Example 1 of §24, the local mean solar time of rise is about $8:41$ and set is about $19:51$. To get the rise and set in eastern standard time we use (5.37) as done in Example 1 in this section above and find we must add 16 minutes to find that Rigel rises at about $8:59$ a.m. and sets at about $20:07$, or $8:07$ p.m.

Exercises §25

1. For each city given in §23, Exercise 1, determine for the date May 6 the local standard time of rise and set for the sun assuming an elliptical orbit for the earth.

2. Repeat Exercise 1 for a date of your choice.

3. For each star given in §23, Exercise 1, determine for the date May 6 the local standard time of rise and set for the star, assuming the earth's orbit is elliptical.

4. Repeat Exercise 3 for a date of your choice.

5. A curious effect popularly called "Manhattanhenge" or the "Manhattan solstice" occurs in the Manhattan borough of New York City on certain days of the year. The numbered east/west streets in Manhattan all point in the same direction: $N61°W$ to the west and $S61°E$ to the east. On certain days of the year, the sun sets (or rises) in such a way that it shines directly down all of these streets simultaneously at the moment of sunset (or sunrise).

   (a) Determine the declination of the sun when this occurs (for sunrise or sunset), assuming the "half sun" effect; that is, the sun's center is directly on the horizon.

(b) What days during the year does this occur at sunrise? At what time?

(c) What days during the year does this occur at sunset? At what time?

(d) Repeat the above, assuming a "full sun" effect; that is, the lower part of the sun just touches the horizon.

Chapter 6

Polyhedra

A *polyhedron* is a region in space bounded by planes. Many theorems in spherical geometry can be understood as statements about associated polyhedra which have one vertex at the center of a sphere and other vertices on the surface of the sphere. We now use this association to study regular polyhedra.

26 Regular solids

Let $\triangle^s ABC$ be a spherical triangle on a sphere in space of center O and radius r. Then the measures of the sides and angles of $\triangle^s ABC$ are closely related to the parts of the tetrahedron $OABC$ in space. As discussed in §5, the measure of the sides $\overset{\frown}{AB}$, $\overset{\frown}{BC}$, and $\overset{\frown}{AC}$ of the triangle are the measures of the angles $\angle AOB$, $\angle BOC$, and $\angle AOC$ about vertex O of the tetrahedron. Furthermore (see Figure 2.4 and Figure 6.1), the measures of $\prec BAC$, $\prec ABC$, and $\prec ACB$ are the measures of dihedral angles between the planes defining the sides of the tetrahedron $OABC$ about vertex O: $m \prec ABC = m\angle A{-}OB{-}C, m \prec BAC = m\angle B-OA-C$ and $m \prec ACB = m\angle A-OC-B$. Thus many theorems about spherical triangles can be rephrased in terms of relationships among the plane angles and dihedral angles of tetrahedron $OABC$ about the vertex O.

Given a vertex of a polyhedron, suppose that the midpoints of the edges containing that vertex are planar. Then the *vertex figure* at that vertex is the polygon defined by those midpoints. A *Platonic* solid $\{p, q\}$ in space is a convex polyhedron whose faces are all the same congruent regular polygon of p sides and which has vertex figures which are all congruent regular polygons of q sides. Five such convex solids[1] have been known since antiquity: the 4-sided tetrahedron $\{3, 3\}$, 6-sided cube $\{4, 3\}$, 8-sided octahedron $\{3, 4\}$, 12-sided dodecahedron $\{5, 3\}$, and 20-sided icosahedron $\{3, 5\}$. (The $\{p, q\}$ notation is called the Schläfli symbol for the polyhedron.)

[1]There are also four non-convex polyhedra of this type; they are called the Kepler–Poinsot solids.

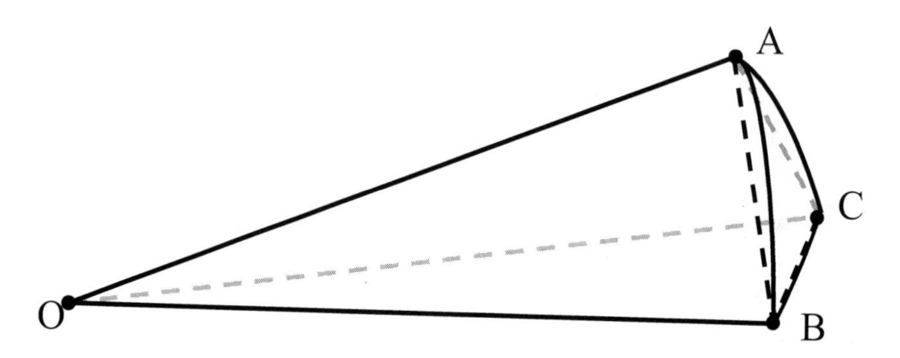

Figure 6.1: Parts of a spherical triangle vs. parts of a tetrahedron: $m \stackrel{\frown}{AB} = m\angle AOB$, $m \stackrel{\frown}{AC} = m\angle AOC$, $m \stackrel{\frown}{BC} = m\angle BOC$, $m \prec ABC = m\angle A - OB - C$, $m \prec BAC = m\angle B - OA - C$, $m \prec ACB = m\angle A - OC - B$.

It is an important fact, which we now proceed to prove, that there are only five Platonic solids. We begin by determining a necessary restriction on the values of p and q. We note that at any vertex X of a convex polyhedron, the sum of the measures of the face angles is less than 2π. (See Figure 6.2.) To see why, consider a sphere with center at X. Because the polyhedron is convex, there exists a plane touching the polyhedron only at X. This plane cuts a great circle Γ in the sphere. The face angles cut arcs of great circles which form a spherical polygon lying on one side of Γ. This polygon is spherically convex because the polyhedron is convex. By §13 Exercise 53, the sum of the measures of the sides of a convex spherical polygon is less than 2π. The measures of these sides are the same as the measures of the face angles at vertex X, so the sum of the latter measures is also less than 2π.

Since each face of the polyhedron is a regular p-gon, the interior angles each have measure $\pi(1 - \frac{2}{p})$. Since there are q such faces meeting at every vertex, the sum of those face angle measures is $q\pi(1 - \frac{2}{p})$. From above we have $q\pi(1 - \frac{2}{p}) < 2\pi$, so $1 - \frac{2}{p} < \frac{2}{q}$, or

$$\frac{1}{p} + \frac{1}{q} > \frac{1}{2}. \tag{6.1}$$

Taking the inequality (6.1) we clear all fractions by multiplying by pq and then we gather terms: $2p+2q-pq > 0$. Then $pq-2p-2q < 0$, so $pq-2p-2q+4 < 4$, and $(p-2)(q-2) < 4$. Now both p and q are whole numbers greater than or equal to 3, so there are only five solutions to this inequality: $\{p, q\}$ can be $\{3,3\}, \{3,4\}, \{4,3\}, \{3,5\}$, or $\{5,3\}$. These are realized as the Platonic solids.

The foregoing discussion shows that there are no more than five Platonic solids. But it does not show that any of them actually can be constructed as solid objects; it only limits the possibilities.

We make use of the sphere to study these and other convex solids in

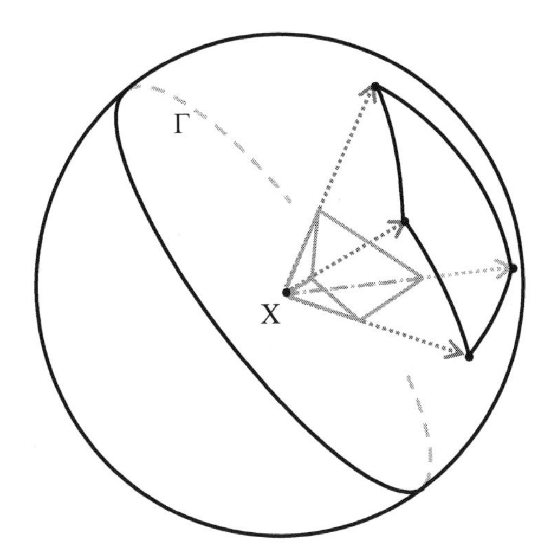

Figure 6.2: The face angles of a convex polyhedron with corner X determine a convex spherical n-gon.

another way as follows. Suppose that O is a point in the interior of a convex polyhedron such that every ray with vertex O intersects the polyhedron in exactly one point. This ray then meets a sphere with center O in a single point. Using this ray, a point of the polyhedron may be associated with a unique point of the sphere. Every vertex of the polyhedron corresponds to a point of the sphere, every edge of the polyhedron corresponds to a great circle arc on the sphere, and every face of the polyhedron corresponds to a spherical polygon. (See Figure 6.3.) The union of all the resulting spherical polygons is the entire sphere, and no two of them overlap (they meet only at edges or vertices.) We call the resulting union of spherical polygons on the sphere the *projection* of the polyhedron to the sphere.

We let V, E, and F be the number of vertices, edges, and faces, respectively, of a Platonic solid. We use an argument due to Legendre to find a relationship among these quantities. Since p is the number of edges per face and each edge lies on two faces, pF gives twice the number of edges in the polyhedron. So $pF = 2E$. In a similar manner, qV gives two times the total number of edges because q edges emanate from every vertex, and every edge is counted twice because an edge has two vertices. Thus $pF = 2E = qV$.

Next we note that the area of the whole sphere ($4\pi r^2$) is the sum of the areas of the faces (since they only overlap at edges). Since there are F congruent faces, we have that $4\pi r^2$ is F times the area of a given face. But the area of a face is found from §14, Exercise 4 to be $(p\theta - (p-2)\pi)r^2$, where θ is the measure of a spherical angle of one of the *projected* polygons. Since there are F of these spherical polygons having non-overlapping union equal to

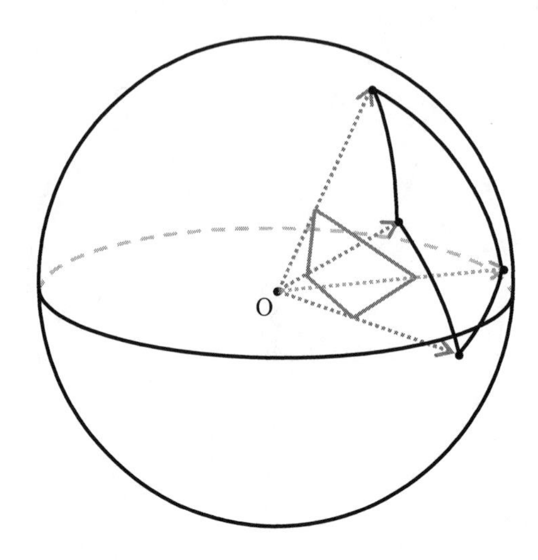

Figure 6.3: Projection of a face of a polyhedron to a sphere with center O.

the whole sphere, the sum of their areas is $4\pi r^2 = (p\theta - (p-2)\pi)r^2 F$. Thus

$$(p\theta - (p-2)\pi)F = 4\pi. \tag{6.2}$$

Since around each vertex of the polyhedron there are q faces, we can conclude that $\theta = 2\pi/q$. Multiplying out the left side of (6.2) and substituting for θ we obtain

$$2\pi pF/q - pF\pi + 2\pi F = 4\pi \tag{6.3}$$

We substitute $pF = qV$ in the first term and $pF = 2E$ in the second term to obtain $2\pi V - 2E\pi + 2\pi F = 4\pi$, or $V - E + F = 2$, as desired. This equation, known as Euler's formula, is more generally true; see Corollary 26.3.

We use Euler's formula $V - E + F = 2$ to determine V, E, and F for each of the Platonic solids. Letting p, q be as above, we take the equation $V - E + F = 2$ and substitute $F = 2E/p$ and $V = 2E/q$: $2E/q - E + 2E/p = 2$, so $E(2/q + 2/p - 1) = 2$ and $2/q + 2/p - 1 = 2/E$. By substituting the values of p and q for each solid into this last equation, we may find the values of E for each. Then we may find V and F for each by using $F = 2E/p$ and $V = 2E/q$. This gives us the following table.

Name	p, q	V	E	F
Tetrahedron	3,3	4	6	4
Cube	4,3	8	12	6
Octahedron	3,4	6	12	8
Dodecahedron	5,3	20	30	12
Icosahedron	3,5	12	30	20

The tetrahedron may be easily formed by taking three equilateral triangles and placing them around a vertex in space so that each pair of triangles has a common side. The third side of each taken together form a fourth equilateral triangle, the fourth face of the tetrahedron.

The octahedron may be formed from a cube by placing a single vertex in the middle of each face of the cube and connecting the vertices of adjacent faces of the cube with an edge. (See Figure 6.4.) This process is called "dualization" of the cube; it results in a solid with as many vertices as the cube has faces, and as many faces as the cube has vertices. If we repeat the process on the octahedron (by placing a vertex at the center of each face and connecting centers of adjacent faces) we obtain another cube. We then say that the cube and octahedron are "dual" to each other via this process. We may similarly "dualize" the tetrahedron, which results in another tetrahedron.

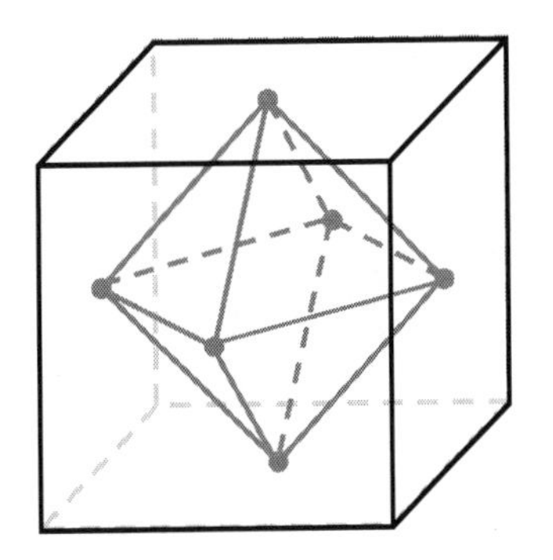

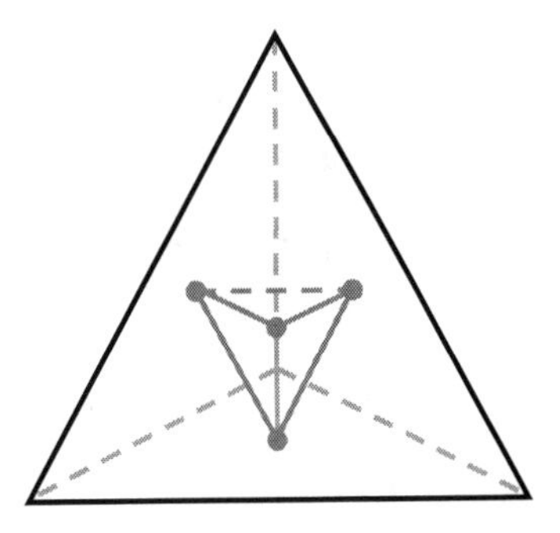

Figure 6.4: Dualization of the cube and tetrahedron to obtain an octahedron and tetrahedron.

We next use some methods of spherical trigonometry to construct the icosahedron.[2] (See Figure 6.5.) According to the above discussion, it should have 12 vertices. We place two of them at a pair of antipodal points of a sphere which for convenience we call the north and south poles. Next, we construct 10 great semicircles (meridians) between the north and south poles N and S, all equally spaced 36 degrees ($\frac{\pi}{5}$ radians) apart. We may divide these into two groups of five meridians each, where in each group the meridians are spaced by 72 degrees ($\frac{2\pi}{5}$ radians). We call one of these the northern group and the other the southern group. On the northern group we place a point on each meridian at a distance c from the north pole, where $c = \arctan(2)$, which is approximately 63.4 degrees. We call these five points N_1, N_2, N_3, N_4 and N_5 (listed in order around the pole). We choose five similar points on the meridians in the southern group and call them S_1, S_2, S_3, S_4 and S_5; choose S_i on the meridian between N_i and N_{i+1} for $i = 1, 2, 3, 4$, and choose S_5 on the meridian between N_5 and N_1. The reason for this choice of c is that the spherical triangles $\triangle^s NN_1N_2$, $\triangle^s NN_2N_3$, $\triangle^s NN_3N_4$, $\triangle^s NN_4N_5$, $\triangle^s NN_5N_1$ are all equiangular spherical triangles, where the angles in each tri-

[2]Our approach is inspired by that found in [Do1965].

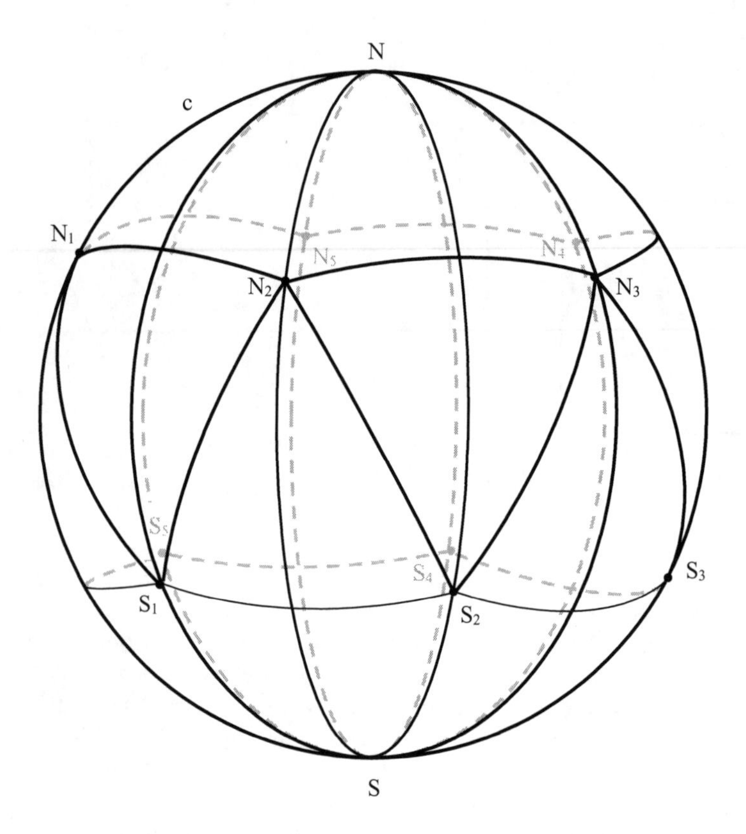

Figure 6.5: Construction of the icosahedron.

angle all have measure $\frac{2\pi}{5}$. A similar conclusion holds for the southern group.[3] To see why, note that if $c = \arctan(2)$ then $\cos(c) = \frac{1}{\sqrt{5}}$ and $\sin(c) = \frac{2}{\sqrt{5}}$. If $m\ \overset{\frown}{NN_1} = m\ \overset{\frown}{NN_2} = \arctan(2)$ then using the spherical law of cosines on $\triangle^s NN_1N_2$, we obtain

$$\begin{aligned}
&\cos(m\ \overset{\frown}{N_1N_2}) \\
= \;& \cos(m\ \overset{\frown}{NN_1})\cos(m\ \overset{\frown}{NN_2}) + \sin(m\ \overset{\frown}{NN_1})\sin(m\ \overset{\frown}{NN_2})\cos(\frac{2\pi}{5}) \\
= \;& \cos^2(c) + \sin^2(c)\frac{\sqrt{5}-1}{4} \\
= \;& \frac{1}{5} + \frac{4}{5}\frac{\sqrt{5}-1}{4} \\
= \;& \frac{1}{\sqrt{5}}.
\end{aligned}$$

(See §3 for calculation of trigonometric functions of multiples of $\frac{\pi}{5}$.) Thus $m\ \overset{\frown}{N_1N_2} = c$, as desired.

We claim that the twelve points N, S, N_i, S_i, $i = 1 \ldots 5$ are the vertices of an icosahedron. We already know that the arcs $\overset{\frown}{NN_1}$, $\overset{\frown}{NN_2}$, $\overset{\frown}{NN_3}$, $\overset{\frown}{NN_4}$, $\overset{\frown}{NN_5}$, $\overset{\frown}{N_1N_2}$, $\overset{\frown}{N_2N_3}$, $\overset{\frown}{N_3N_4}$, $\overset{\frown}{N_4N_5}$, and $\overset{\frown}{N_5N_1}$ are all congruent and these arcs are all congruent to the corresponding lengths in the southern group. We next show that $m\ \overset{\frown}{N_1S_1} = c$. For this, we consider the spherical triangle $\triangle^s NN_1S_1$: we have $m\ \overset{\frown}{NN_1} = c$, $m\ \overset{\frown}{NS_1} = \pi - c$, and $m \prec N_1NS_1 = \frac{\pi}{5}$. Using the spherical law of cosines,

$$\begin{aligned}
&\cos(m\ \overset{\frown}{N_1S_1}) \\
= \;& \cos(m\ \overset{\frown}{NN_1})\cos(m\ \overset{\frown}{NS_1}) + \sin(m\ \overset{\frown}{NN_1})\sin(m\ \overset{\frown}{NS_1})\cos(m \prec N_1NS_1) \\
= \;& \cos(c)\cos(\pi - c) + \sin(c)\sin(\pi - c)\cos(\frac{\pi}{5}) \\
= \;& -\cos^2(c) + \sin^2(c)\cos(\frac{\pi}{5}) \\
= \;& -\frac{1}{5} + \frac{4}{5}\frac{\sqrt{5}+1}{4} \\
= \;& \frac{1}{\sqrt{5}}.
\end{aligned}$$

[3] Some readers may be satisfied with the following manner of choosing the points N_i and S_i. On two meridians separated by $\frac{2\pi}{5}$, suppose two points X and Y are each chosen at spherical distance c from N. If c is close to zero then the spherical triangle NXY is isosceles and nearly planar. Since the vertex angle is $\frac{2\pi}{5}$, the base angles are close to $\frac{3\pi}{10}$, so $XY > NX$ (as longer sides are opposite larger angles.) On the other hand, if $m\ \overset{\frown}{NX} = m\ \overset{\frown}{NY} = \frac{\pi}{2}$, then $m\ \overset{\frown}{XY} = \frac{2\pi}{5} < m\ \overset{\frown}{NX}$. Since XY depends continuously on the position of X and Y, there must be some value of $NX = NY$ between 0 and $\frac{\pi}{2}$ such that $m\ \overset{\frown}{NX} = m\ \overset{\frown}{NY} = m\ \overset{\frown}{XY}$.

Thus $m\ \widehat{N_1S_1} = c$. A similar argument shows that the arcs $\widehat{N_2S_1}$, $\widehat{N_2S_2}$, $\widehat{N_3S_2}$, $\widehat{N_3S_3}$, $\widehat{N_4S_3}$, $\widehat{N_4S_4}$, $\widehat{N_5S_4}$, $\widehat{N_5S_5}$, and $\widehat{N_1S_5}$ all have measure c. Thus with twelve vertices we have cut up the entire sphere into twenty spherical triangles all of which are equilateral of side c and angles of $\frac{2\pi}{5}$. Now we take any pair of vertices connected by an arc on the sphere and instead connect them with a segment in space. The result is a regular polyhedron with 20 faces, 12 vertices, and (since $V - E + F = 2$) 30 edges. This is the icosahedron. We leave it to Exercise 2 to check that the vertex figure of the resulting polyhedron around every vertex is a regular pentagon.

We may easily construct the dodecahedron from the icosahedron by dualization. We merely take the center of each of the 20 triangles and connect them in pairs if the faces from which they are chosen have an edge in common. The result is a polyhedron with 20 vertices, 12 faces, and 30 edges. This is the dodecahedron.

We next show how to use a spherical triangle to find the measures of angles between faces of a Platonic solid.

Proposition 26.1 *Suppose that a Platonic solid has p sides on every face and q faces around every vertex. Suppose that I is the interior measure of the angle between two adjacent faces of the polyhedron. Then*

$$\sin(\frac{I}{2}) = \csc(\frac{\pi}{p})\cos(\frac{\pi}{q}). \tag{6.4}$$

Proof.

(See Figure 6.6.) We assume the points of the solid lie on a sphere with center O. Suppose two adjacent faces have a side $\overline{AE}$ with midpoint B. Let C be the center of one of these faces and let D be the center of the other. Then since (planar) $\triangle CAE$ and $\triangle DAE$ are isosceles with vertex angles at C and D, respectively, the line segment $\overline{BC}$ is perpendicular to $\overline{AB}$ and similarly $\overline{BD} \perp \overline{AB}$. Thus $\angle CBD$ is a plane angle of the dihedral angle $\angle C - AB - D$, and $I = m\angle CBD$. By symmetry $CB = DB$ and $\angle OBC \cong \angle OBD$, so $I = 2m\angle OBC$. Since C is the center of a face, $\overleftrightarrow{OC} \perp \overleftrightarrow{BC}$, so $\alpha \equiv m\angle COB = \frac{\pi}{2} - m\angle OBC = \frac{\pi}{2} - \frac{I}{2}$.

Next we let F and G be the points where $\overrightarrow{OB}$ and $\overrightarrow{OC}$, respectively, meet the sphere with center O and radius OA. Consider $\triangle^s AFG$. Since $\angle FOG$ is a central angle for $\widehat{FG}$, $m\ \widehat{FG} = \alpha = \frac{\pi}{2} - \frac{I}{2}$. Note that $\widehat{GA} \cong \widehat{GE}$ and $\angle AOB \cong \angle EOB$ so $m\ \widehat{AF} = m\ \widehat{FE}$. By Exercise 3 of §12, $\widehat{AF} \perp \widehat{FG}$, so $\prec AFG$ is a right angle. The face with center C projects from center O out to the sphere to create a regular p-sided spherical polygon so that $m \prec AGE = \frac{2\pi}{p}$. By Exercise 4 of §12, $m \prec AGF = m \prec EGF = \frac{1}{2} m \prec AGE = \frac{\pi}{p}$. Since there are q polygons around every vertex of the polyhedron, $m \prec GAF = \frac{\pi}{q}$. By

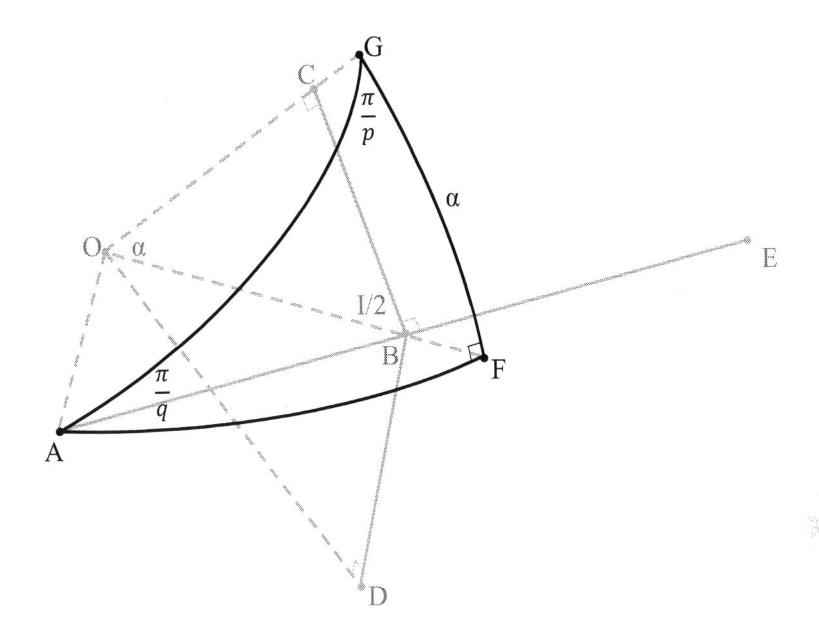

Figure 6.6: Proposition 26.1.

Geber's theorem (or (4.36) of Theorem 17.1),

$$\sin(\sphericalangle AGF) = \frac{\cos(\sphericalangle GAF)}{\cos(\widehat{GF})},$$

so $\cos(\widehat{GF}) = \cos(\sphericalangle GAF)\csc(\sphericalangle AGF) = \cos(\frac{\pi}{q})\csc(\frac{\pi}{p})$. But $\cos(\widehat{GF}) = \cos(\alpha) = \cos(\frac{\pi}{2} - \frac{I}{2}) = \sin(\frac{I}{2})$, and we obtain (6.4). ◊

Earlier we noted that the sum of the measures of the angles around a vertex of a convex polyhedron is less than 2π. But how great is the deficit from 2π? Descartes[4] characterized it as follows.

Theorem 26.2 *Suppose that a convex polyhedron has V vertices and that for $i = 1, 2, \ldots V$, the sum of the measures of the face angles around the ith vertex is $2\pi - \delta_i$. Then*

$$\sum_{i=1}^{V} \delta_i = 4\pi \tag{6.5}$$

Proof. We choose a point O in the interior of the polyhedron and choose a sphere centered at O large enough to contain the polyhedron in its interior. By dilation we may assume the radius of the sphere to be 1, so its surface

[4]See [Co1973, p. 23].

area is 4π. Then we project the faces of the polyhedron out to the sphere to obtain a division of the sphere into spherical polygons. The sum of the areas of these spherical polygons is 4π. Suppose that a face in the original polyhedron has angles $\theta_1, \theta_2, \ldots, \theta_n$ and the projected angles in the sphere have measures $\phi_1, \phi_2, \ldots, \phi_n$. The area of the projected face is $(\sum_{i=1}^{n} \phi_i) - (n-2)\pi$. But in the planar face, $\sum_{i=1}^{n} \theta_i = (n-2)\pi$. So the area of the projected face is $\sum_{i=1}^{n} \phi_i - \sum_{i=1}^{n} \theta_i = \sum_{i=1}^{n}(\phi_i - \theta_i)$. Thus the area of the projected polygon is the sum of the differences between each projected angle and the original plane angle. Adding these over all faces gives the left side of (6.5). Since this is the area of the sphere, it must be 4π. $\diamondsuit$

Corollary 26.3 *A convex solid with V vertices, E edges, and F faces satisfies $V - E + F = 2$.*

The proof is left as Exercise 1.

We conclude this section by using Proposition 26.1 to study the generalization of the Platonic solids in four-dimensional space. A *polytope* in four-dimensional space is a set bounded by three-dimensional hyperplanes. We here consider convex polytopes. The portions of the hyperplanes which lie on the boundary of the solid are the analogues of the faces of polyhedra in three dimensions; these are called *cells* of the polytope. These cells have vertices, edges, and faces which comprise the vertices, edges, and faces of the polytope. To define what is meant by a "regular" polytope, it will help to have the notion of a vertex figure. Suppose that the midpoints of the edges containing a vertex of the polytope lie in a single hyperplane. Then we define the *vertex figure* at the given vertex to be the polytope whose vertices are these midpoints.

A polytope is said to be regular if each of its boundary cells is a three-dimensional Platonic solid $\{p, q\}$, and the vertex figure at every vertex is also a Platonic solid (the same at every vertex). Because a (two-dimensional) face of the vertex figure is a vertex figure of a $\{p, q\}$ cell (and the latter vertex figure is a regular polygon of q sides), the vertex figure of the polytope must have the form $\{q, r\}$, where r is the number of $\{p, q\}$ which contain a particular edge. Such a polytope in four dimensions is denoted by the Schläfli symbol $\{p, q, r\}$. We proceed to determine the possibilities for the values of $\{p, q, r\}$.

Let us take a point P on some edge e of the polytope and let H be a (three-dimensional) hyperplane perpendicular (in four dimensions) to e at P. The intersection of the hyperplane with the polytope is a three-dimensional convex subset of the hyperplane. Furthermore, the hyperplane meets each of the r copies of $\{p, q\}$ which contain e. The hyperplane does not contain any of these $\{p, q\}$ because it is perpendicular to the edge e of each. Thus its intersection with each $\{p, q\}$ is a plane angle of the dihedral angle formed by the faces whose common edge is e. The measure of this plane angle is the value I of Proposition 26.1. Thus the intersection of H with the boundary of the polytope consists of r angles of measure I forming a convex corner about point P. As noted earlier (§13 Exercise 53), the sum of the measures of these angles must be less than 2π. Thus $rI < 2\pi$, or $\frac{I}{2} < \frac{\pi}{r}$. Note that $0 < I < \pi$ and $r \geq 3$ so $0 < \frac{I}{2} < \frac{\pi}{2}$ and $0 < \frac{\pi}{r} < \frac{\pi}{2}$. Since $\sin(\cdot)$ is an increasing function for angles between 0 and $\frac{\pi}{2}$, $\sin(\frac{I}{2}) < \sin(\frac{\pi}{r})$. By Proposition 26.1, we obtain $\csc(\frac{\pi}{p})\cos(\frac{\pi}{q}) < \sin(\frac{\pi}{r})$, or

$$\cos(\frac{\pi}{q}) < \sin(\frac{\pi}{r})\sin(\frac{\pi}{p}). \tag{6.6}$$

The only values permissible for p, q, and r are those where $\{p, q\}$ and $\{q, r\}$ are Platonic solids. We leave it to the reader to verify that this inequality is satisfied for only six polytopes:

Proposition 26.4 *The only possibilities for regular polytopes in four dimensions are* $\{3,3,3\}, \{3,3,4\}, \{4,3,3\}, \{3,4,3\}, \{3,3,5\}$ *and* $\{5,3,3\}$.

The first three of these are easy to construct. Suppose that we let (x, y, z, w) be coordinates on pairwise perpendicular axes in four dimensions.

The solid $\{3,3,4\}$ is found by taking vertices at the eight points

$$(\pm 1, 0, 0, 0), (0, \pm 1, 0, 0), (0, 0, \pm 1, 0), (0, 0, 0, \pm 1).$$

This is called the *cross polytope* and is the four-dimensional analogue of the octahedron.

The solid $\{4,3,3\}$ is found by taking vertices at the 16 points $(\pm 1, \pm 1, \pm 1, \pm 1)$. This is called the 4-cube and is the analogue of the three-dimensional cube.

The solid $\{3,3,3\}$ is constructed most easily as a polytope in five dimensions. We find it by taking vertices at the five points

$$(1,0,0,0,0), (0,1,0,0,0), (0,0,1,0,0), (0,0,0,1,0), (0,0,0,0,1).$$

These points lie on the four-dimensional hyperplane consisting of points whose coordinates have sum equal to 1, so can be seen as lying in four dimensions. This polytope is called the 4-*simplex*, the four-dimensional analogue of the tetrahedron.

The other three polytopes in Proposition 26.4 do exist but their construction is difficult. We refer the reader to [Co1973] or [So1929] for details.

Exercises §26

1. Prove Corollary 26.3.

2. Show that the vertex figure of the icosahedron is a regular pentagon.

3. Construct an icosahedron by cutting it out of a cube of side 1 as follows. (See Figure 6.7.) Choose a face of the cube and draw a segment of length $2x$ parallel to two of the sides and having midpoint at the center of the face. Repeat this successively on all faces, making sure that the

segment in any face has direction perpendicular to the segments in the neighboring faces. Then pick an endpoint of one of these segments and connect it to the four nearest endpoints of segments in neighboring faces. Repeat this process for all the 12 endpoints. Then determine the value of x which makes these connecting segments all have the same length.

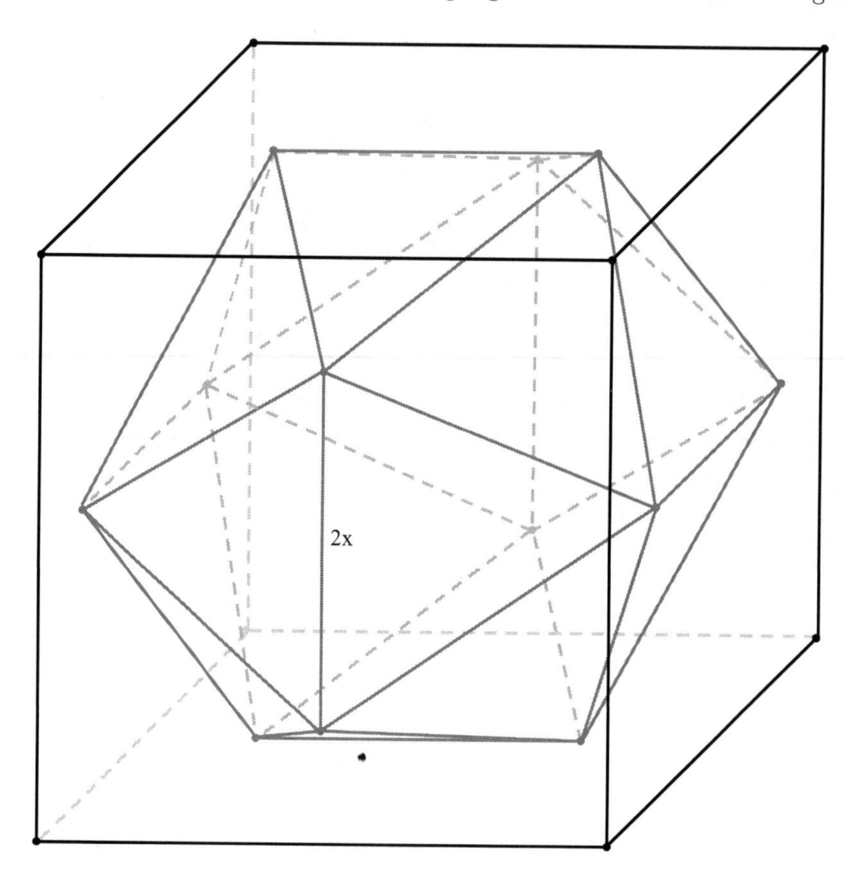

Figure 6.7: Exercise 3.

4. Imitate the argument used with Figure 6.5 for the construction of an icosahedron inside a sphere to construct a cube inside a sphere.

5. If I, p, q are as in this section, s is the length of the side of the Platonic solid and r is the radius of the inscribed sphere of this polyhedron, then $r = \frac{s}{2}\cot(\frac{\pi}{p})\tan(\frac{I}{2})$.

6. If I, p, q, s are as in the previous exercise, show that the angle β formed at O by the rays to the vertex of the Platonic solid and the center of a face satisfies $\cot(\beta) = \cot(\frac{\pi}{p})\cot(\frac{\pi}{q})$.

7. If I, p, q, s are as in the previous exercise, show that the interior angle γ between a radius and an edge satisfies $\sin(\gamma) = \cos(\frac{\pi}{p}) \csc(\frac{\pi}{q})$.

8. If α, β, γ are as above, prove that $\alpha + \beta + \gamma = (10 - p - q)\pi/4$.

9. If I, p, q, s are as above and E is the number of edges of the polyhedron then the surface area is $\frac{1}{2}s^2 E \cot(\frac{\pi}{p})$.

10. If I, p, q, s, E are as above, then the volume of the solid is

$$\frac{1}{12}s^3 E \cot^2(\frac{\pi}{p}) \tan(\frac{I}{2}).$$

11. If I, p, q, s are as above, then the radius of the circumscribed sphere of the solid is

$$R = \frac{s}{2}\tan(\frac{\pi}{q}) \tan(\frac{I}{2}).$$

12. Let I, p, q, s be as above. The so-called mid-sphere has center O but passes through the midpoints of the sides of the solid. Show that the radius of the mid-sphere is

$$\frac{s}{2}\cos(\frac{\pi}{p}) \tan(\frac{I}{2}) \sec(\frac{\pi}{q}).$$

13. For the Platonic solid $\{p, q\}$ take its vertices, the centers of its faces and project them out to the circumscribed sphere. We construct a polyhedron with $V + E + F$ vertices and $pF = 2E$ faces. For each projected face of $\{p, q\}$ create p faces as follows. Let O be the center of the sphere, C the center of a face of $\{p, q\}$, A, E adjacent vertices of that face, B the midpoint of $\overline{AE}$, and F, G the projections of B, C to the circumscribed sphere of $\{p, q\}$. Let M be the midpoint of $\overline{AG}$. Then construct the plane through M perpendicular to $\overline{OM}$ and consider the half-space on the same side of that plane as O. For each of the F faces of $\{p, q\}$ we obtain p such half-spaces. The new polyhedron is the intersection of the resulting $pF = 2E$ half-spaces. Prove that the angle δ between two of these faces satisfies either $\cos(\delta) = \cos^2(\frac{\pi}{p}) + \sin(\frac{\pi}{p})\cos(\frac{\pi}{p})\cot(\frac{\pi}{q})$ or $\cos(\delta) = \cos^2(\frac{\pi}{q}) + \sin(\frac{\pi}{q})\cos(\frac{\pi}{q})\cot(\frac{\pi}{p})$.

14. Suppose that $\alpha, \gamma, p, q, s, E, r$ above are taken for the dodecahedron, so $p = 3$, $q = 5$, and $E = 30$. Suppose that we create a new polyhedron as follows. If O is the center, A a vertex then we slice the icosahedron with a plane perpendicular to $\overline{OA}$ which passes through the midpoints of all the edges with vertex A. We repeat this for all V vertices, and remove the resulting caps around each vertex. The result is a polyhedron with $V + F$ faces: F regular p-gons and V regular q-gons all having side $s/2$. It is called the icosidodecahedron. Show that its volume is

$\frac{1}{48}s^3 E \cot(\frac{\pi}{q}) \cos(\gamma)$ less than the volume of the icosahedron. (The same solid may be obtained by cutting the dodecahedron at its vertices.)

15. The rhombic dodecahedron is a regular solid with 12 faces formed as follows. Take the cube with center O, and edge $\overline{AB}$ with midpoint M. Take the plane containing $\overline{AB}$ which is perpendicular to $\overline{OM}$. Repeat for all edges of the cube, obtaining 12 planes. The rhombic dodecahedron is the intersection of the half-spaces on the same side of the planes as O. The 12 faces are all rhombuses. Show that the (obtuse) angle between each pair of adjacent faces is $\frac{2\pi}{3}$.

16. The triacontahedron is a regular solid with 30 faces formed as follows. Take the icosahedron with center O, and edge $\overline{AB}$ with midpoint M. Take the plane containing $\overline{AB}$ which is perpendicular to $\overline{OM}$. Repeat for all edges of the cube, obtaining 30 planes. The triacontahedron is the intersection of the half-spaces on the same side of the planes as O. The 30 faces are all rhombuses. Show that the (obtuse) angle between each pair of adjacent faces is $\frac{4\pi}{5}$.

17. Prove Proposition 26.4.

27 Crystals

A crystal is an object which is composed of atoms or molecules arranged in a regular pattern which repeats in various directions in three dimensions. The small scale atomic pattern in the object generally is reflected in its external appearance to an observer. The surface of a crystal tends to be composed of planes. Then the crystal has the shape of a polyhedron, and the relationships among angles in that polyhedron can often be studied with the tools of spherical geometry. In this section we briefly introduce the reader to a few principles of crystallography.

We can begin study of crystals by discussing the notion of a lattice. A lattice may be informally described as an arrangement of points in space which appears the same when viewed from any point of the arrangement. Here we define this notion more precisely with algebraic tools. Suppose that $\mathbf{a}$, $\mathbf{b}$, and $\mathbf{c}$ are three nonzero vectors in space which do not lie in the same plane when given the same initial point. (The algebraic condition on $\mathbf{a}$, $\mathbf{b}$, and $\mathbf{c}$ is that of *linear independence*, which means that if a "linear combination" $z_1\mathbf{a} + z_2\mathbf{b} + z_3\mathbf{c}$ is zero for real z_1, z_2 and z_3 then $z_1 = z_2 = z_3 = 0$.) Then the (three-dimensional) *lattice* M determined by $\mathbf{a}$, $\mathbf{b}$, and $\mathbf{c}$ consists of all the vectors of the form $z_1\mathbf{a} + z_2\mathbf{b} + z_3\mathbf{c}$, where z_1, z_2 and z_3 belong to the set $\mathbf{Z}$ of integers $\ldots, -3, -2, -1, 0, 1, 2, 3, \ldots$. We think of each of these vectors as having initial point $(0, 0, 0)$, so the vector may be identified with its terminal point. If we think of the terminal point of each of these vectors as being an atom or molecule of a crystal, then we have imposed an algebraic structure on that crystal. Of course, the algebraic structure has infinitely many vectors in it where the crystal has only finitely many atoms, but the crystal has so many atoms that it does not hurt to assume it is infinite in extent.

The set of all vectors which form the lattice has the property that the sum of two vectors in the lattice is in the lattice, and an integer multiple of a vector in the lattice is in the lattice. Such a set is called a *module* over the integers.[5] The integers are called the *scalars* of the module. The vectors **a**, **b**, and **c** are said to be a *basis* of the module.

The vectors **a**, **b**, and **c** form a corner of a polyhedron called a parallelepiped. (See Figure 6.8 and recall Figure 1.6.) The parallelepiped has three pairs of parallel plane faces. This parallelepiped we shall call a cell. By translation of the cell in directions **a**, **b**, and **c** the space occupied by the crystal may be seen as the union of cells.

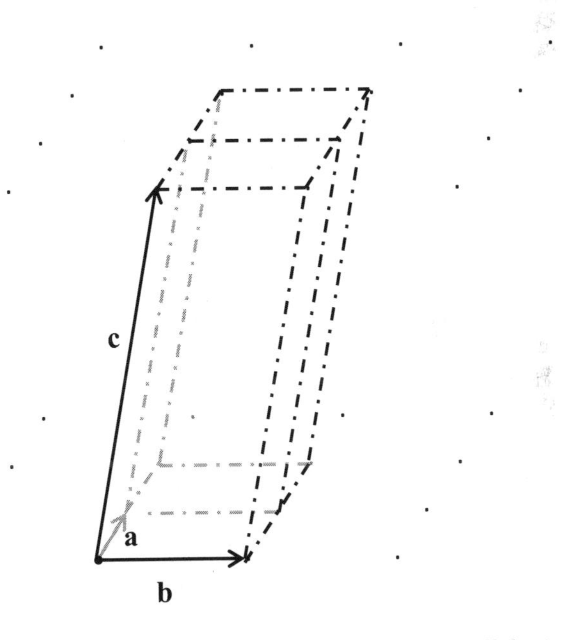

Figure 6.8: A portion of a lattice; basis $\mathbf{a}, \mathbf{b}, \mathbf{c}$; and parallelepiped cells.

Having a common initial point, the three vectors **a**, **b**, and **c** point to three points A, B, and C on a sphere centered at $(0, 0, 0)$ (assuming the radius is large enough to contain the endpoints of the three vectors.) The points A, B, and C form a spherical triangle since the vectors **a**, **b**, and **c** do not lie in a plane. The measures of the sides of that triangle are the measures of the angles between pairs of **a**, **b**, and **c**. As noted in §26 (see Figure 6.1), the measures

[5]A number of other properties need to hold for a module, including that the addition be associative and communtative; see a text on abstract algebra for more details.

of the angles of the spherical triangle are the measures of the dihedral angles between the planes formed by pairs of $\mathbf{a}$, $\mathbf{b}$, and $\mathbf{c}$.

It will turn out that many of the points in the lattice form planes in space. In fact, sometimes these planes of points form the faces of the solid crystal.

Let $\mathbf{a}^*$ be the vector perpendicular to $\mathbf{b}$ and $\mathbf{c}$ such that $\mathbf{a}^* \cdot \mathbf{a} = 1$. (Specifically, $\mathbf{a}^* = (\mathbf{b} \times \mathbf{c})/(\mathbf{a} \cdot \mathbf{b} \times \mathbf{c})$. The fact that $\mathbf{a}$, $\mathbf{b}$, and $\mathbf{c}$ are linearly independent guarantees that the denominator is nonzero.) Similarly let $\mathbf{b}^*$ be the vector perpendicular to $\mathbf{a}$ and $\mathbf{c}$ such that $\mathbf{b}^* \cdot \mathbf{b} = 1$, and let $\mathbf{c}^*$ be the vector perpendicular to $\mathbf{a}$ and $\mathbf{b}$ such that $\mathbf{c}^* \cdot \mathbf{c} = 1$. Then $\mathbf{a}^*$, $\mathbf{b}^*$, and $\mathbf{c}^*$ are also linearly independent (for reasons that will become clear.) Thus the set of all the vectors of the form $z_1\mathbf{a}^* + z_2\mathbf{b}^* + z_3\mathbf{c}^*$, where z_1, z_2 and z_3 are integers, forms a lattice called the *reciprocal lattice* M^*. The vectors $\mathbf{a}^*, \mathbf{b}^*$ and $\mathbf{c}^*$ are said to be the basis of the reciprocal lattice.

If we select an element $\mathbf{v}^* = v_1\mathbf{a}^* + v_2\mathbf{b}^* + v_3\mathbf{c}^*$ of the reciprocal lattice, we may take its dot product with every element of the original lattice (which we refer to as the *direct lattice* to distinguish it from the reciprocal lattice). This dot product associates to every element $\mathbf{v} = z_1\mathbf{a} + z_2\mathbf{b} + z_3\mathbf{c}$ of the direct lattice the integer $z_1v_1 + z_2v_2 + z_3v_3$. Thus the element $\mathbf{v}^*$ may be associated with a mapping $f : M \to \mathbf{Z}$. The mapping f has the property that for all $\mathbf{v}, \mathbf{w}$ in the lattice and any integer z, $f(\mathbf{v} + \mathbf{w}) = f(\mathbf{v}) + f(\mathbf{w})$ and $f(z\mathbf{v}) = zf(\mathbf{v})$. Such a mapping f is said to be *linear* in $\mathbf{Z}$. Thus from a mathematical point of view the elements of the reciprocal lattice may be seen as belonging to the so-called *dual space* $L(M : \mathbf{Z})$ of the lattice M: this is the set of all mappings f from M to $\mathbf{Z}$ which are linear in $\mathbf{Z}$.

The reciprocal lattice is an important lattice in crystallography and solid state physics. It is a useful tool in understanding the manner in which x-ray diffraction occurs for the crystal of the direct lattice. If an x-ray beam is fired into a crystal, the atoms of the crystal lattice scatter the waves of the x-ray in various directions depending on the molecular structure of the lattice, the angle at which the x-ray strikes the crystal, and the wavelength of the x-ray. This scattering is called *diffraction.* For our purposes the most important principle governing diffraction is the so-called Bragg law of diffraction. Bragg's law says that under certain conditions some of the wave scatterings of an x-ray from the atoms of a lattice will have their wavelengths match up to produce what is called "constructive interference" in a certain specific direction. In such a direction an x-ray will diffract through a set of atoms/molecules in the crystal lattice provided those atoms/molecules lie in a plane. The x-ray will seem to "reflect" as light reflects off a mirror (the angle of incidence equals the angle of reflection) but only for certain special angles of incidence.

The reciprocal lattice is helpful here. Let $\mathbf{v}^* = v_1\mathbf{a}^* + v_2\mathbf{b}^* + v_3\mathbf{c}^*$ be an element of the reciprocal lattice. Given an integer u, let us ask: what is the set of all $\mathbf{v}$ in the direct lattice such that $\mathbf{v}^* \cdot \mathbf{v} = u$? The set of all such $\mathbf{v}$ must lie in a plane in space because such a $\mathbf{v} = z_1\mathbf{a} + z_2\mathbf{b} + z_3\mathbf{c}$ satisfies $z_1v_1 + z_2v_2 + z_3v_3 = u$, which is the equation of a plane. The vector $\mathbf{v}^*$ is perpendicular to that plane. Now that plane might contain no points of the lattice. But if it contains any points, it contains infinitely many points — in fact, a whole two-dimensional lattice of points. (See Exercise 2.) Then the direct lattice is the union of parallel planes of lattice points. When that happens, Bragg's law says that an x-ray in an appropriate direction interacts the same way with all of the atoms/molecules at the points of the parallel plane lattice to produce a diffraction. If d is the smallest distance between a pair of the lattice planes, and λ is the wavelength of the x-ray, then θ is said to be an angle of diffraction for that set of lattice planes if for some integer $n \geq 0$, $n\lambda = 2d\sin(\theta)$. The angle θ is both the incidence and reflection angle for the x-ray from one of the parallel planes. However, this does not mean that any x-ray incident at angle θ will produce a diffraction. In fact, whether diffraction occurs depends on the direction of approach of the x-ray, which can be determined by experimentation, or by use of a device called the Ewald sphere.[6]

Let us now connect the reciprocal lattice with spherical geometry. Recall that $\mathbf{a}$, $\mathbf{b}$, and $\mathbf{c}$ denote a basis of the direct lattice. The element $\mathbf{a}^*$ of the reciprocal lattice satisfies $\mathbf{a}^* \cdot \mathbf{a} = 1$, and $\mathbf{a}^* \cdot \mathbf{b} = \mathbf{a}^* \cdot \mathbf{c} = 0$. Thus $\mathbf{a}^*$ is a vector perpendicular to both $\mathbf{b}$ and $\mathbf{c}$ but on the same side of the plane of $\mathbf{b}$ and $\mathbf{c}$ as $\mathbf{a}$ (because $\mathbf{a}^* \cdot \mathbf{a} = 1 > 0$ means that the angle between $\mathbf{a}^*$ and $\mathbf{a}$ is acute). Now suppose we imagine a sphere centered at $(0,0,0)$ with radius large enough to contain the terminal points of $\mathbf{a}$, $\mathbf{b}$, $\mathbf{c}$, $\mathbf{a}^*$, $\mathbf{b}^*$, and $\mathbf{c}^*$. As noted earlier, the vectors $\mathbf{a}$, $\mathbf{b}$, and $\mathbf{c}$ point to three points A, B, and C on that sphere. Because $\mathbf{a}^*$ is perpendicular to both $\mathbf{b}$ and $\mathbf{c}$ and the angle between $\mathbf{a}^*$ and $\mathbf{a}$ is acute, $\mathbf{a}^*$ must point to the point A', the pole of $\bigcirc BC$ on the same side of $\bigcirc BC$ as A. Similarly, if we consider the vectors $\mathbf{b}^*$ and $\mathbf{c}^*$ of the reciprocal lattice such that $\mathbf{b}^* \cdot \mathbf{b} = 1$, $\mathbf{b}^* \cdot \mathbf{a} = \mathbf{b}^* \cdot \mathbf{c} = 0$, $\mathbf{c}^* \cdot \mathbf{c} = 1$, and $\mathbf{c}^* \cdot \mathbf{b} = \mathbf{c}^* \cdot \mathbf{a} = 0$, then $\mathbf{b}^*$ points to B' and $\mathbf{c}^*$ points to C'. Since we assumed $\mathbf{a}$, $\mathbf{b}$, and $\mathbf{c}$ do not lie in a plane, points A, B, and C form a spherical triangle. Thus the polar triangle is well-defined, which means that $\mathbf{a}^*$, $\mathbf{b}^*$, and $\mathbf{c}^*$ do not lie in a plane, so are linearly independent, as noted earlier. So the vectors $\mathbf{a}^*, \mathbf{b}^*$, and $\mathbf{c}^*$ form a basis of the reciprocal lattice (see Exercise 3) which points to the vertices of the polar triangle $\triangle^s A'B'C'$.

From Theorem 11.18, we know that the polar triangle of the polar triangle is the original triangle. We then may conclude that the reciprocal lattice of the reciprocal lattice is the direct lattice. (See Exercise 4.) This justifies the term "reciprocal."

Let d_{bc} be the distance between the nearest parallel lattice planes in the directions parallel to $\mathbf{b}$ and $\mathbf{c}$. We claim that $1/d_{bc}$ is the length of the vector $\mathbf{a}^*$. This will give some more explanation for the introduction of the term "reciprocal." We know that $\mathbf{a}^* \cdot \mathbf{a} = 1$. The distance between the lattice plane containing $\mathbf{b}$ and $\mathbf{c}$ and its nearest parallel is the length of the projection of $\mathbf{a}$ onto $\mathbf{a}^*$, which is the absolute value of the dot product of $\mathbf{a}$ with $\mathbf{a}^*/|\mathbf{a}^*|$.

[6]For more details on the Ewald sphere, see any text on crystallography or solid state physics.

Thus $d_{bc} = \frac{\mathbf{a}^*}{|\mathbf{a}^*|} \cdot \mathbf{a} = \frac{1}{|\mathbf{a}^*|}\mathbf{a}^* \cdot \mathbf{a} = \frac{1}{|\mathbf{a}^*|}$. So $|\mathbf{a}^*| = 1/d_{bc}$.

This observation also gives a geometric way of understanding the construction of the reciprocal lattice: $\mathbf{a}^*$ is a vector perpendicular to the plane containing $\mathbf{b}$ and $\mathbf{c}$ whose length is equal to the reciprocal of the shortest distance between a pair of lattice planes parallel to $\mathbf{b}$ and $\mathbf{c}$ in the direct lattice.

We conclude this section by offering an explanation for the Bragg law from more general ideas. Suppose we first consider the interaction of an x-ray with the single line of atoms lined up in the direction of $\mathbf{a}$. We assume the x-ray has wavelength λ and approaches this line of atoms along the direction of the vector $\mathbf{v}$. (See Figure 6.9.) It then interacts the same way with each atom along the line and emerges in the direction of the vector $\mathbf{w}$. Assume that $|\mathbf{v}| = |\mathbf{w}| = \lambda$. We assume that the angle of incidence of $\mathbf{v}$ with $\mathbf{a}$ is θ_1 and the angle of emergence of $\mathbf{w}$ with $\mathbf{a}$ is θ_2. The central principle of diffraction is that in comparing the interaction of the x-ray with adjacent atoms, the x-ray interacting with the atom on the left should travel a distance equal to an integer number of wavelengths more (or less) than the x-ray beam on the right. Then the two waves emerge in a sense "reinforcing" the same phase of the wave to the right and produce an x-ray in the direction of $\mathbf{w}$. (If they emerge at different phases from the interaction with the atoms then no clear wave is detectable.) This is an angle of incidence producing the so-called constructive interference of x-ray diffraction. It should be understood that an angle of incidence where this occurs is rare. Let us see what algebraic conditions must occur for this to happen.

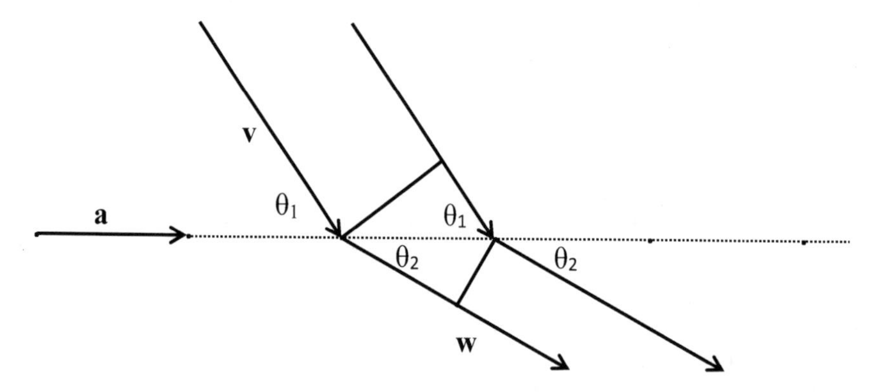

Figure 6.9: Wave $\mathbf{v}$ incident at angle θ_1 deflected by atoms separated by $\mathbf{a}$ into wave $\mathbf{w}$ at angle θ_2. Vectors $\mathbf{v}$, $\mathbf{w}$, and $\mathbf{a}$ need not be planar.

As seen in Figure 6.9, the wave on the left travels $|\mathbf{a}|\cos(\theta_2) - |\mathbf{a}|\cos(\theta_1)$ further than the wave on the right. In order to obtain constructive interference, this quantity needs to be an integer multiple of λ, say $n_a\lambda$. Note that the dot product $\mathbf{a}\cdot\mathbf{w} = |\mathbf{a}||\mathbf{w}|\cos(\theta_2) = |\mathbf{a}|\lambda\cos(\theta_2)$. Similarly, $\mathbf{a}\cdot\mathbf{v} = |\mathbf{a}|\lambda\cos(\theta_1)$. Thus the quantity $|\mathbf{a}|\cos(\theta_2) - |\mathbf{a}|\cos(\theta_1)$ may be calculated via the dot products $\mathbf{a}\cdot\mathbf{w}/\lambda - \mathbf{a}\cdot\mathbf{v}/\lambda = \mathbf{a}\cdot(\mathbf{w}-\mathbf{v})/\lambda = n_a\lambda$. So we have $\mathbf{a}\cdot(\mathbf{w}-\mathbf{v}) = n_a\lambda^2$.

This analysis may be done in each of the directions $\mathbf{a}$, $\mathbf{b}$, and $\mathbf{c}$ which form the lattice, from which we obtain what are known as the three Laue equations:

$$\begin{aligned} \mathbf{a}\cdot(\mathbf{w}-\mathbf{v}) &= n_a\lambda^2 \\ \mathbf{b}\cdot(\mathbf{w}-\mathbf{v}) &= n_b\lambda^2 \\ \mathbf{c}\cdot(\mathbf{w}-\mathbf{v}) &= n_c\lambda^2 \end{aligned} \tag{6.7}$$

Observe that

$$\begin{aligned} (n_b\mathbf{a} - n_a\mathbf{b})\cdot(\mathbf{w}-\mathbf{v}) &= 0 \\ (n_c\mathbf{b} - n_b\mathbf{c})\cdot(\mathbf{w}-\mathbf{v}) &= 0. \end{aligned}$$

If we let $\mathbf{p} = n_b\mathbf{a} - n_a\mathbf{b}$ and $\mathbf{q} = n_c\mathbf{b} - n_b\mathbf{c}$ then we also have that for any integers z_1, z_2, $(z_1\mathbf{p} + z_2\mathbf{q})\cdot(\mathbf{w}-\mathbf{v}) = 0$. The vectors $z_1\mathbf{p} + z_2\mathbf{q}$ (z_1 and z_2 ranging over the integers) generally form a two-dimensional lattice plane L of vectors in the direct lattice. (By adding an integer multiple of $\mathbf{a}$ to all the points in L we can obtain another lattice plane parallel to L.) The vector $\mathbf{w}-\mathbf{v}$ is then perpendicular to L. An appropriate scalar multiple of $\mathbf{w}-\mathbf{v}$ belongs to the reciprocal lattice. (In fact, $\frac{1}{\lambda^2}(\mathbf{w}-\mathbf{v}) = n_a\mathbf{a}^* + n_b\mathbf{b}^* + n_c\mathbf{c}^*$ by (6.7).) But because $|\mathbf{v}| = |\mathbf{w}|$, the vector $\mathbf{w}-\mathbf{v}$ bisects the angle between $\mathbf{w}$ and $-\mathbf{v}$. (This can be seen in Figure 6.10: the parallelogram formed by $\mathbf{v}$ and $\mathbf{w}$ is a rhombus, and the diagonals of a rhombus bisect the angles. See §1, Exercise 3, or just note that the dot products $-\mathbf{v}\cdot(\mathbf{w}-\mathbf{v})$ and $\mathbf{w}\cdot(\mathbf{w}-\mathbf{v})$ are the same, so the cosine of the angle in each pair is the same.) Thus the incident x-ray along vector $\mathbf{v}$ forms the same angle with $\mathbf{w}-\mathbf{v}$ as $-\mathbf{w}$. Then the angle of incidence of $\mathbf{v}$ with the above two-dimensional lattice plane of atoms is the same as the angle of reflection. (Figure 6.11.) This is part of Bragg's law. Referring to equations (6.7), note that $\mathbf{a}$, $\mathbf{b}$, and $\mathbf{c}$ are linearly independent in space, so if $\mathbf{w}-\mathbf{v}$ is nonzero, the right-hand sides of these three equations cannot all be zero. For simplicity suppose that $n_a \neq 0$. If $n_a > 0$ then $\mathbf{a}\cdot(\mathbf{w}-\mathbf{v}) = d|\mathbf{w}-\mathbf{v}|$, where d is the length of the projection of $\mathbf{a}$ onto the unit vector $(\mathbf{w}-\mathbf{v})/|\mathbf{w}-\mathbf{v}|$. Since $\mathbf{a}$ points between two lattice planes parallel to L, d is the perpendicular distance between these lattice planes. This gives us $n_a\lambda^2 = d|\mathbf{w}-\mathbf{v}|$. Now $|\mathbf{w}-\mathbf{v}|$ is the length of one of the diagonals of the rhombus in Figure 6.12. Since the diagonals of a rhombus are perpendicular (§1, Exercise 3), $|\mathbf{w}-\mathbf{v}| = 2|\mathbf{w}|\sin(\theta) = 2\lambda\sin(\theta)$. (See Figure 6.12, or see Exercise 5.) We conclude $n_a\lambda^2 = 2d\lambda\sin(\theta)$, so $n_a\lambda = 2d\sin(\theta)$, as stated in Bragg's law.

If $n_a < 0$ a similar argument produces the same conclusion if we reverse the role of $\mathbf{v}$ and $\mathbf{w}$.

Hence the principle of Bragg's law may be derived from the more general principle of Laue's equations. A lattice that causes constructive interference

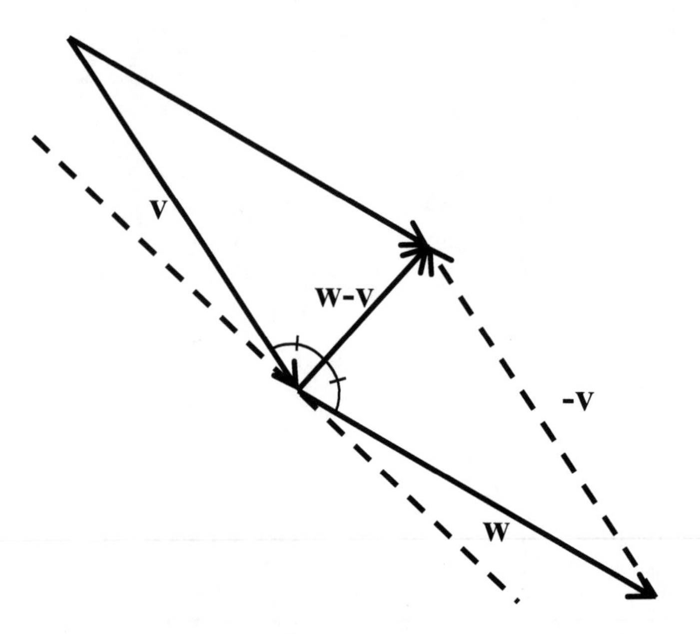

Figure 6.10: If $|\mathbf{v}| = |\mathbf{w}|$ then in the rhombus formed by $\mathbf{v}$ and $\mathbf{w}$, $\mathbf{w} - \mathbf{v}$ bisects the angle between $\mathbf{w}$ and $-\mathbf{v}$.

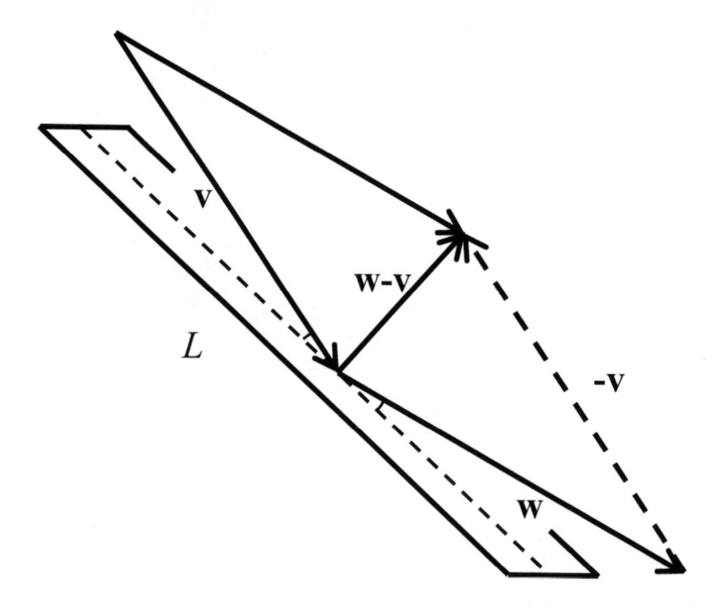

Figure 6.11: The angle of incidence of $\mathbf{v}$ with the lattice plane L is the same as the angle of reflection that $\mathbf{w}$ has with L. The vector $\mathbf{w} - \mathbf{v}$ is normal to the plane.

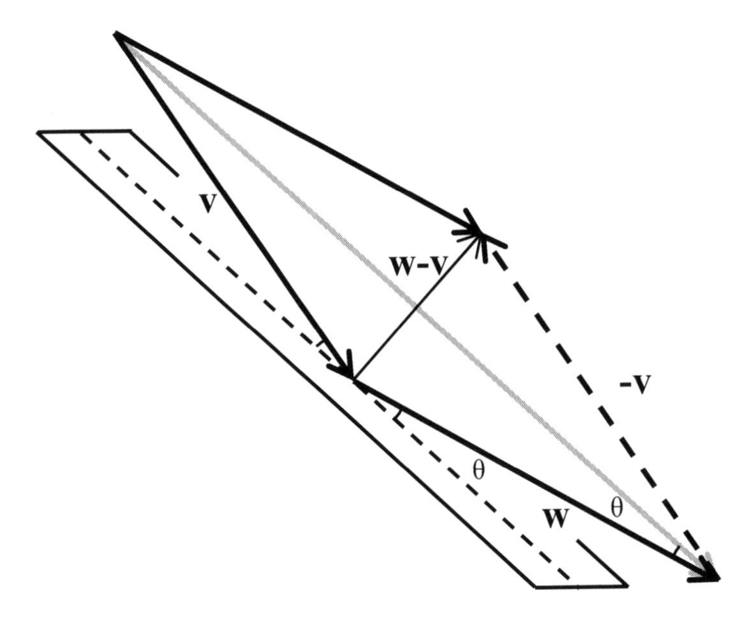

Figure 6.12: $|\mathbf{w} - \mathbf{v}| = 2|\mathbf{w}|\sin(\theta) = 2\lambda\sin(\theta)$.

in an x-ray may be seen as causing the x-ray to reflect off two-dimensional lattice planes of the three-dimensional lattice.

It is difficult to delve much further into crystallography without expertise in chemistry and we leave further study of the matter to the references.

Exercises §27.

1. Prove that if $f_1, f_2 \in L(M : \mathbf{Z})$ and u is an integer then $f_1 + f_2 \in L(M : \mathbf{Z})$ and $uf \in L(M : \mathbf{Z})$.

2. Prove that if $f : M \to \mathbf{Z}$ is linear then for an integer z, $f(\mathbf{v}) = z$ for either infinitely many $\mathbf{v}$ in M or no $\mathbf{v}$ in M.

3. Prove that if $\mathbf{a}$, $\mathbf{b}$, and $\mathbf{c}$ are a basis of a lattice then $\mathbf{a}^*$, $\mathbf{b}^*$, and $\mathbf{c}^*$ form a basis of the reciprocal lattice.

4. Justify the statement that the reciprocal lattice of the reciprocal lattice is the original direct lattice.

5. Expand $(\mathbf{w} - \mathbf{v}) \cdot (\mathbf{w} - \mathbf{v})$ to show that it equals $2\lambda^2 - 2\lambda^2 \cos(2\theta)$. Conclude that $|\mathbf{w} - \mathbf{v}| = 2\lambda\sin(\theta)$.

6. Consider a parallelopiped determined by three vectors $\mathbf{a} = \vec{OA}$, $\mathbf{b} = \vec{OB}$, and $\mathbf{c} = \vec{OC}$ with the same initial point O. Recall that $|\mathbf{a}|$, $|\mathbf{b}|$, and $|\mathbf{c}|$ are the lengths of the three edges from a vertex. Let $\alpha = m\angle BOC$, $\beta =$

$m\angle AOC$, and $\gamma = m\angle AOB$. Then the volume V of the parallelopiped is given by

$$V = |\mathbf{a}||\mathbf{b}||\mathbf{c}|\sqrt{1 - \cos^2(\alpha) - \cos^2(\beta) - \cos^2(\gamma) + 2\cos(\alpha)\cos(\beta)\cos(\gamma)}. \tag{6.8}$$

Hint. Take a sphere centered at O and suppose that **a**, **b**, and **c** point to points X, Y, and Z on that sphere. Use both plane and spherical trigonometry to show that the volume is

$$|\mathbf{a}||\mathbf{b}||\mathbf{c}|\sin(\alpha)\sin(\gamma)\sin(\prec XYZ),$$

which is the same as:

$$|\mathbf{a}||\mathbf{b}||\mathbf{c}|\sin(\beta)\sin(\gamma)\sin(\prec YXZ)$$
$$|\mathbf{a}||\mathbf{b}||\mathbf{c}|\sin(\alpha)\sin(\beta)\sin(\prec XZY)$$

Then use (4.55) and (4.56).

7. Let V, α, β, and γ be as in Exercise 6. Then show:

$$|\mathbf{a}^*| = \frac{|\mathbf{b}||\mathbf{c}|\sin(\alpha)}{V}, |\mathbf{b}^*| = \frac{|\mathbf{a}||\mathbf{c}|\sin(\beta)}{V}, |\mathbf{c}^*| = \frac{|\mathbf{a}||\mathbf{b}|\sin(\gamma)}{V}.$$

Conclude similar dual formulas of the form

$$|\mathbf{a}| = \frac{|\mathbf{b}^*||\mathbf{c}^*|\sin(\alpha^*)}{V^*},$$

where α^* is the angle between $\mathbf{b}^*$ and $\mathbf{c}^*$, and V^* is the volume of the parallelopiped determined by $\mathbf{a}^*, \mathbf{b}^*$ and $\mathbf{c}^*$.

8. Let V^* be the volume of the *reciprocal cell*, i.e., the parallelopiped determined by the vectors $\mathbf{a}^*, \mathbf{b}^*$ and $\mathbf{c}^*$. Prove that $V^* = 1/V$.

Chapter 7

Spherical mappings

We briefly review the general notion of mapping. Let A and B be sets. A *mapping* f from A to B, denoted $f : A \to B$, is a mechanism which associates to every element a in A a single element b in B. We write $b = f(a)$ and say that b is the *image* of a under f. We will say that f maps set A to set B, and that f maps element a to element b. The set A is known as the *domain* of f and B is the *target*. If whenever $f(a_1) = f(a_2)$ we must have $a_1 = a_2$, we say that f is *one-one* or *injective*. (That is, if f maps two elements of A to the same element in B, then in fact the two elements in A must have been the same.) The set of all b in B such that there exists an a in A with $f(a) = b$ is called the *range* or *image* of f. If the range of f is the same as the target of f, then we say that f is *onto* or *surjective*. If f is both injective and surjective, then we say that f is *bijective* or a *one-one correspondence*.

If $f : A \to B$ and $g : B \to C$ then $g \circ f : A \to C$ is the mapping such that $g \circ f(x) = g(f(x))$. The mapping $g \circ f$ is referred to as the *composition* of g and f. The *identity* mapping on a set A is the mapping $I : A \to A$ such that $I(a) = a$ for all a in A. If $f : A \to B$ is bijective then the *inverse* of f, $f^{-1} : B \to A$ is the mapping satisfying $f \circ f^{-1}(b) = b$ for all b in B and $f^{-1} \circ f(a) = a$ for all a in A.

In this chapter we will be concerned with mappings where the domain is a portion of sphere (perhaps all of the sphere) and f is bijective. First we shall study mappings where the target is also a sphere, and then we shall study the case where the target is a (two-dimensional) plane.

28 Rotations and reflections

Let σ be a sphere, P a point of σ, Γ the polar circle of P, Q a point of Γ and $\ell : \Gamma \to \mathbf{R}/2\pi$ a labelling of Γ such that $\ell(Q) = 0$. In this section we regard spherical polar coordinates as a mapping from σ to an ordered pair (ϕ, θ) such that $0 \leq \phi \leq \pi$ and $\theta \in \mathbf{R}/2\pi$. (See Figure 7.1.) As in §8, for a point X on σ, ϕ is the spherical distance from P to X. But here we define θ slightly differently. Let Y be the (unique) point where $\overrightarrow{PX}$ meets Γ. Then θ is defined to be $\ell(Y)$.

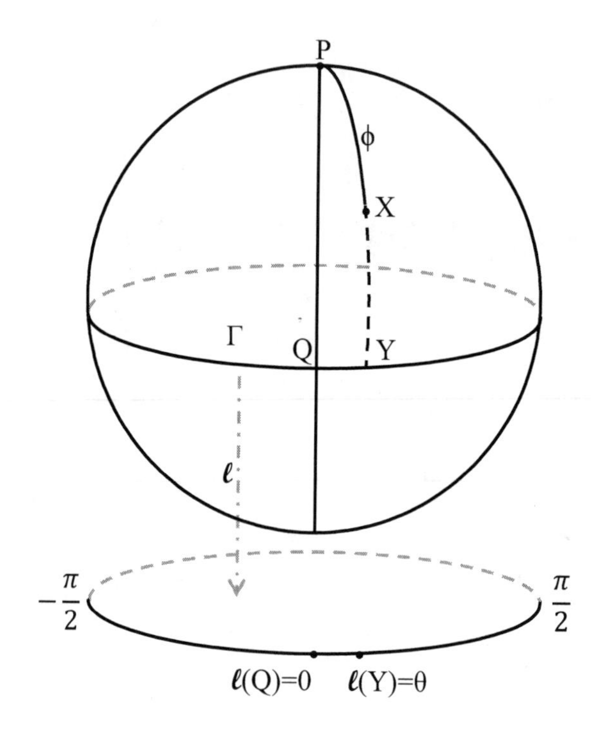

Figure 7.1: Spherical polar coordinates.

We can see the similarity to the coordinates as explained in §8 as follows. Suppose $\ell(Y)$ is represented by a number between 0 and π. Then since $m\ \widehat{PY} = m\ \widehat{PQ} = \frac{\pi}{2}$, $m \prec XPQ = m \prec YPQ = m\ \widehat{QY} = \ell(Y) \pmod{2\pi}$. Because $\ell(Y)$ is well-defined up to an integer multiple of 2π, it makes sense to speak of a rotation by angle $\ell(Y)$. That is, rotations by angles $\alpha, \alpha + 2\pi, \alpha + 4\pi, \alpha + 6\pi$, etc., all represent the same rotation because a rotation by an angle which is an integer multiple of 2π is the identity mapping. Rotation by angle $\ell(Y)$ is the same as rotation by the angle $\prec XPQ$ in the counterclockwise direction so a rotation of $\ell(Y)$ about P takes PQP^a into PXP^a. This is how the coordinates were described in §8.

Then a *rotation* with pole P through angle τ is a mapping $T : \sigma \to \sigma$ such that every point on the sphere with coordinates (ϕ, θ) is mapped to a point with coordinates $(\phi, \theta + \tau (\text{mod}\, 2\pi))$. Note that $T(P) = P$ and $T(P^a) = P^a$, where P^a is the antipode of P.

If Γ is the great circle whose pole is P, then the *reflection* in Γ is the mapping $F : \sigma \to \sigma$ such that every point with coordinates (ϕ, θ) is mapped to a point with coordinates $(\pi - \phi, \theta)$. Note that $F(P) = P^a$, $F(P^a) = P$ and for every X on Γ, $F(X) = X$. We say that Γ is the axis of reflection of F.

We leave it as an exercise to check that $T, F : \sigma \to \sigma$ are bijections. (See Exercise 3.)

Suppose U is a subset of σ and $f : U \to \sigma$, Then we say that f is an *isometry* if given two points X and Y in U, the spherical distance from $f(X)$ to $f(Y)$ is the same as the distance from X to Y. An isometry must be a one-one mapping (see Exercise 1.)

Then $T, F : \sigma \to \sigma$ are isometries and we say that T and F are "isometries of the sphere σ." To see this, suppose our rotation pole is P and we consider two points A_1 and A_2 with spherical coordinates (ϕ_1, θ_1) and (ϕ_2, θ_2). The spherical distance d between A_1 and A_2 satisfies (4.19). After rotation by angle τ, the distance d' between the rotated points satisfies $\cos(d') = \cos(\phi_1)\cos(\phi_2) + \sin(\phi_1)\sin(\phi_2)\cos((\theta_2 + \tau) - (\theta_1 + \tau)) = \cos(\phi_1)\cos(\phi_2) + \sin(\phi_1)\sin(\phi_2)\cos(\theta_2 - \theta_1)$. Then $\cos(d) = \cos(d')$, so $d = d'$ since $0 \le d, d' \le \pi$. We leave it as an exercise to check that a reflection mapping is also an isometry of σ. (See Exercise 4.)

We say that $f : \sigma \to \sigma$ has a *fixed point* X if $f(X) = X$, and say that f *fixes* X. We leave it to Exercise 5 that a rotation T fixes only P and P^a unless $T = I$ (in which case all points are fixed points). A reflection F fixes X if and only if X is on the axis of reflection.

Rotations and reflections differ in another important respect: rotations preserve the so-called *orientation* of subsets of the sphere, while reflections reverse orientation. A few examples of orientation will illustrate the idea of it. If a spherical triangle $\triangle^s ABC$ is viewed by an observer outside the sphere, its vertices A, B, and C appear to trace the perimeter of the triangle in either a clockwise direction or a counterclockwise direction. Suppose this direction is counterclockwise. Under rotations and reflections, the images of the three points A, B, and C again form a spherical triangle. (See Figure 7.2.) Under a rotation T, these vertices $T(A)$, $T(B)$, and $T(C)$ trace the triangle counterclockwise. Under a reflection F, these vertices $F(A)$, $F(B)$, and $F(C)$ trace the triangle clockwise. A similar property holds for any region where a direction for tracing its perimeter is specified. Note that if a reflection is followed by another reflection, the orientation of an object is reversed twice, so the orientation is preserved.

We now proceed to a definition of orientation as follows. We first prove that any isometry of the sphere is the composition of at most three reflections. Then an isometry will be said to be orientation preserving depending on whether the number of reflections required is even or odd. Two congruent objects (e.g., triangles) on the sphere will be said to have the same orientation if one can be mapped to the other with an even number of reflections. They will have different orientations if the number of reflections is odd.

Lemma 28.1 *Suppose U is a subset of a sphere σ which contains at least three points A, B, and C not lying on any great circle. Let $f : U \to \sigma$ be an*

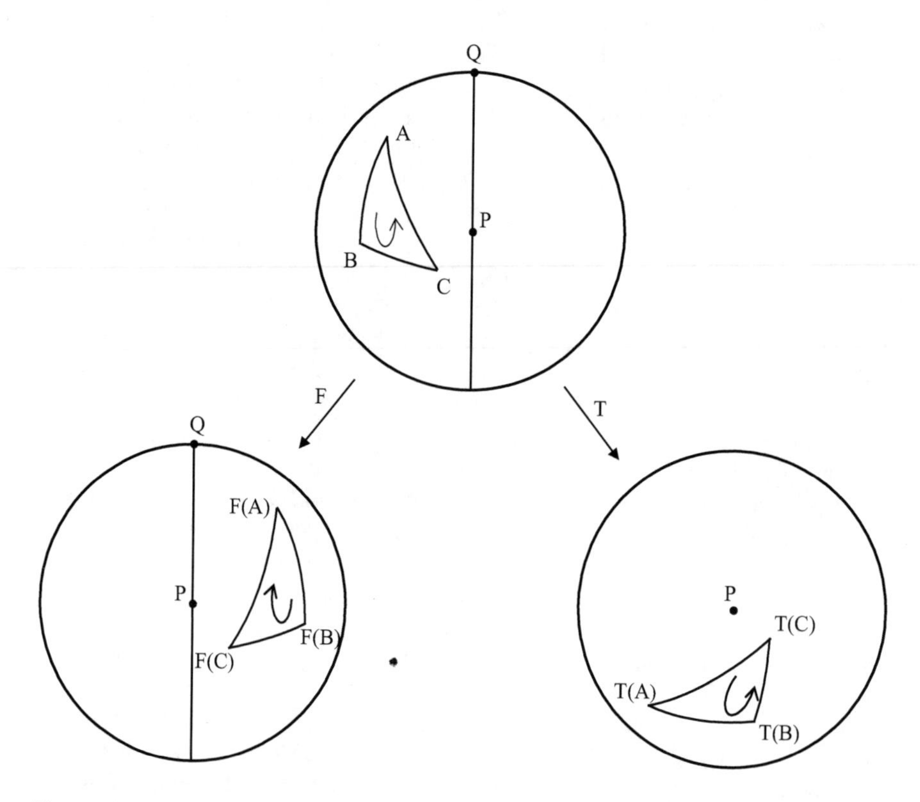

Figure 7.2: Rotation and reflection, top view. Left: reflection F in $\bigcirc PQ$ reverses orientation of $\triangle^s ABC$. Right: rotation T about P preserves orientation of $\triangle^s ABC$.

isometry. If $f(A) = A$, $f(B) = (B)$, and $f(C) = C$, then $f(X) = X$ for any X in U.

Proof. Suppose that $f(X) \neq X$. Then X is different from A, B, and C. Since f is an isometry, $f(X)$ and $f(A)(= A)$ are at the same spherical distance from each other as X and A. So $f(X)$ and X are at the same distance from A. Since $f(X) \neq X$, A lies on the great circle which is the set of points spherically equidistant from X and $f(X)$. (See §12, Exercise 10.) We similarly conclude that B and C are on this great circle. This is a contradiction because A, B, and C do not all lie on the same great circle. So $f(X) = X$. ♢

Lemma 28.2 *An isometry of the sphere is determined by the images of three distinct points not lying on a single great circle. That is, if A, B, and C are distinct points of the sphere not lying on a single great circle and f_1 and f_2 are isometries of the sphere such that $f_1(A) = f_2(A)$, $f_1(B) = f_2(B)$ and $f_1(C) = f_2(C)$, then $f_1(X) = f_2(X)$ for all X on the sphere.*

Proof. If f_2 is the identity mapping, then the conclusion follows from Lemma 28.2.

In general, let f_2^{-1} denote the inverse of f_2. Then the mapping $f_2^{-1} \circ f_1$ is an isometry of the sphere which maps A to A, B to B, and C to C. By the above argument, it must send X to X for all X of the sphere, so $f_1(X) = f_2(X)$ for all X of the sphere. ♢

Theorem 28.3 *Suppose that we are given three points A, B, and C on a sphere and three corresponding points D, E, and F on the same sphere so that distances are equal: $d(A, B) = d(D, E)$, $d(A, C) = d(D, F)$, and $d(B, C) = d(E, F)$. Then there exists a composition of no more than three successive reflections which maps A to D, B to E, and C to F.*

Proof. The idea of the proof is to use one reflection to map A to D, another reflection to map B to E, and a third to map C to F, if necessary.

First, suppose that $A = D$, $B = E$, and $C = F$. Then the identity is the desired mapping; this can be understood as the composition of zero reflections.

Second, suppose there are exactly two pairs of common points. Then without loss of generality we may assume that $A = D$, $B = E$, and $C \neq F$. Let Γ be the set of all points on the sphere at the same spherical distance from C and F. The spherical distance between A and C is the same as the distance between $D(= A)$ and F. Thus A is on Γ. Similarly, B is on Γ. By §12, Exercise 10, Γ is the great circle of all points equidistant from C and F. A reflection R in Γ maps each of A and B to itself, and maps C to F (see Exercise 7) so a single reflection suffices.

Third, suppose that the triangles have one pair of corresponding vertices which consists of the same point, but the other two pair consist of distinct points. Then we may assume that $A = D$, $B \neq E$, and $C \neq F$. Let Γ be the great circle of points spherically equidistant from B and E. We claim that reflection in Γ maps A to itself, and switches B and E. Once we know this, composition with another reflection as above produces an isometry which maps A, B, C to D, E, F, respectively, so two successive reflections are sufficient. By assumption $\overset{\frown}{AB} \cong \overset{\frown}{DE}$; since $A = D$, $\overset{\frown}{AB} \cong \overset{\frown}{AE}$. Thus A is on Γ, so reflection in Γ maps A to itself. By Exercise 7, reflection in Γ switches B and E.

Fourth, suppose that $A \neq D$, $B \neq E$, and $C \neq F$. Then a reflection R in the great circle Γ of points equidistant between A and D switches A and D by Exercise 7. Composing R with mappings found above, we find that the composition of no more than three successive reflections with map A to D, B to E, and C to F. $\diamondsuit$

Theorem 28.4 *Any isometry of the sphere can be written as the composition of at most three successive reflections.*

Proof. Let f be an isometry of the sphere. Let A, B, and C be three points of the sphere not lying on the same great circle, and let $D = f(A)$, $E = f(B)$, and $F = f(C)$. By Theorem 28.3, there exists a composition R of three successive reflections such that $D = R(A)$, $E = R(B)$, and $F = R(C)$. Since f and R are isometries which coincide at three points not lying on a single great circle, f and R are identical mappings on the sphere by Lemma 28.2. $\diamondsuit$

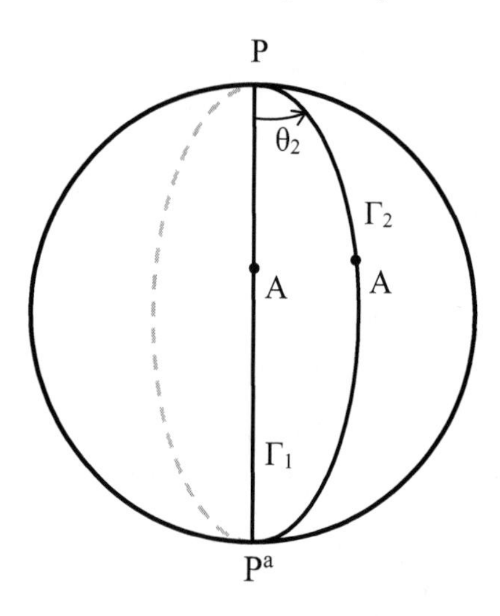

Figure 7.3: Reflection in Γ_1 followed by reflection in Γ_2.

We now show that the composition of two reflections is a rotation. (See Figure 7.3.) Suppose that F_1 and F_2 are reflections of a sphere in great circles Γ_1 and Γ_2, respectively. Assume that P is one of the intersection points of Γ_1 and Γ_2. We claim that the composition $F_2 \circ F_1$ is a rotation about P. To see this, suppose we introduce spherical polar coordinates about P as follows. Let P^a be the antipode of P, and suppose A_1 and A_2 are points of Γ_1 and Γ_2, respectively, other than P and P^a. We let (ϕ, θ) be spherical coordinates about P with zero meridian PA_1P^a. Suppose that the meridian PA_2P^a is the meridian where $\theta = \theta_2$. Then by Exercises 8 and 9 reflection in Γ_1 is the mapping $(\phi, \theta) \mapsto (\phi, -\theta (\operatorname{mod} 2\pi))$ and reflection in Γ_2 is $(\phi, \theta) \mapsto (\phi, 2\theta_2 - \theta (\operatorname{mod} 2\pi))$. The composition of reflection in Γ_1 followed by reflection in Γ_2 is then $(\phi, \theta) \mapsto (\phi, \theta + 2\theta_2 (\operatorname{mod} 2\pi))$. This is a rotation about P in angle $2\theta_2$.

Next we show that the composition of two successive rotations is also a rotation, even if the axis of rotation is different. This surprising result was proven by Euler in 1776 with an argument substantially similar to the one we use.

Theorem 28.5 *The composition of two successive rotations of a sphere is also a rotation.*

Proof. (See Figure 7.4.) Let us call the two rotations T_1 and T_2. Assume that T_1 is a rotation about point A_1 by angle θ_1 and that T_2 is a rotation about point A_2 by angle θ_2. Let S_1 be rotation about point A_1 by angle $\theta_1/2$ (in the same direction as T_1) and S_2 be rotation about point A_2 by angle $\theta_2/2$ (in the same direction as T_2). Let $\overleftrightarrow{r_1}$ be the spherical ray $S_1^{-1}(\overleftrightarrow{A_1A_2})$ and $\overleftrightarrow{r_2}$ be the spherical ray $S_2(\overleftrightarrow{A_2A_1})$. Let F be reflection in $\bigcirc A_1A_2$, F_1 be reflection in the great circle containing $\overleftrightarrow{r_1}$, and F_2 be reflection in the great circle containing $\overleftrightarrow{r_2}$. Then $T_1 = F \circ F_1$ and $T_2 = F_2 \circ F$, so $T_2 \circ T_1 = F_2 \circ F \circ F \circ F_1 = F_2 \circ F_1$, and since the composition of successive reflections is a rotation, we conclude that $T_2 \circ T_1$ is a rotation. ◊

Note that the proof of Theorem 28.5 gives an explicit way of determining the axis of rotation of $T_2 \circ T_1$: it is the intersection of the great circles containing $\overleftrightarrow{r_1}$ and $\overleftrightarrow{r_2}$.

We will now define the distinction between orientation-preserving and orientation reversing isometries of the sphere based on the number of reflections required to achieve the isometry. But before doing so we must ask: is it possible for an isometry to be expressed as both an odd number and an even number of reflections?

Proposition 28.6 *If the identity isometry is expressed as a composition of successive reflections, the number of reflections used is even.*

Proof. Suppose the identity can be expressed as the composition of an odd number of reflections (say $2n+1$ reflections: $I = F_1 \circ F_2 \circ F_3 \circ \ldots \circ F_{2n+1}$). Then pairing the reflections: $I = (F_1 \circ F_2) \circ (F_3 \circ F_4) \circ (F_5 \circ F_6) \circ \ldots \circ (F_{2n-1} \circ F_{2n}) \circ F_{2n+1}$ we conclude that each composed pair of reflections is a rotation, so $I = T_1 \circ T_2 \circ \ldots \circ T_n \circ F_{2n+1}$. Then applying Theorem 28.5, we conclude that $T_1 \circ T_2 \circ \ldots \circ T_n$ is a rotation T, so $I = T \circ F_{2n+1}$. But this means that the rotation T is the same as the reflection F_{2n+1}, since the latter is its

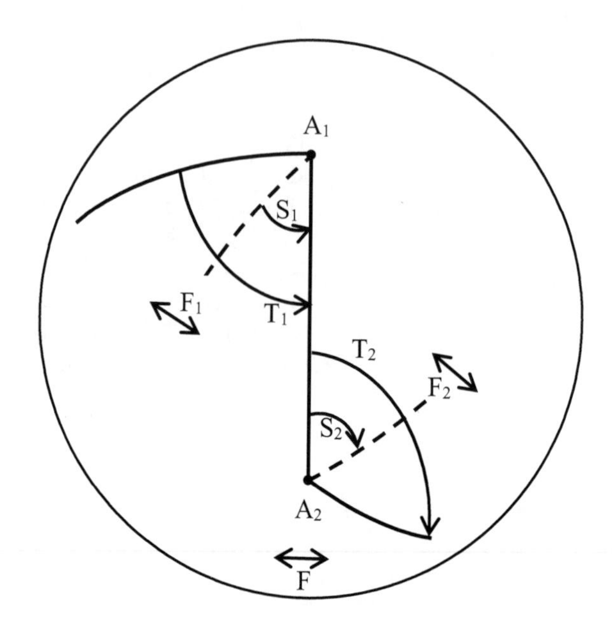

Figure 7.4: Theorem 28.5.

own inverse. But a rotation cannot be a reflection because a reflection fixes a great circle, whereas a rotation fixes either two antipodal points or the whole sphere. This being a contradiction, we conclude that the identity must be the product of an even number of reflections. ◊

Theorem 28.7 *An isometry of the sphere can be written as the composition of an odd number or an even number of reflections, but not both.*

Proof. Theorem 28.3 shows that any isometry can be expressed as the composition of no more than three reflections. If it can be expressed as both a composition of an even number of reflections $F_1 \circ F_2 \circ \ldots \circ F_n$ and in another way as a composition of an odd number of reflections $G_1 \circ G_2 \circ \ldots \circ G_m$, then the composition $G_m^{-1} \circ G_{m-1}^{-1} \circ \ldots \circ G_1^{-1} \circ F_1 \circ F_2 \circ \ldots \circ F_n$ is the identity, and is the composition of an odd number of reflections, which contradicts Theorem 28.6. ◊

We now may make the following definition.

Definition 28.8 *An isometry of a sphere is said to be* orientation preserving *or* orientation reversing *if it can be expressed as the successive composition of and even or odd number of reflections, respectively.*

Theorem 28.7 shows that the notion of orientation preserving or reversing is well-defined.

Theorem 28.9 *An orientation preserving isometry of a sphere must be a rotation.*

Proof. Such an isometry can be written as the composition of an even number $2n$ of rotations. These can be grouped and composed pairwise so the isometry can be written as the composition of n rotations. But by Theorem 28.5, this composition is a rotation. $\diamondsuit$

We note that Theorem 28.9 can be extended to include the case of an isometry which is only defined on a portion of the sphere. This is important in applications in plate tectonics which we consider shortly.

If U is a subset of the sphere and f is a mapping from U to the sphere such that for any X, Y in U, the spherical distance from X to Y is equal to the spherical distance from $f(X)$ to $f(Y)$, then we say that f is an isometry on U.

Theorem 28.10 *Suppose U is a subset of a sphere containing at least three points not on a single great circle. A isometry from U to the sphere extends to a unique isometry from the whole sphere to itself.*

That is, if $f : U \to \sigma$ is an isometry, then there exists a mapping $\overline{f} : \sigma \to \sigma$ which is an isometry and $\overline{f}(X) = f(X)$ for all X in U.

Proof. Let A, B, and C be three points of U not lying on a single great circle. By Theorem 28.3, there exists an isometry $R : \sigma \to \sigma$ equal to the composition of at most three reflections such that $R(A) = f(A)$, $R(B) = f(B)$, and $R(C) = f(C)$. Then $R^{-1} \circ f$ is an isometry such that $R^{-1} \circ f(A) = A$, $R^{-1} \circ f(B) = B$, and $R^{-1} \circ f(C) = C$. By Lemma 28.1, $R^{-1} \circ f(X) = X$ for all X in U so $f(X) = R(X)$ for all X in U. $\diamondsuit$

We now apply the results above to the theory of plate tectonics. The surface of the earth is covered with a series of fairly rigid plates which move about on the surface of the earth. Each such plate may be treated as a region U whose occasional movements produce an isometry of the initial position of the plate to a new position. Intuitively this change of position preserves orientation. This isometry, by Theorem 28.10 induces an isometry of the entire surface of the earth, which must be a rotation by Theorem 28.9.

Suppose that we consider two plates which meet. From the point of view of an observer on one plate, the other plate appears to be moving relative to the first plate, thinking of the first plate as being stationary. This motion induces an orientation-preserving isometry of the earth's surface, hence a rotation about an axis. The axis of this rotation has poles on the earth's surfaces. The rotation is entirely determined by a pole, the angular rate of the rotation, and the direction of that rotation. This, along with the geometry of the intersection of the two plates, determines the nature of the interaction between the two adjacent plates. Figure 7.5 illustrates various possibilites. In each of them, plate B rotates about P in the direction of the arrows when viewed from Plate A. In the first case (upper left in the figure), the plate boundary crosses small circles centered at a pole P of the rotation. This causes plate B to

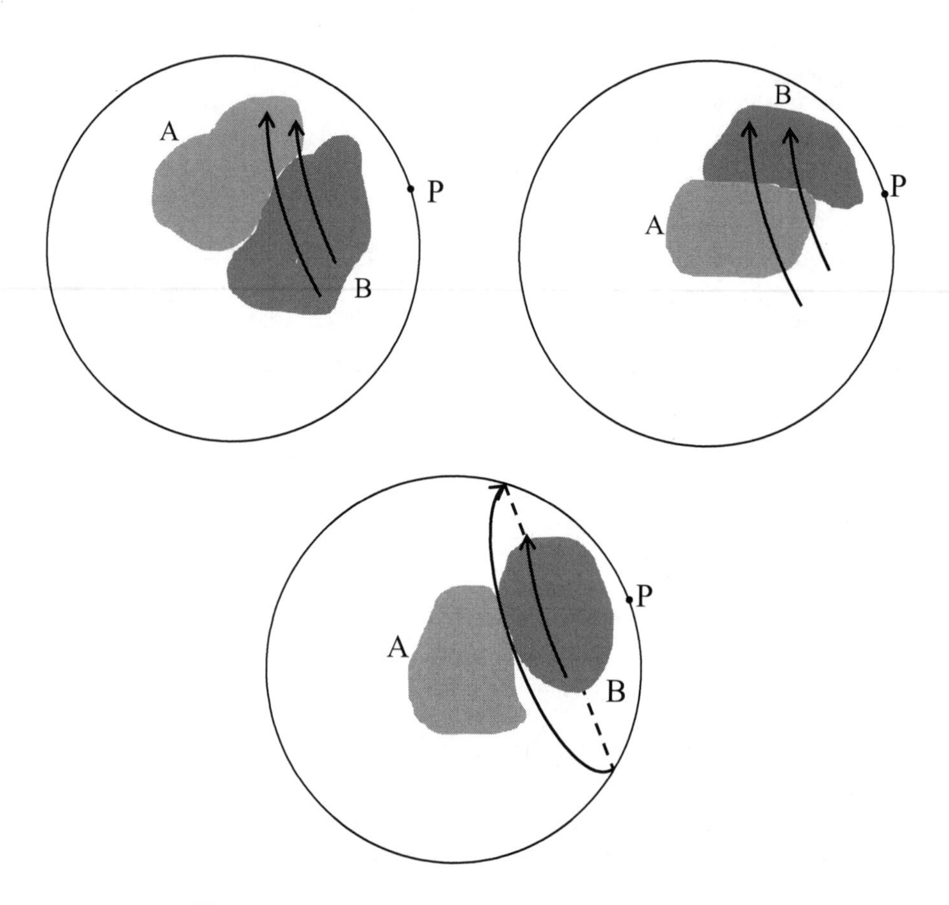

Figure 7.5: Motion of plate B relative to plate A.

smash into plate A. The result is generally that one plate slides underneath the other in what is called a "subduction zone." This results in the creation of mountains on the plate boundary. In the second case (upper right in the figure), plate B is moving away from plate A, resulting in a rift valley between the plates. In the third case, plate B moves along the plate boundary relative to plate A forming a fault. The famous San Andreas Fault in California is of this type.

In the case of such a fault, the rate at which the fault slides can be related to the rate of the rotation around the axis P. Suppose the rate of rotation is δ radians per unit time, the radius of the earth is R, and the spherical distance from P to point X on the fault is ρ. Then the rate of slippage along the small circle is a distance of $R\delta\sin(\rho)$ per unit time. (See Exercise 10.) Knowing the parameters R, δ, and ρ then allows us to determine the slippage rate of the fault.

In reality, not all portions of a fault slip at the same constant rate; the slippage rate is an average over time. A fault will slip in fits and starts, sometimes steadily and sometimes only in great amounts suddenly, resulting in great earthquakes. In the case of the San Andreas Fault, certain portions of the fault are slipping steadily, but two portions of the fault (a northern section around San Francisco and a southern section around Los Angeles) are currently "locked." That is, no slippage is detected using instruments on the ground. But knowing the average rate of slippage gives valuable information about the behavior of the fault. Knowing the average slippage rate gives the scientist an idea of how much locked fault needs to slide in order to "catch up" with the average slippage. This in turn gives a sense of how serious an earthquake might be when that slippage actually occurs.

For further details on applications in plate tectonics, see [TS2014] or [Fo2004].

Exercises §28

1. Show that an isometry must be a one-one mapping.

2. Show that a rotation of angle τ about the axis of P is the same mapping, regardless of what zero meridian is chosen in the coordinate system used.

3. Show that rotations and reflections must be bijections.

4. Show that a reflection is an isometry.

5. Let T be a rotation as defined above which is not the identity. Show that X is a fixed point of T if and only if $X = P$ or $X = P^a$.

6. Let F be a reflection as defined above. Show that X is a fixed point of F if and only if X is on the axis of reflection Γ.

7. Prove that if the great circle Γ is the set of all points on the sphere equidistant from points A and B, then reflection in Γ maps A to B.

8. If spherical coordinates (ϕ, θ) are given with equator Γ_1 then the mapping $(\phi, \theta) \mapsto (\phi, -\theta)$ is reflection across the great circle of the zero meridian. (Hint: Show that this mapping is an isometry which maps the great circle containing the zero meridian to itself and interchanges its poles. Then apply Lemma 28.2.)

9. If spherical coordinates (ϕ, θ) are given with equator Γ_1 then the mapping $(\phi, \theta) \mapsto (\phi, 2\theta_2 - \theta)$ is reflection across the great circle of the meridian where $\theta = \theta_2$.

10. Show that if the radius of the earth is R, the spherical distance from the fault to the pole of rotation is ρ radians and the rotation rate around the pole is δ radians per unit time, the fault slips a distance of $R\delta \sin(\rho)$ per unit time. (Hint: show that the radius in space of the small circle of the fault is $R \sin(\rho)$.)

11. Assume that the motion of the North American plate about the Pacific plate has rotation center $48.7°N$, $78.2°W$ and that the rate of angular rotation is 7.5×10^{-7} degrees per year[1]. Use this to approximate the average annual rate of slippage of the San Andreas Fault at San Bernardino, California.

12. Referring to Exercise 11, determine the average annual rate of slippage in San Francisco. In fact, there has been no slippage along the fault in San Francisco since 1906 — so what distance would the fault have slipped during that period of time if it had been slipping at the average rate?

29 Spherical projections

In this section we consider the old problem of how best to represent a spherical surface on a flat map via a mapping from the sphere to a plane region. This kind of mapping we call a *projection*. The central difficulty is how to do this without causing some kind of distortion. The kinds of distortion we are talking about include distortion of distance, angles, shapes, and areas. The reader is probably familiar with standard flat maps of the earth. These maps allow a great deal of distortion of the poles because most household uses of maps do not require any detail, or even any accuracy, of the poles. Land masses and distances near the poles typically appear much larger than they should be, relative to the land masses near the equator. We will see that it is possible to produce maps which avoid some kinds of distortions, but not all of them at once.

We first consider the *cylindrical projection*. We suppose that we have a sphere with designated poles and equator. We place it inside a cylinder of

[1] See [TS2014].

the same radius, making sure that the line through the poles is vertical and the equator is horizontal. We project all points on the sphere radially to the cylinder. (See Figure 7.6.) The cylinder can then be unwrapped without any further distortion to produce a flat map of the sphere. This projection has some pleasant qualities. The image of a meridian on a sphere is a vertical line, and the image of a circle of constant latitude is a horizontal (radial) circle on the cylinder. When the cylinder is unwrapped, the parallels are horizontal lines, the meridians are vertical lines and longitude and latitude are x and y coordinates on the map. Note that near the equator distortion is small because the sphere is close to the cylinder there, and is nearly vertical like the cylinder. But for regions near the poles, large regions get compressed vertically while being spread around a distant radius, when in fact these points are rather close to each other. (In fact, we have not even defined the mapping for the poles.) The cylindrical projection thus distorts distance and shapes.

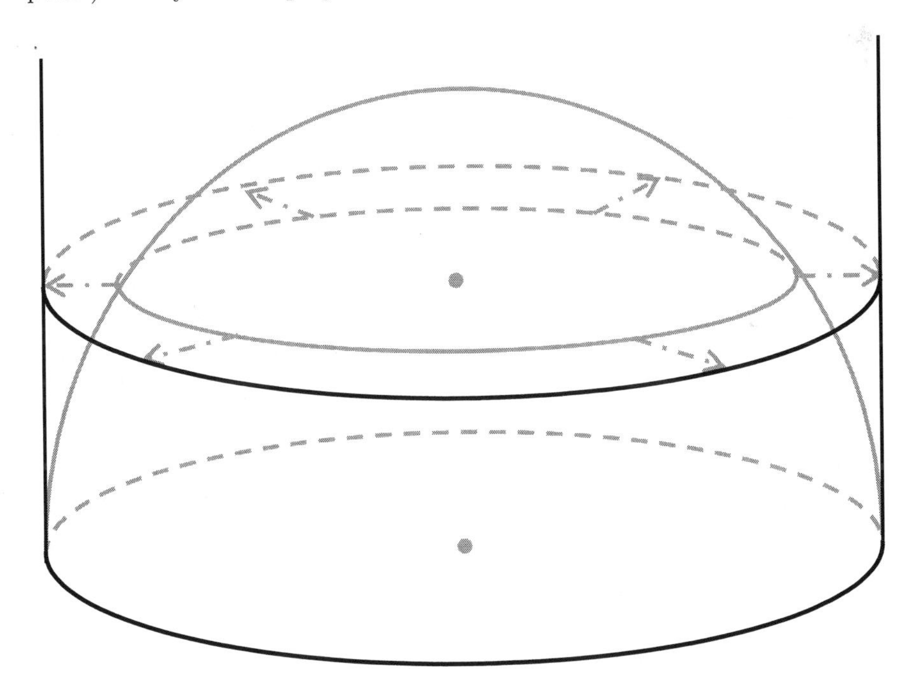

Figure 7.6: Cylindrical projection.

It may come as a surprise that cylindrical projection preserves areas. That is, given any region on the sphere, its area is the same as its image on the cylinder. Suppose we take a region on the sphere bounded between two parallels of latitude and two meridians at angle θ. (See Figure 7.7.) Then what we have is called a "zone of a lune." The projection to the cylinder is then a region bounded vertically by parallel cross sections of the cylinder and horizontally by vertical lines along the cylinder. The result is a rectangle when

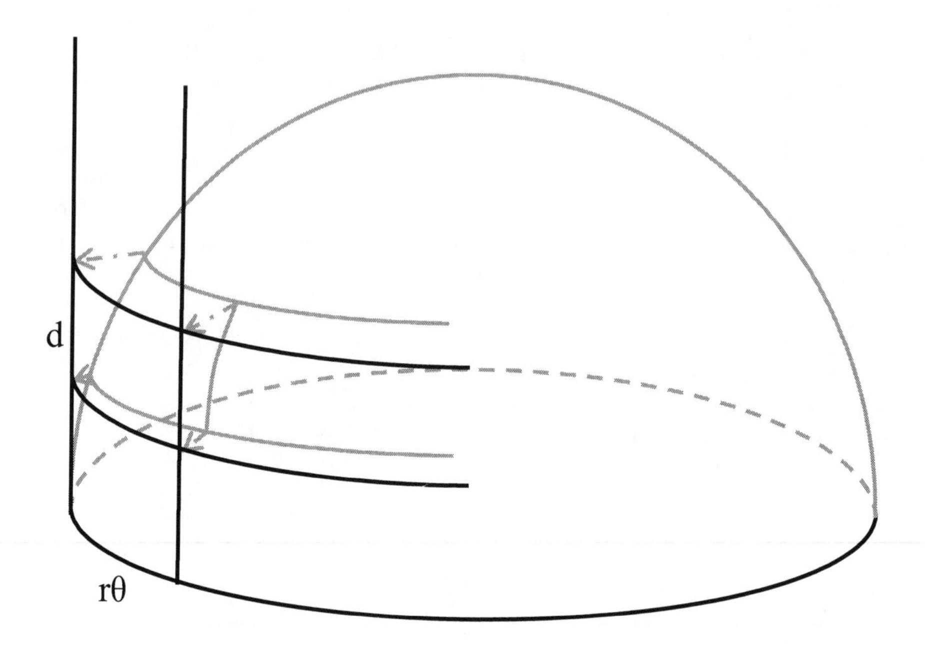

Figure 7.7: Cylindrical projection preserves area.

unrolled. Suppose the projection to the cylinder is of height d. Since the lune has angle θ, the two meridians project to two vertical lines on the cylinder forming an angle of θ with the central axis of the cylinder. The two circular arcs bounding the top and bottom of the projection thus have angle θ and radius r, so have length $r\theta$. Thus the width of the rectangle on the unrolled map is the $r\theta$ and its height is d. The area of the rectangle is then $rd\theta$. According to §7, Exercise 1, $rd\theta$ is also the area of the zone of the lune on the sphere. Thus the projection preserves areas of any region bounded horizontally by meridians and vertically by parallels of latitude. This fact can be extended to a general region by expressing it as the non-overlapping union of such regions. (See Exercise 1 for another justification.)

The idea of a projection of the sphere to a cylinder so that meridians are projected to vertical lines on the cylinder can be done in another useful way. We can preserve this property by stretching the vertical direction of the projection. The distortion of regions toward the poles would suggest that stretching the image on the cylinder more toward the poles would produce a good map. This is the idea behind the *Mercator projection*, discovered by Gerard Kremer (also known as Mercator) in the sixteenth century.

The Mercator projection is designed to preserve another property of the original sphere: it preserves angles between curves. A mapping which preserves angles between curves where they intersect is called a *conformal mapping*. A dilation of the plane is perhaps the simplest example of a conformal mapping.

In plane geometry we see that dilation by a factor of $c > 0$ takes a triangle and makes it c times as big, but the angles of the triangle are the same. One way of thinking about what a conformal mapping means is that the stretching that occurs is the same in every direction. Thus given an infinitesimally small rectangle, its image under the projection is a similar rectangle.

Suppose that 1 is the radius of the sphere and ℓ_1, ℓ_2 are latitude and longitude of a point in radians. Then the radius of the ℓ_1 latitude parallel is $\cos(\ell_1)$ and $s = \cos(\ell_1)\ell_2$ measures signed distance around the ℓ_1 parallel circle from the zero meridian. Let (x, y) be (circular, vertical) coordinates on the cylinder. Then let us think of cylindrical projection as a composition mapping $(\ell_1, \ell_2) \mapsto (\ell_1, s) \mapsto (y, x)$. Then $x = \ell_2 = s/(\cos(\ell_1))$ and $y = \sin(\ell_1)$. The derivative $\frac{dy}{d\ell_1} = \cos(\ell_1)$ measures the stretching factor of the mapping in the latitudinal direction. The derivative $\frac{dx}{ds} = 1/\cos(\ell_1)$ measures the stretching factor in the longitudinal direction. We seek a function $z = f(\ell_1)$ such that the vertical stretching is $\frac{dz}{d\ell_1} = \frac{dx}{ds} = 1/\cos(\ell_1) = \sec(\ell_1)$. This will make a mapping where the dilation of the vertical and horizontal distances is the same, and the result will be a mapping which is conformal. To find z we must integrate the secant, and the integral can be written as $z = \ln(\sec(\ell_1) + \tan(\ell_1)) + C = \ln(\tan(\frac{\pi}{4} + \frac{\ell_1}{2})) + C$. (See Exercise 2.) Since $z = 0$ when $\ell_1 = 0$, $C = 0$. So the Mercator projection is given by

$$(\ell_1, \ell_2) \mapsto (z, x) = (\ln(\tan(\frac{\pi}{4} + \frac{\ell_1}{2}), \ell_2). \qquad (7.1)$$

The coordinates may be multiplied by the same constant to dilate the image to make a map which is large enough for personal use.

The result is a map where the local shapes are preserved — but we pay a significant price for this advantage by the dramatic size distortion away from the equator. There is another interesting property of the Mercator projection that arises from preserving angles. A *loxodrome* or *rhumb line* is a curve on the sphere which makes the same angle with all meridians (and parallels). This is the curve that would be traced following a constant course of navigation. (That is, the direction of travel on the compass is constant.) Because the meridians and parallels map to vertical and horizontal lines under the Mercator projection, a loxodrome projects to a curve which maintains the same angle with a set of parallel lines. That is, the loxodrome projects to a straight line. Thus a ship travelling by constant course could see its path on a Mercator projection simply by plotting a straight line with appropriate slope.

We next consider a projection whose roots date to the ancient world, the *stereographic projection.* Suppose that P is a point of a sphere, and L is a plane tangent to the antipode of P. Let X be any point on the sphere other than P, and let Y be the point where $\overleftrightarrow{XP}$ meets L. Then Y is the stereographic projection of X to L.

We can see that for a great or small circle with pole P, its image must be a circle in L. (See Figure 7.8.) Suppose that R is the radius of the sphere, X is chosen on a small circle parallel to L of radius r_X (in space), the center of the circle on the sphere is distance d from P, and the distance P^aY is r_Y. Then by similar triangles, $r_Y/r_X = 2R/d$. Thus P^aY is the same regardless of the choice of X on the given small circle and the image of the circle on the sphere is a circle in L of radius $r_Y = 2Rr_X/d$. Furthermore, for X near P, the line $\overleftrightarrow{PX}$ is nearly parallel to L, so the radius r_Y is very large. Thus we can think of the image of P under the projection as being a "point at infinity" in L.

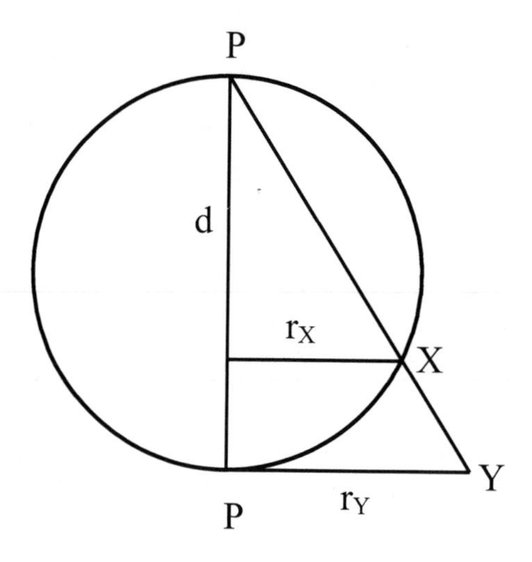

Figure 7.8: Stereographic projection of horizontal circle of radius r_X.

There are two significant properties of the stereographic projection that we proceed to prove. First, the projection is conformal, and second, the image of a circle on the sphere is a line in L (if it passes through P) or a circle in L (if it does not pass through P).

Proposition 29.1 *Stereographic projection is conformal.*

Proof. Suppose that two tangent lines to the sphere form an angle at point X different from P. Let L_1 and L_2 be the planes passing through P which contain the lines. These planes intersect the sphere in small circles which pass through both X and P. Let L_3 be the perpendicular bisector of $\overline{PX}$. Then L_3 contains the center of the sphere, so the reflection of the sphere in L_3 is itself. We have that $\overline{PX}$ is contained in both L_1 and L_2. By Theorem 2.18, L_3 is perpendicular to L_1 and L_2, so the reflections of L_1 and L_2 in L_3 are themselves. Reflection makes the angle between the small circles have the same measure at P as at X (which has the same measure as the angle between the original two lines). Furthermore, the tangent lines to the small circles at P are the intersection of L_1 and L_2 with the tangent plane to the sphere at P. Since L is parallel to the tangent plane at P, the tangent lines to the small circles at P are parallel to the stereographic projection of the original angle, which is the intersection of L_1 and L_2 with L. Thus the measure of the projected angle is the same as the measure of the original angle. $\diamondsuit$

Before proceeding, we consider some properties of cones in space. Suppose that Γ is a circle, and P a point not in the plane of Γ. (See Figure 7.9.) Then the union of all lines passing through P and a point of Γ form a circular cone Λ_1 with vertex P. Let A be the point on the circle nearest to P and let B be the point on the circle furthest from P. Then the plane PAB is perpendicular to the plane of the circle base. Let the angle bisector of $\angle APB$ meet the plane of Γ at point E. Let L_1 be the plane perpendicular to $\overleftrightarrow{PE}$ passing through B. We claim that Λ_1 is also a right elliptical cone with elliptical base equal to the intersection of the cone with L_1. This elliptical base has center on $\overrightarrow{PE}$.

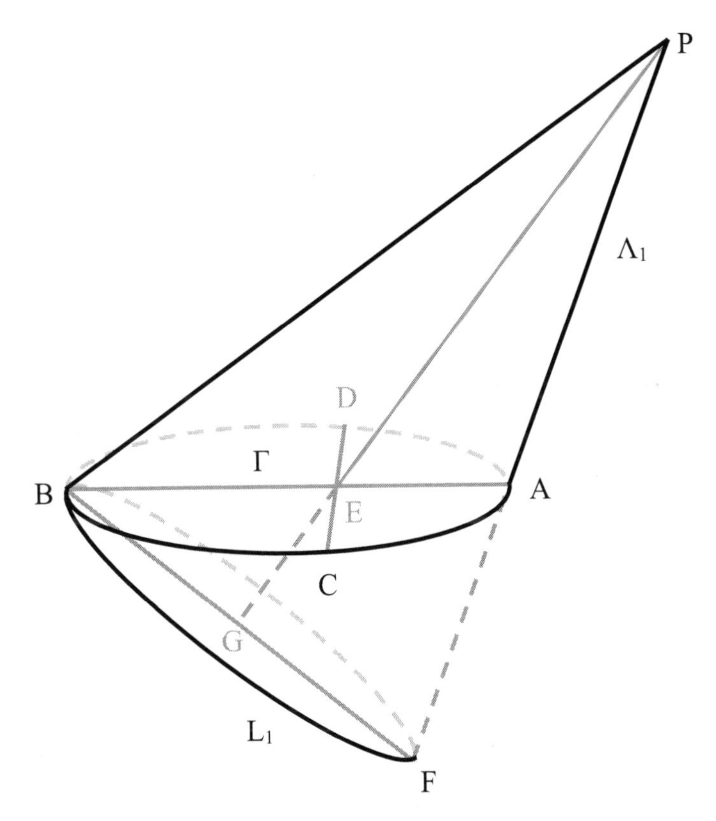

Figure 7.9: A circular cone is also a right elliptical cone.

To see this, we first note that Λ_1 is the image under a linear mapping of some right circular cone Λ_2 which fixes the circle but translates the vertex in a plane parallel to the circular base. (See Exercise 5 and Figure 7.13.) Under this mapping the image of a plane is a plane and the image of an ellipse is an ellipse. (The same is true of the inverse of the mapping.) The bounded planar sections of the right circular cone are known to be ellipses, so their images under the mapping are also ellipses. This shows that the plane L_1 must intersect Λ_1 in an ellipse. Let $\overrightarrow{PA}$ meet L_1 in F, and $\overrightarrow{PE}$ meet L_1 in G. Now Λ_1 is its own reflection in plane PAB, and every point of $\overleftrightarrow{FB}$ is reflected to itself since $\overleftrightarrow{FB}$ is the intersection of planes PAB and L_1. Since $\overleftrightarrow{PE}$ is contained in plane PAB and is perpendicular to plane L_1, planes L_1 and PAB are perpendicular (by Theorem 2.18). Thus the reflection of L_1 in plane PAB is itself. Since both Λ_1 and L_1 are symmetric in plane PAB, so is their intersection (the ellipse). Thus $\overline{FB}$ is the major or minor axis of the ellipse $L_1 \cap \Lambda_1$. The midpoint of $\overline{FB}$ must be G because $\overrightarrow{PE}$ is the angle bisector of $\angle APB$ and $\overleftrightarrow{FB}$ is perpendicular to it at G. Thus G is the center of the ellipse. Let L_2 be the plane containing $\overleftrightarrow{PE}$ perpendicular to $\overline{FB}$. Then this plane cuts the ellipse in its other axis. Then the ellipse is its own reflection in the plane L_2, so cone Λ_1 is also. Thus Λ_1 is a right elliptical cone with base $L_1 \cap \Lambda_1$.

Because cone Λ_1 is its own reflection in plane PCD, the reflection of its original circular base is also a circle in the cone. Also, a section of the cone parallel to either circle is another circle. (See Figure 7.10.)

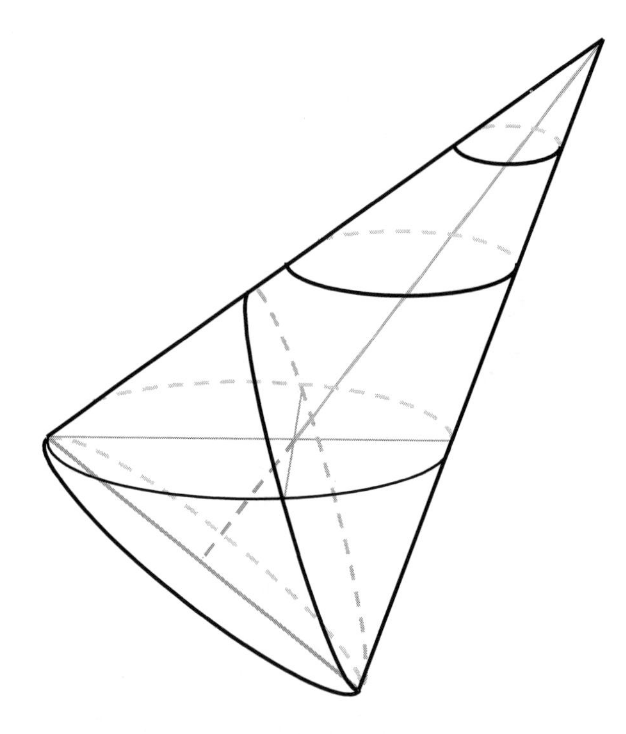

Figure 7.10: Parallel cuts and reflections of the original circular base.

Proposition 29.2 *Stereographic projection maps a circle on the sphere to a line (if the circle on the sphere passes through the projection point P) or to a circle (if the circle on the sphere does not contain P).*

Proof. Suppose that a circle on the sphere is contained in plane L_1. If the circle contains P, then its image under stereographic projection is $L_1 \cap L$, which is a line.

Suppose that the circle does not contain P. Then the set of all lines passing through P and X (as X varies on the circle) is a circular cone. The intersection of this cone with L is the image of the circle under stereographic projection. As noted earlier, this cone is a circular cone which can also be seen as a right elliptical cone where the reflection of the circular base is also a section of the cone, as are the sections of the cone parallel to it. Let A through G be as chosen above in Figure 7.9.

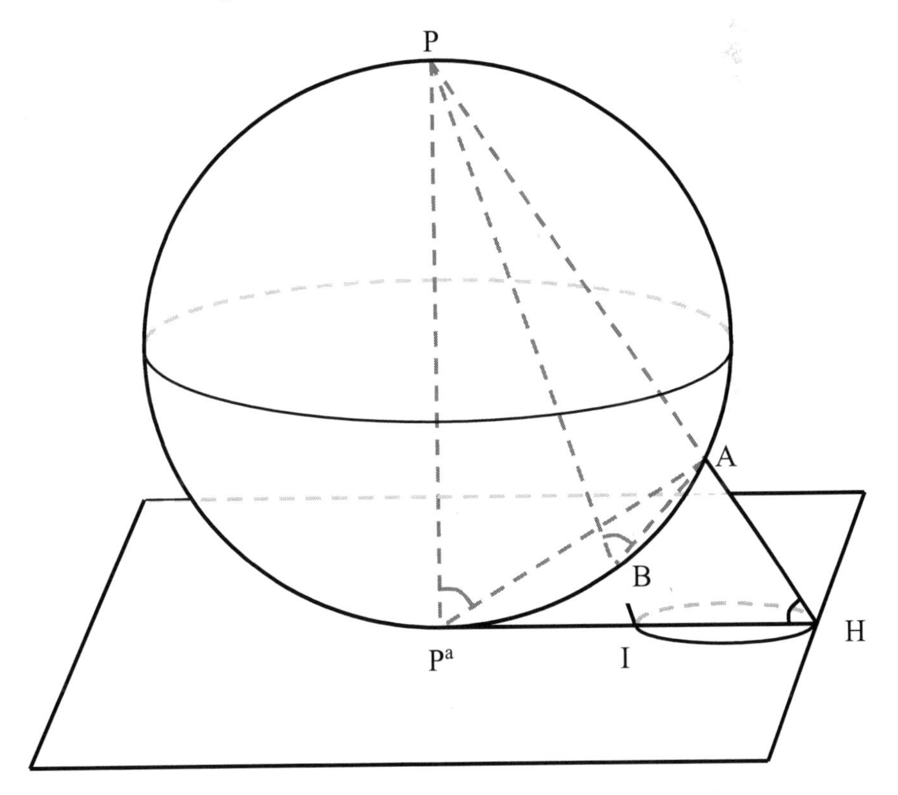

Figure 7.11: Stereographic projection of a circle is a circle.

We claim that the image of the original circular base under stereographic projection can be obtained as follows. Take the reflection of the original circular base in plane PCD. We claim that the section of the cone parallel to this reflection which passes through P^a is the stereographic projection. (See

Figure 7.11.) To see this, let $\overrightarrow{PA}$ meet L at H and $\overrightarrow{PB}$ meet L at I. Then since the measure of an arc inscribed in a circle is half the measure of the arc it cuts on the circle, $m\angle PBA = \frac{1}{2}m\overset{\frown}{PA} = m\angle PP^aA$. Now $\angle PAP^a$ is a right angle since it cuts a diameter $\overline{PP^a}$ of the sphere. Thus $\angle PP^aA$ is complementary to $\angle APP^a$. Also, $\angle PP^aH$ is a right angle since $\overleftrightarrow{P^aH}$ is tangent to the sphere at P^a. Thus $\angle PHP^a$ is complementary to $\angle APP^a$. Since $\angle PHP^a$ and $\angle PP^aA$ are both complementary to $\angle APP^a$, $\angle PHP^a \cong \angle PP^aA$, so $\angle PHI \cong \angle PBA$. This means that the reflection of the original circular base of the cone in plane PCD must be parallel to its stereographic image. Thus the stereographic image is a circle. ◇

Instead of having an image plane tangent to the point P^a, the image plane could be any parallel plane, but the result is merely a dilation of the stereographic projection defined here so Theorem 29.1 and Theorem 29.2 still hold. Thus it is not uncommon to define stereographic projection as having image to the parallel plane through the center of the sphere.

We lastly prove an important theorem about what is not possible for a projection of the sphere: that it is not possible to project any portion of the sphere to a plane in a way that preserves distances. That is, distance distortion is always unavoidable.

Theorem 29.3 *Given a spherically convex subset of the sphere, there is no isometric mapping of it to a region of a plane.*

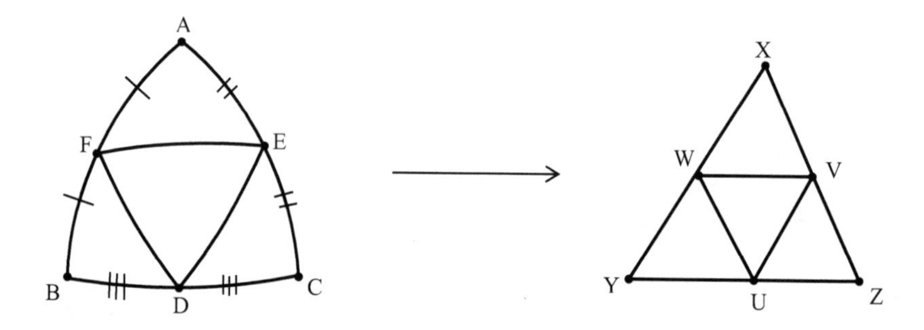

Figure 7.12: Theorem 29.3.

Proof. (See Figure 7.12.) To the contrary, suppose that such an isometric mapping from the subset to the plane does exist. Let A, B, and C be three points in the spherically convex set, and let X, Y, and Z be their images under the isometry. If C is between A and B at distance d from A, then $m\overset{\frown}{XZ} = d$ and $m\overset{\frown}{AC} + m\overset{\frown}{CB} = m\overset{\frown}{AB}$, so $XZ + ZY = XY$ and $XZ = d$ and we conclude that Z is between X and Y at distance d from X. Thus the image of $\overset{\frown}{XY}$ is segment $\overline{AB}$. By a similar argument (see Exercise 8) the image of a spherical $\triangle^s ABC$ is a plane triangle $\triangle XYZ$. Let D, E, and F be the midpoints of $\stackrel{\frown}{BC}$, $\stackrel{\frown}{AC}$, and $\stackrel{\frown}{AB}$, respectively. Then their images are the midpoints U, V, and W of $\overline{YZ}$, $\overline{XZ}$, and $\overline{XY}$, respectively. In the plane, we know that these midpoints produce four congruent triangles: $\triangle XWV \cong \triangle WYU \cong \triangle VUZ \cong \triangle UVW$. By the isometry, we have spherical SSS correspondences — hence congruences — between the corresponding spherical triangles: $\triangle^s AFE \cong \triangle^s FBD \cong \triangle^s EDC \cong \triangle^s DEF$. Then corresponding angles are congruent: $\prec AFE \cong\prec FBD$ and $\prec AEF \cong\prec ECD$. But then we have an AAA correspondence between $\triangle^s AFE$ and $\triangle^s ABC$, so they are congruent by spherical AAA congruence. But this is impossible since the measures of the sides of $\triangle^s AFE$ are half those of $\triangle^s ABC$. Thus the assumption is false, and the isometric mapping cannot exist. ♢

Exercises §29.

1. Show that at a point of latitude λ_1, the cylindrical projection stretches lengths of arcs on parallels by a factor $1/\cos(\lambda_1)$ and stretches small distances along meridians by an approximate factor of $\cos(\lambda_1)$. Conclude that cylindrical projection multiplies the areas of small rectangles by a factor of approximately 1.

2. Check that for $-\frac{\pi}{2} < \ell_1 < \frac{\pi}{2}$ the quantity $\sec(\ell_1)+\tan(\ell_1)$ is positive (so has a logarithm) and check the identity $\sec(\ell_1)+\tan(\ell_1) = \tan(\frac{\pi}{4} + \frac{\ell_1}{2})$.

3. Suppose that a ship travels along a loxodrome forming an angle of θ with the meridians in such a way that its latitude changes by $\Delta\ell_1$. Argue that the distance traveled along the loxodrome is $\Delta\ell_1 \sec(\theta)$. (Hint: Think of the loxodrome as the union of many small segments all of which form angle θ with the meridians.)

4. Suppose that on a Mercator projection two points are separated by a longitudinal difference of $\Delta\ell_2$ and the vertical positions are $L_0 < L_1$. Show that the angle formed with the meridians satisfies $\tan(\theta) = \Delta\ell_2/(L_1 - L_0)$ and use Exercise 3 to find the actual spherical distance traveled along the loxodrome.

5. Suppose that (x, y, z) are real coordinates for space and we are given a cone whose base is a circle in the xy plane and apex $(0, b, c)$ for $c \neq 0$. Show that the mapping $(x, y, z) \mapsto (x, y - \frac{b}{c}z, z)$ maps the cone to a right circular cone with the same base and apex $(0, 0, c)$. Also show that the mapping is invertible with inverse $(x, y, z) \mapsto (x, y + \frac{b}{c}z, z)$. Show that it maps planes in space to planes, and ellipses to ellipses. (See Figure 7.13.)

6. Show that for the stereographic projection, the radius r_Y of the image point is given by $r_Y = 2R\tan(\frac{\pi}{4} + \frac{\ell_1}{2})$.[2]

[2]Note the similarity to the Mercator projection.

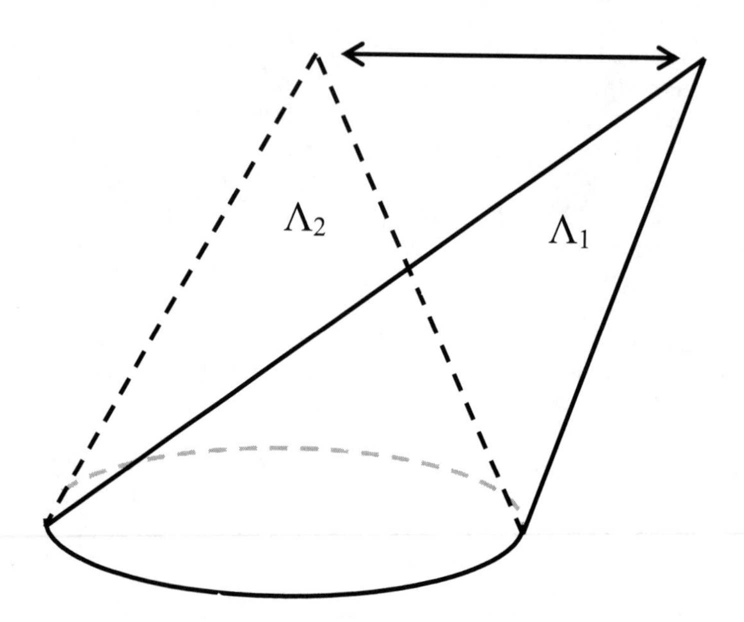

Figure 7.13: Transformation of non-right circular cone to right circular cone.

7. Prove that the center of the stereographic projection of a circle is the stereographic projection of the point which is the intersection of the tangent planes to the points on the given circle.

8. Prove the detail left out of Theorem 29.3 that the image of a spherical triangle under an isometry to the plane must be a plane triangle.

Chapter 8

Quaternions

In this chapter we see how to use the coordinate algebraic structure of the quaternions to organize the derivation of formulas in spherical geometry. Essentially this is done by showing how to describe the rotations and reflections in space with quaternions.

30 Review of complex numbers

Let us review how the complex numbers can be used to do geometric operations in plane geometry. Recall that the imaginary number $\mathbf{i}$ is introduced in algebra as a new number corresponding to $\sqrt{-1}$. We then form the set of so-called complex numbers $a + b\mathbf{i}$ which can be added by the rule $(a + b\mathbf{i}) + (c + d\mathbf{i}) = (a + c) + (b + d)\mathbf{i}$. They can be multiplied by use of the identity $\mathbf{i}^2 = -1$, the distributive laws, and the fact that $r\mathbf{i} = \mathbf{i}r$ for all real numbers r: $(a + b\mathbf{i})(c + d\mathbf{i}) = ac + ad\mathbf{i} + b\mathbf{i}c + b\mathbf{i}d\mathbf{i} = ac + ad\mathbf{i} + bc\mathbf{i} + bd\mathbf{i}^2 = ac + ad\mathbf{i} + bc\mathbf{i} + bd(-1) = ac - bd + (ad + bc)\mathbf{i}$. If the reader is concerned about just what exactly a complex number "is," we can understand the expression $a + b\mathbf{i}$ to be simply a notation for the ordered pair (a, b). Then the multiplication rule just given could be understood as a product defined on the set of ordered pairs of real numbers and written $(a, b)(c, d) = (ac - bd, ad + bc)$. The expression of (a, b) in the notation $a + b\mathbf{i}$ aids in remembering the product rule, and in seeing the set of complex numbers as being an extension of the real numbers.

Take any complex number $a + b\mathbf{i}$ and write it as

$$\sqrt{a^2 + b^2}\left(\frac{a}{\sqrt{a^2 + b^2}} + \frac{b}{\sqrt{a^2 + b^2}}\mathbf{i}\right) = \sqrt{a^2 + b^2}(\cos(\phi) + \sin(\phi)\mathbf{i})$$

for some real ϕ, $0 \leq \phi < 2\pi$. This is called the *polar form* of the complex number. The number $r = |a + b\mathbf{i}| = \sqrt{a^2 + b^2}$ is called the *absolute value* or *norm* of the complex number. The number ϕ is called the *argument* of the complex number. It is the angle required to rotate the positive x-axis counterclockwise into the direction of $a+b\mathbf{i}$. (See Figure 8.1.) Recall that the

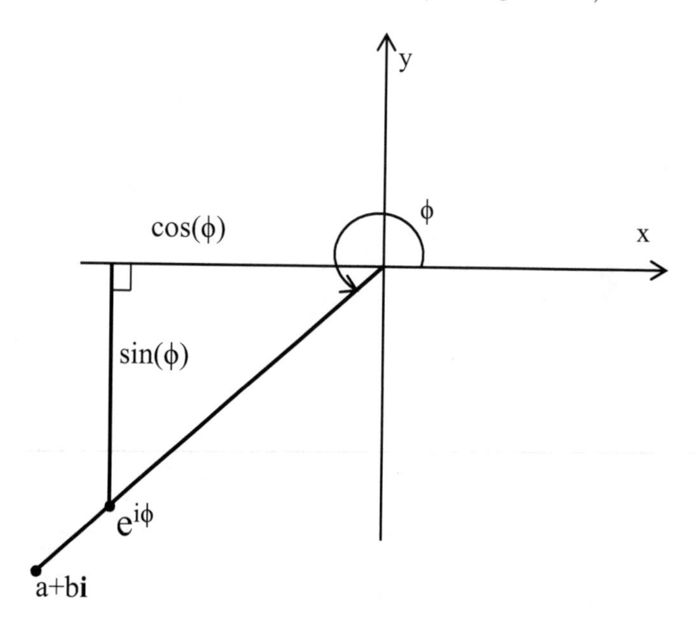

Figure 8.1: $a+b\mathbf{i}$ and its argument ϕ.

expression $e^{\mathbf{i}\theta} = \cos(\theta) + \mathbf{i}\sin(\theta)$. Then θ is the argument of $e^{\mathbf{i}\theta}$. If we multiply $a+b\mathbf{i}$ by $e^{\mathbf{i}\theta}$ we obtain

$$\begin{aligned}
& (a+b\mathbf{i})e^{\mathbf{i}\theta} \\
= \; & [\cos(\phi)\cos(\theta) - \sin(\phi)\sin(\theta) \\
& +\mathbf{i}(\cos(\phi)\sin(\theta) + \sin(\phi)\cos(\theta))]\sqrt{a^2+b^2} \\
= \; & \sqrt{a^2+b^2}(\cos(\phi+\theta) + \sin(\phi+\theta)\mathbf{i}),
\end{aligned}$$

by (1.11) and (1.10). This product forms a complex number with argument $\phi+\theta$. Thus the multiplication of any complex number $a+b\mathbf{i}$ by $e^{\mathbf{i}\theta}$ results in a complex number whose argument is θ larger than the argument of $a+b\mathbf{i}$. Hence multiplication by $e^{\mathbf{i}\theta}$ results in rotation of the complex plane by angle θ counterclockwise. We say that the mapping $w \mapsto e^{\mathbf{i}\theta}w$ is rotation by angle θ about the origin. (See Figure 8.2.)

We can see how this construction can be partly extended to space as follows. (See Figure 8.3.) Given a point (a,b,c) in space, view the first two coordinates (a,b) as a complex number $a+b\mathbf{i}$ so the point in space is the ordered pair $(a+b\mathbf{i},c)$. Then multiplication of $a+b\mathbf{i}$ by $e^{\mathbf{i}\theta}$ causes space to be rotated by angle θ counterclockwise in the z axis when viewed from above the positive z axis.

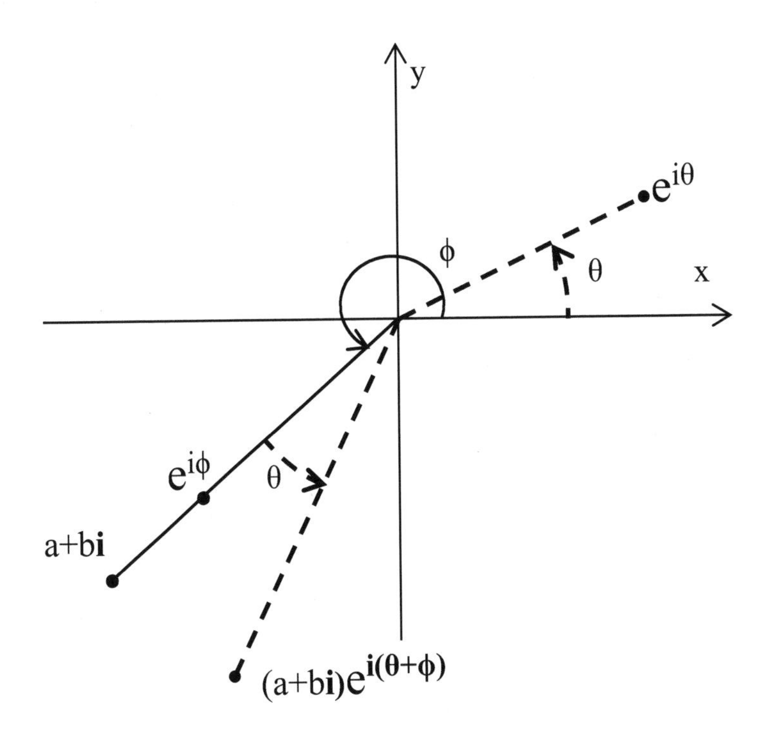

Figure 8.2: Multiplication by $e^{i\theta}$ is counterclockwise rotation by θ.

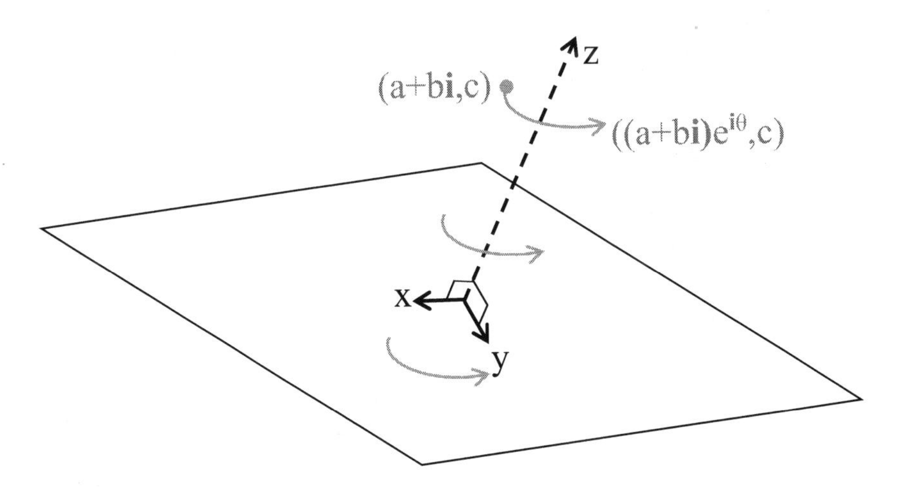

Figure 8.3: Rotation by angle θ counterclockwise around the z axis.

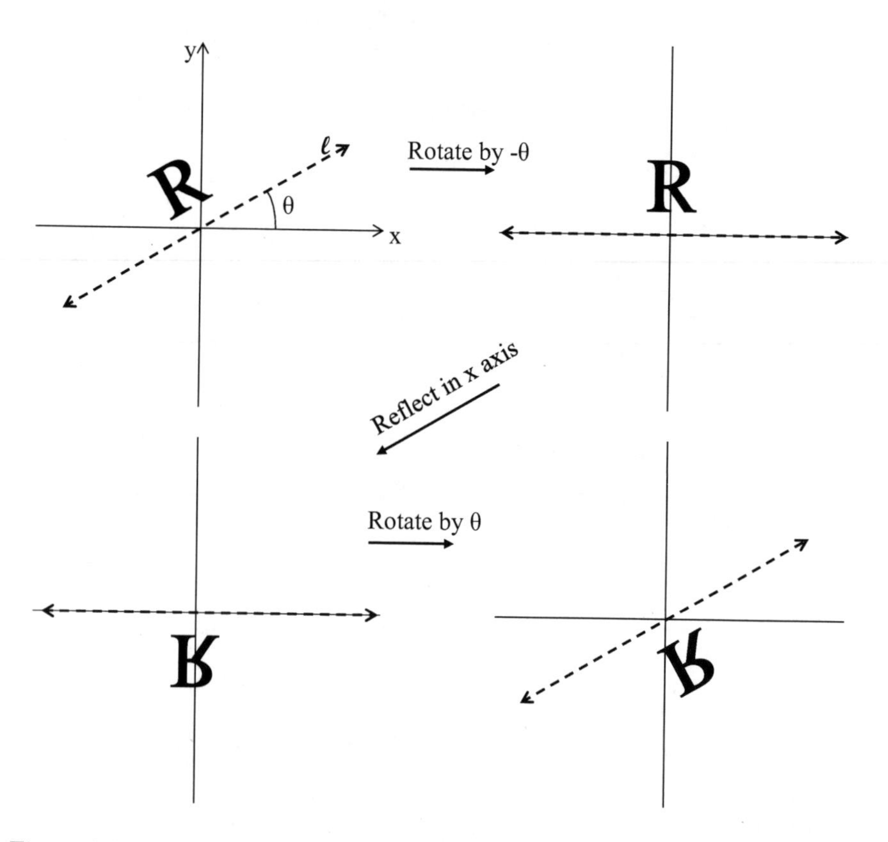

Figure 8.4: Reflection in line ℓ forming angle θ with the x axis, where "rotation by $-\theta$" means "clockwise rotation by angle θ."

Next we consider reflections in the complex plane. (See Figure 8.4.) Given a point $w = x+y\mathbf{i}$, its reflection in the x axis is $x-y\mathbf{i}$, which is the *conjugate* $\overline{w}$ of w. So we say that the mapping $w \mapsto \overline{w}$ is reflection in the x axis. To reflect a point in an arbitrary line ℓ through the origin, suppose that ℓ forms angle θ with the positive x axis. Then we can achieve reflection in line ℓ by (1) rotating the plane clockwise by angle θ first (which moves the line ℓ to the x axis,) then (2) reflecting in the x axis, and then (3) rotating back counterclockwise by angle θ. This composition of three mappings is

$$w \mapsto we^{-\mathbf{i}\theta} \mapsto \overline{w}e^{\mathbf{i}\theta} \mapsto \overline{w}e^{\mathbf{i}2\theta}$$

so we can say that $w \mapsto \overline{w}e^{\mathbf{i}2\theta}$ is reflection in the line ℓ.

Now suppose we have two lines ℓ_1 and ℓ_2 which make angles of θ_1 and θ_2, respectively, $\theta_1 \leq \theta_2$ with the positive x axis. (See Figure 8.5.) We consider what happens when we reflect in ℓ_1 and then reflect in ℓ_2. This composition of mappings is

$$w \mapsto \overline{w}e^{\mathbf{i}2\theta_1} \mapsto \overline{\overline{w}e^{\mathbf{i}2\theta_1}}e^{\mathbf{i}2\theta_2} = we^{2(\theta_2-\theta_1)\mathbf{i}}.$$

This mapping is rotation about the origin by angle $2(\theta_2 - \theta_1)$, which is twice the angle between ℓ_1 and ℓ_2.

These ideas can be extended to reflections in space. (See Figure 8.6.) Suppose that two distinct planes P_1 and P_2 in space meet in a line Z. Let plane P_3 be perpendicular to Z. Let $\ell_1 = P_1 \cap P_3$ and $\ell_2 = P_2 \cap P_3$. We create a coordinate system by letting the line Z be the z axis and setting up x and y axes in P_3 perpendicular to the z axis. Reflection in plane P_1 maps P_3 to itself. In fact, such a reflection maps P_3 to itself by reflection in ℓ_1. Similarly, reflection in P_2 maps P_3 to itself, and such a mapping acts on P_3 by reflection in ℓ_2. Then if reflection in P_1 is composed with reflection in P_2 we obtain a rotation of P_3 by twice the angle θ between ℓ_1 and ℓ_2 indicated in Figure 8.6. The same argument holds for any plane perpendicular to Z: reflection in P_1 or P_2 maps such a plane to itself and the composition of reflection in P_1 and P_2 is a rotation of the plane of angle 2θ. Thus reflection in P_1 followed by reflection in P_2 is rotation about the z axis by angle 2θ.

31 Quaternions: Definitions and basic properties

The set $\mathbf{H}$ of *quaternions* may be formally understood as the set of all ordered quadruples (a, b, c, d) under an addition and multiplication that we will define presently. To define the addition and multiplication in $\mathbf{H}$, it will be helpful to identify the quadruple (a, b, c, d) with an expression of the form $q = a + b\mathbf{i} + c\mathbf{j} + d\mathbf{k}$, where $\mathbf{i}$, $\mathbf{j}$, and $\mathbf{k}$ are thought of as "imaginary" numbers or vectors. The number a is the *real part* $\mathrm{Re}(q)$ of q and the expression $b\mathbf{i}+c\mathbf{j}+d\mathbf{k}$ is the *vector part* $\mathrm{Ve}(q)$ of q. If a quaternion has a real part which is zero, we say

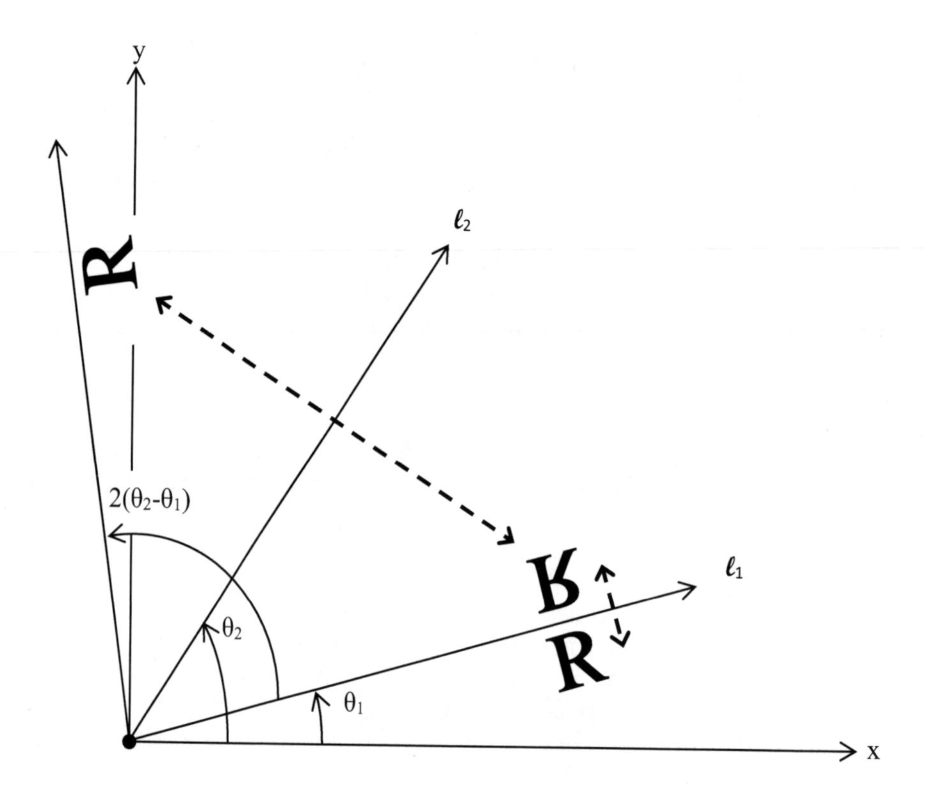

Figure 8.5: Reflection in ℓ_1 followed by reflection in ℓ_2 results in rotation by $2(\theta_2 - \theta_1)$.

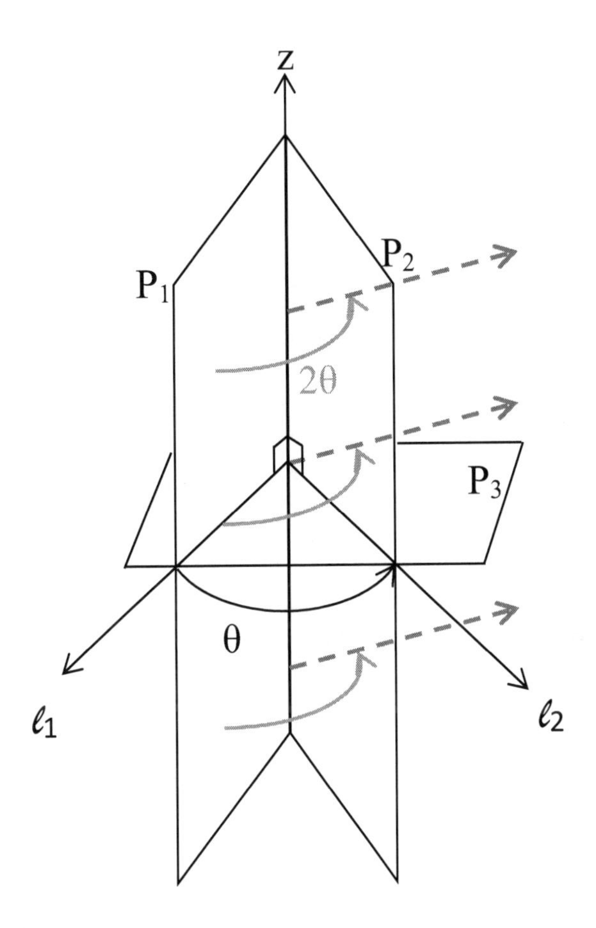

Figure 8.6: Reflection in P_1 followed by reflection in P_2 results in rotation by 2θ about the z axis.

that the quaternion is a "pure quaternion" or a "vector." If a quaternion has a vector part which is zero, we say that the quaternion is "real."

If $q_1 = a_1 + b_1\mathbf{i} + c_1\mathbf{j} + d_1\mathbf{k}$ and $q_2 = a_2 + b_2\mathbf{i} + c_2\mathbf{j} + d_2\mathbf{k}$, we define the sum of q_1 and q_2 to be $q_1 + q_2 = (a_1 + a_2) + (b_1 + b_2)\mathbf{i} + (c_1 + c_2)\mathbf{j} + (d_1 + d_2)\mathbf{k}$. Then we define their product as follows: let $\mathbf{ii} = \mathbf{i}^2 = -1$ and similarly $\mathbf{j}^2 = \mathbf{k}^2 = -1$. We also define

$$\mathbf{ij} = \mathbf{k}, \mathbf{jk} = \mathbf{i}, \mathbf{ki} = \mathbf{j}$$
$$\mathbf{ji} = -\mathbf{k}, \mathbf{kj} = -\mathbf{i}, \mathbf{ik} = -\mathbf{j}$$

We then extend these rules to a product of q_1 and q_2 by requiring the quaternion product to distribute over addition from the right and the left (that is, $q_1(q_2 + q_3) = q_1q_2 + q_1q_3$ and $(q_2 + q_3)q_1 = q_2q_1 + q_3q_1$ for quaternions q_1, q_2, q_3) and that the product of a real quaternion r and any other quaternion q must commute (that is, $rq = qr$).

We look at the product q_1q_2 by first looking at the case where $a_1 = a_2 = 0$:

$$\begin{aligned}
& (b_1\mathbf{i} + c_1\mathbf{j} + d_1\mathbf{k})(b_2\mathbf{i} + c_2\mathbf{j} + d_2\mathbf{k}) \\
= \ & b_1b_2\mathbf{i}^2 + c_1c_2\mathbf{j}^2 + d_1d_2\mathbf{k}^2 \\
= \ & b_1c_2\mathbf{ij} + b_2c_1\mathbf{ji} + b_1d_2\mathbf{ik} + b_2d_1\mathbf{ki} + c_1d_2\mathbf{jk} + c_2d_1\mathbf{kj} \\
= \ & -(b_1b_2 + c_1c_2 + d_1d_2) + (c_1d_2 - c_2d_1)\mathbf{i} + (b_2d_1 - b_1d_2)\mathbf{j} + (b_1c_2 - b_2c_1)\mathbf{k} \\
= \ & -\mathrm{Ve}(q_1) \cdot \mathrm{Ve}(q_2) + \mathrm{Ve}(q_1) \times \mathrm{Ve}(q_2).
\end{aligned}$$

This fact deserves to be featured:

Proposition 31.1 *If* $\mathrm{Re}(q_1) = \mathrm{Re}(q_2) = 0$ *then*

$$q_1q_2 = -q_1 \cdot q_2 + q_1 \times q_2 \tag{8.1}$$

where $(\cdot)$ *and* $(\times)$ *denote the dot and cross product, respectively.*

Thus the quaternion product combines the two types of products on vectors into a single operation!

Let us examine the real and vector parts of the quaternion product. Note that

$$\begin{aligned}
pq &= (\mathrm{Re}(p) + \mathrm{Ve}(p))(\mathrm{Re}(q) + \mathrm{Ve}(q)) \\
&= \mathrm{Re}(p)\mathrm{Re}(q) + \mathrm{Re}(p)\mathrm{Ve}(q) + \mathrm{Re}(q)\mathrm{Ve}(p) + \mathrm{Ve}(p)\mathrm{Ve}(q).
\end{aligned}$$

Of these four terms, the first is real, the next two have real part zero, and from (8.1) the last term is $-\mathrm{Ve}(p) \cdot \mathrm{Ve}(q) + \mathrm{Ve}(p) \times \mathrm{Ve}(q)$. So

$$pq = \mathrm{Re}(p)\mathrm{Re}(q) - \mathrm{Ve}(p) \cdot \mathrm{Ve}(q) + \mathrm{Re}(p)\mathrm{Ve}(q) + \mathrm{Re}(q)\mathrm{Ve}(p) + \mathrm{Ve}(p) \times \mathrm{Ve}(q)$$

and we conclude that

$$\mathrm{Re}(pq) = \mathrm{Re}(p)\mathrm{Re}(q) - \mathrm{Ve}(p) \cdot \mathrm{Ve}(q) \tag{8.2}$$

and

$$\mathrm{Ve}(pq) = \mathrm{Re}(p)\mathrm{Ve}(q) + \mathrm{Re}(q)\mathrm{Ve}(p) + \mathrm{Ve}(p) \times \mathrm{Ve}(q). \tag{8.3}$$

Because the right side of (8.2) is the same if p and q are switched we obtain the useful

$$\mathrm{Re}(pq) = \mathrm{Re}(qp) \tag{8.4}$$

for all quaternions p and q.

Note that the real quaternion 0 is the additive identity and the real quaternion 1 is the multiplicative identity.

If $q = a+b\mathbf{i}+c\mathbf{j}+d\mathbf{k}$ then $\overline{q} = a-b\mathbf{i}-c\mathbf{j}-d\mathbf{k}$ is the *conjugate* of q. The *norm* or *absolute value* of q is is $|q| = \sqrt{a^2+b^2+c^2+d^2}$. Note that $q\overline{q} = \overline{q}q = |q|^2$. A *unit* quaternion is a quaternion whose absolute value is 1. The conjugate of a quaternion is useful in the same way it is with complex numbers: it allows us to calculate a multiplicative *inverse* for a nonzero quaternion. If $q \neq 0$, then let $q^{-1} = \frac{1}{|q|^2}\overline{q}$. Then we easily verify that $qq^{-1} = q^{-1}q = 1$.

It is easy to check that $\overline{p+q} = \overline{p}+\overline{q}$ for any quaternions p and q, and if r is real, $\overline{rp} = r\overline{p}$. (See Exercise 2.) If q is real, then $\overline{q} = q$. If q is a pure quaternion, $\overline{q} = -q$. An important property relating the conjugate to the product is that

$$\overline{pq} = (\overline{q})(\overline{p}). \tag{8.5}$$

To see why this is true, simply check that the real and vector parts of both sides are the same, using (8.2) and (8.3). We check (8.2) and leave (8.3) to Exercise 8. We have $\mathrm{Re}((\overline{q})(\overline{p})) = \mathrm{Re}(\overline{q})\mathrm{Re}(\overline{p}) - \mathrm{Ve}(\overline{q}) \cdot \mathrm{Ve}(\overline{p}) = \mathrm{Re}(p)\mathrm{Re}(q) - (-\mathrm{Ve}(q)) \cdot (-\mathrm{Ve}(p)) = \mathrm{Re}(p)\mathrm{Re}(q) - \mathrm{Ve}(p) \cdot \mathrm{Ve}(q) = \mathrm{Re}(pq) = \mathrm{Re}(\overline{pq})$, as desired.

We now consider the products of three quaternions. The quaternion product is associative: $p(qr) = (pq)r$; we leave this as an exercise to the reader (see Exercise 5). Property (8.5) and the associative property lead immediately to the property

$$|pq| = |p||q|. \tag{8.6}$$

To see this, note that $|pq|^2 = (pq)(\overline{pq}) = pq(\overline{q})(\overline{p}) = p(q\overline{q})\overline{p} = p|q|^2\overline{p}$. Since $|q|^2$ is real, this equals $|q|^2 p\overline{p} = |q|^2|p|^2 = |p|^2|q|^2$. Thus $|pq|^2 = |p|^2|q|^2$ and taking the square root gives us (8.6).

We next see how the associativity of the quaternion product can be used to discover the two key identities for triple products of vectors. We begin with that fact that for quaternions p, q, and r, $(pq)r = p(qr)$. We take real parts of both sides of the equation, using (8.2) to find

$$\mathrm{Re}(pq)\mathrm{Re}(r) - \mathrm{Ve}(pq) \cdot \mathrm{Ve}(r) = \mathrm{Re}(p)\mathrm{Re}(qr) - \mathrm{Ve}(p) \cdot \mathrm{Ve}(qr),$$

and we may write that these both equal $\mathrm{Re}(pqr)$. If p, q and r are vectors then the real part of each is zero, so $\mathrm{Ve}(pq) \cdot \mathrm{Ve}(r) = \mathrm{Ve}(p) \cdot \mathrm{Ve}(qr)$; but using (8.3), we find $p \times q \cdot r = p \cdot q \times r$, i.e., $r \cdot p \times q = p \cdot q \times r$. Permuting the letters, we find that for vectors (pure quaternions) p, q, r,

$$-\mathrm{Re}(pqr) = p \cdot q \times r = q \cdot r \times p = r \cdot p \times q. \tag{8.7}$$

One may show that the absolute value of this quantity is the volume in space of the parallelepiped formed by the vectors p, q, and r. (See Exercise 11.)

Next we look at the vector part of the associative law, using (8.3). We find

$$\begin{aligned} & \mathrm{Ve}(pq)\mathrm{Re}(r) + \mathrm{Re}(pq)\mathrm{Ve}(r) + \mathrm{Ve}(pq) \times \mathrm{Ve}(r) \\ = \; & \mathrm{Ve}(p)\mathrm{Re}(qr) + \mathrm{Ve}(qr)\mathrm{Re}(p) + \mathrm{Ve}(p) \times \mathrm{Ve}(qr); \end{aligned}$$

for vectors p, q, and r, this reduces to

$$-(p \cdot q)r + (p \times q) \times r = -(q \cdot r)p + p \times (q \times r). \tag{8.8}$$

We obtain two other identities by permuting the letters and using basic properties of the dot and cross product:

$$-(p \cdot r)q + (p \times r) \times q = -(q \cdot r)p - p \times (q \times r) \tag{8.9}$$

and

$$-(q \cdot p)r - (p \times q) \times r = -(p \cdot r)q - (p \times r) \times q. \tag{8.10}$$

By taking (8.8)-(8.9)+(8.10), we find $2p \times (q \times r) = -2(q \cdot p)r + 2(p \cdot r)q$, or

$$p \times (q \times r) = (p \cdot r)q - (q \cdot p)r. \tag{8.11}$$

It should now be clear how the quaternion product combines the dot and cross product together, and that many properties relating the dot and cross product can be seen as consequences of general properties of the quaternion product.

We conclude this section by illustrating how to use vector methods to change coordinates in spherical astronomy. Let us consider the example of changing coordinates between declination-hour angle and altitude-azimuth coordinates as done in Example 1 of §22. Referring to Figure 8.7, we assume $\mathbf{i}$ and $\mathbf{j}$ point to the north (N) and west (W) horizon points on the celestial sphere, respectively, and assume $\mathbf{k}$ points to the zenith. Then we have spherical polar coordinates of the type set up in §8 where Z is the pole (and the distance to Z is $90° - a$). At the zenith the zenith distance $90° - a$ is zero, and the azimuth is $0°$ and $90°$ at the north and west points, respectively. Suppose $x\mathbf{i} + y\mathbf{j} + z\mathbf{k}$ is a unit vector pointing to a point X on the celestial sphere. Then according to §8, (2.3),

$$\begin{aligned} (x, y, z) &= (\sin(90° - a)\cos(A), \sin(90° - a)\sin(A), \cos(90° - a)) \\ &= (\cos(a)\cos(A), \cos(a)\sin(A), \sin(a)). \end{aligned}$$

We similarly introduce three vectors $\mathbf{e}_1, \mathbf{e}_2$ and $\mathbf{e}_3$ such that $\mathbf{e}_3$ points to the north celestial pole P, $\mathbf{e}_2 = \mathbf{j}$ and $\mathbf{e}_1$ points to the point M on the celestial equator where $h = 0$. Then we have a spherical polar coordinate system where P is the pole (with pole distance $90° - \delta$ and the hour angle h is 0 and $90°$ at the indicated points M and W, respectively.) The vectors $\mathbf{e}_1, \mathbf{e}_2$ and $\mathbf{e}_3$ are

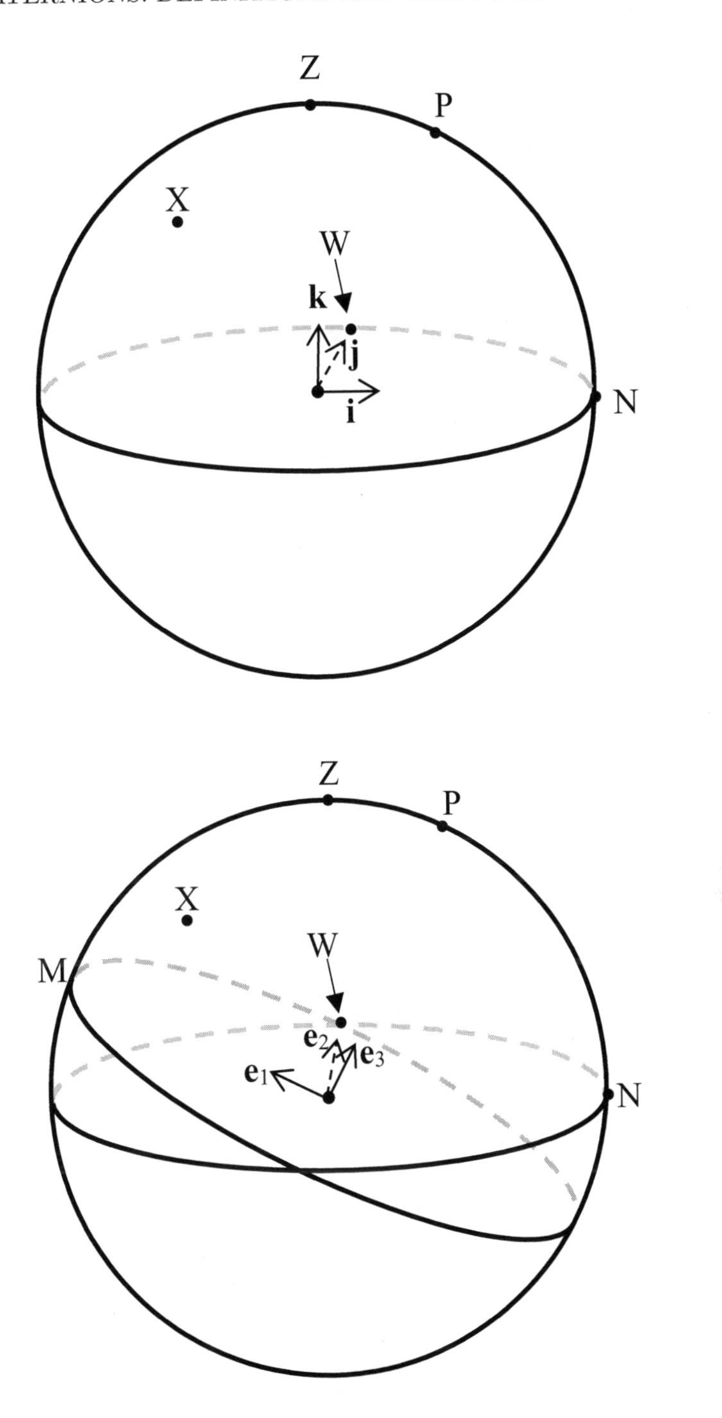

Figure 8.7: Changing coordinates between altitude-azimuth coordinates and declination-hour angle coordinates.

pairwise perpendicular. Suppose $\tilde{x}\mathbf{e}_1 + \tilde{y}\mathbf{e}_2 + \tilde{z}\mathbf{e}_3$ is a unit vector pointing to the above point X on the celestial sphere. Again applying §8, (2.3),

$$\begin{aligned}(\tilde{x}, \tilde{y}, \tilde{z}) &= (\sin(90^\circ - \delta)\cos(h), \sin(90^\circ - \delta)\sin(h), \cos(90^\circ - \delta)) \\ &= (\cos(\delta)\cos(h), \cos(\delta)\sin(h), \sin(\delta))\end{aligned}$$

So we have $x\mathbf{i} + y\mathbf{j} + z\mathbf{k} = \tilde{x}\mathbf{e}_1 + \tilde{y}\mathbf{e}_2 + \tilde{z}\mathbf{e}_3$, or

$$\begin{aligned}&\cos(\delta)\cos(h)\mathbf{e}_1 + \cos(\delta)\sin(h)\mathbf{e}_2 + \sin(\delta)\mathbf{e}_3 \\ =\ &\cos(a)\cos(A)\mathbf{i} + \cos(a)\sin(A)\mathbf{j} + \sin(a)\mathbf{k}.\end{aligned} \tag{8.12}$$

Considering Figure 8.7, we note that the angle between Z and P is $90^\circ - \phi$, where ϕ is the latitude of the observer. So $\mathbf{e}_1 \cdot \mathbf{i} = -\cos(90^\circ - \phi) = -\sin(\phi)$, $\mathbf{e}_1 \cdot \mathbf{j} = 0$, $\mathbf{e}_1 \cdot \mathbf{k} = \cos(\phi)$, $\mathbf{e}_2 \cdot \mathbf{i} = \mathbf{e}_2 \cdot \mathbf{k} = 0$, $\mathbf{e}_2 \cdot \mathbf{j} = 1$, $\mathbf{e}_3 \cdot \mathbf{i} = \cos(\phi)$, $\mathbf{e}_3 \cdot \mathbf{j} = 0$ and $\mathbf{e}_3 \cdot \mathbf{k} = \sin(\phi)$. If we take the dot product of both sides of (8.12) with $\mathbf{e}_1$, we get $\cos(\delta)\cos(h) + 0 + 0 = -\cos(a)\cos(A)\sin(\phi) + \sin(a)\cos(\phi)$, which is (5.7) of §22. If we take the dot product of both sides of (8.12) with $\mathbf{e}_2$, we get $0 + \cos(\delta)\sin(h) + 0 = 0 + \cos(a)\sin(A) + 0$, which is (5.8) of §22. If we take the dot product of both sides of (8.12) with $\mathbf{e}_3$, we get $0 + 0 + \sin(\delta) = \cos(a)\cos(A)\cos(\phi) + 0 + \sin(a)\sin(\phi)$, which is (5.6) of §22. If we take the dot product of both sides of (8.12) with **i**, **j**, and **k** then we obtain the equations of §22, (5.9).

A similar treatment of the change of coordinates between equatorial and ecliptic coordinates is left to Exercise 12.

Exercises §31

1. If $p = a_1 + b_1\mathbf{i} + c_1\mathbf{j} + d_1\mathbf{k}$ and $q = a_2 + b_2\mathbf{i} + c_2\mathbf{j} + d_2\mathbf{k}$ show that

$$\begin{aligned}pq &= (a_1a_2 - b_1b_2 - c_1c_2 - d_1d_2) + (a_1b_2 + a_2b_1 + c_1d_2 - c_2d_1)\mathbf{i} \\ &+ (a_1c_2 - b_1d_2 + a_2c_1 + b_2d_1)\mathbf{j} + (a_1d_2 + b_1c_2 - b_2c_1 + a_2d_1)\mathbf{k}.\end{aligned}$$

2. Prove that if p and q are quaternions and r is real then $\overline{p+q} = \overline{p} + \overline{q}$ and $\overline{rp} = r\overline{p}$.

3. Show that the multiplicative inverse of a pure unit quaternion is the same as its negative.

4. If p and q are vectors (pure quaternions), p is perpendicular to q, and p has absolute value 1, then $pqp = q$.

5. Prove that the quaternions are associative as follows. First show that $q_1(q_2q_3) = (q_1q_2)q_3$ if each of q_1, q_2, q_3 is either 1, **i**, **j**, or **k**. Then extend this to all quaternions.

6. Show that if p, q, and r are quaternions,

$$\mathrm{Re}(pqr) = \mathrm{Re}(qrp) = \mathrm{Re}(rpq). \tag{8.13}$$

What about a similar statement for a product of n quaternions?

7. Show that three vectors p, q, and r lie in the same plane if and only if $\text{Re}(pqr) = 0$.

8. Complete the proof of (8.5) by showing that the vector part of each side is the same.

9. Show that for quaternions p, q, $\text{Re}(pq) + \text{Re}(p\overline{q}) = 2\text{Re}(p)\text{Re}(q)$ and $\text{Re}(p\overline{q}) - \text{Re}(pq) = 2\text{Ve}(p) \cdot \text{Ve}(q)$.

10. Show that $\text{Re}(pqp^{-1}) = \text{Re}(q)$ and $\text{Ve}(pqp^{-1}) = p\text{Ve}(q)p^{-1}$.

11. Argue that $|p \cdot q \times r|$ is the volume of the parallelepiped formed by p, q, and r.

12. Derive the change of variable equations §22, (5.10), (5.11), (5.12), (5.13), (5.14), and (5.15) for switching between equatorial and ecliptic coordinates by using vectors.

32 Application to the sphere

We recall the following fact from complex numbers: that if the complex number a is multiplied by the complex number $\cos(\theta) + \sin(\theta)\mathbf{i}$, to obtain the complex number b, then b is found geometrically by rotating a counterclockwise by θ in the complex plane. We may write $b = (\cos(\theta) + \sin(\theta)\mathbf{i})a$, or $ba^{-1} = \cos(\theta) + \sin(\theta)\mathbf{i}$.

It turns out that an analogous property holds for the quaternions, but concerning rotations in three-dimensional space. Recall the fact that $\mathbf{a}$ and $\mathbf{b}$ are vectors than $\mathbf{a} \times \mathbf{b} = -\mathbf{b} \times \mathbf{a}$. If $\mathbf{a}$ and $\mathbf{b}$ are noncollinear then $\mathbf{a} \times \mathbf{b}$ and $\mathbf{b} \times \mathbf{a}$ are two nonzero opposite vectors perpendicular to both $\mathbf{a}$ and $\mathbf{b}$. Suppose we view the plane of $\mathbf{a}$ and $\mathbf{b}$ from above $\mathbf{a} \times \mathbf{b}$. By the right-hand rule for cross products, the direction around the origin from $\mathbf{a}$ to $\mathbf{b}$ is counterclockwise. (See Figure 8.8.)

Theorem 32.1 *Suppose that* $\mathbf{a}$ *and* $\mathbf{b}$ *are nonzero vectors in space (i.e., pure quaternions) having the same norm and forming an angle of measure* θ. *Let* $\mathbf{e}$ *be* $(\mathbf{a} \times \mathbf{b})/|\mathbf{a} \times \mathbf{b}|$. *Then* $\mathbf{ba}^{-1} = \cos(\theta) + \sin(\theta)\mathbf{e}$. *If* $\mathbf{a}$ *and* $\mathbf{b}$ *are collinear vectors then the same equation holds for* $\mathbf{e}$ *arbitrary, and* $\theta = 0$ *if* $\mathbf{a} = \mathbf{b}$ *and* $\theta = \pi$ *if* $\mathbf{a} = -\mathbf{b}$.

Proof. (See Figure 8.9.) Suppose first that $\mathbf{a}$ and $\mathbf{b}$ are noncollinear and that $|\mathbf{a}| = |\mathbf{b}| = 1$. Then $\mathbf{a}^{-1} = \overline{\mathbf{a}}/|\mathbf{a}|^2 = -\mathbf{a}$ since $|\mathbf{a}| = 1$ and $\mathbf{a}$ is a vector. Then $\mathbf{ba}^{-1} = \mathbf{b}(-\mathbf{a}) = -\mathbf{ba} = -(-\mathbf{b} \cdot \mathbf{a} + \mathbf{b} \times \mathbf{a}) = \mathbf{a} \cdot \mathbf{b} + \mathbf{a} \times \mathbf{b}$. By Theorem 4.1 and (1.36), this equals $|\mathbf{a}||\mathbf{b}|\cos(\theta) + |\mathbf{a}||\mathbf{b}|\sin(\theta)\mathbf{e}$, or $\cos(\theta) + \sin(\theta)\mathbf{e}$, as desired. The case where $|\mathbf{a}| = |\mathbf{b}|$ is different from 1 is easily derived from the case where $|\mathbf{a}| = |\mathbf{b}| = 1$ and is left to the exercises.

In the degenerate case where $\mathbf{a} = \mathbf{b}$ (and $\theta = 0$), the equation reads $1 = 1$. In the case $\mathbf{a} = -\mathbf{b}$ (and $\theta = \pi$), the equation reads $-1 = -1$. ♢

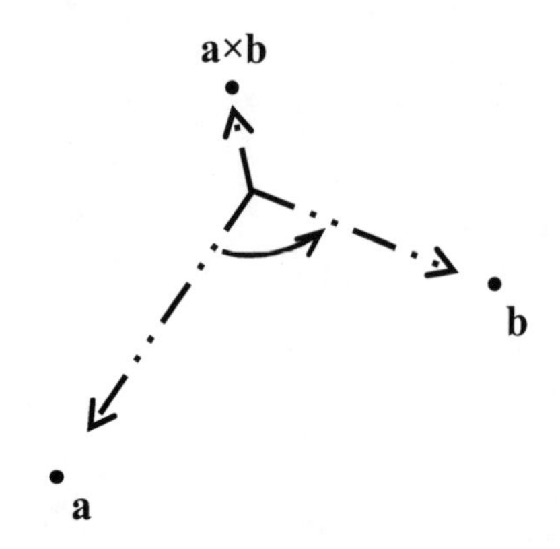

Figure 8.8: The view of the plane of $\mathbf{a}, \mathbf{b}$ from above $\mathbf{a} \times \mathbf{b}$.

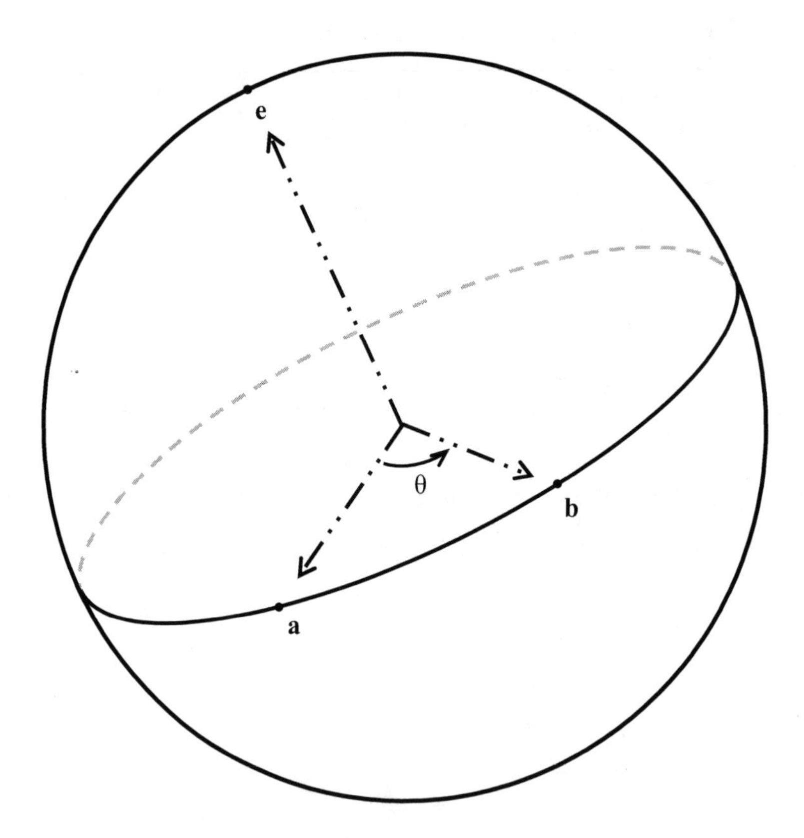

Figure 8.9: $\mathbf{b}\mathbf{a}^{-1} = \cos(\theta) + \sin(\theta)\mathbf{e}$.

Theorem 32.1 has an interesting implication for our understanding of the sphere. Suppose that we have a sphere of radius 1, and $\mathbf{a}$ and $\mathbf{b}$ are vectors associated with two non-antipodal points. Let Γ be the great circle passing through $\mathbf{a}$ and $\mathbf{b}$, and let $\mathbf{e}$ be as defined in Theorem 32.1. Then $\mathbf{e}$ is a unit vector perpendicular to both $\mathbf{a}$ and $\mathbf{b}$. That is, $\mathbf{e}$ is a pole of Γ.

The theorem then says that $\mathbf{ba}^{-1} = \cos(\theta) + \sin(\theta)\mathbf{e}$. But it says even more: it says that if $\mathbf{v}$ and $\mathbf{w}$ are *any* two position vectors of points on Γ separated by an angle of θ, where the angle θ is traced from $\mathbf{v}$ to $\mathbf{w}$ in the same direction on the circle as the angle θ from $\mathbf{a}$ to $\mathbf{b}$, then we also have $\mathbf{wv}^{-1} = \cos(\theta) + \sin(\theta)\mathbf{e}$. The reason is that if the angle is traced in the same direction for both pairs, then $\mathbf{v} \times \mathbf{w} = \mathbf{a} \times \mathbf{b}$, and both point in the direction of $\mathbf{e}$, using the right-hand rule for cross product.

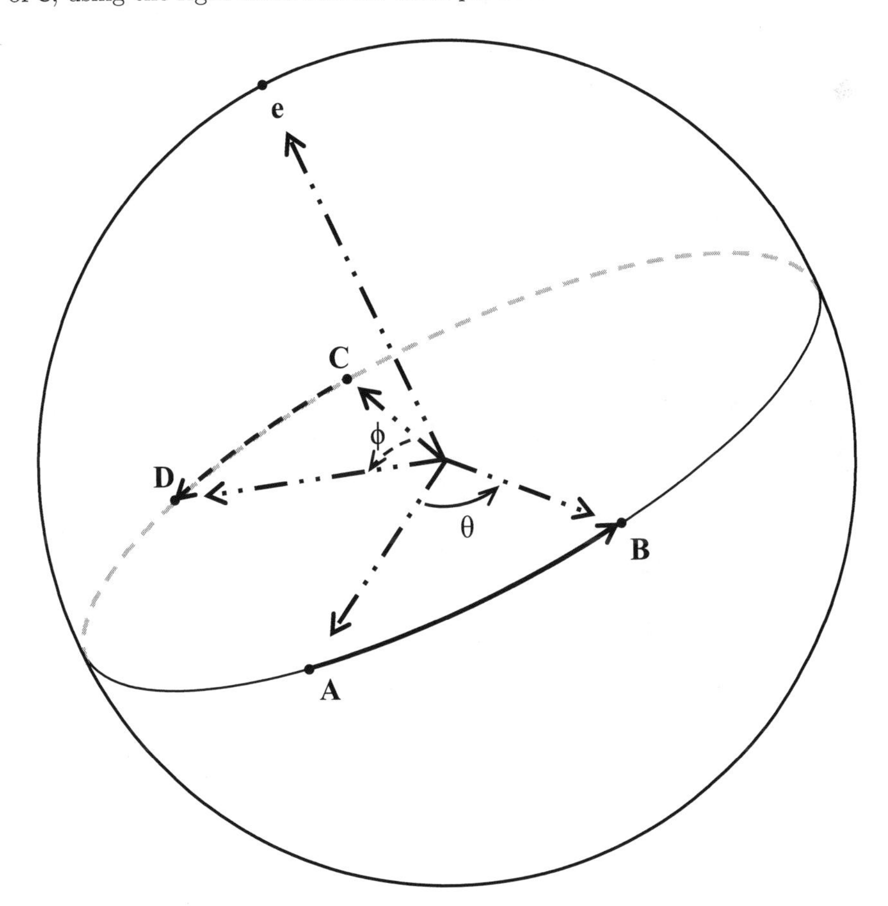

Figure 8.10: $\vec{AB} = \mathbf{BA}^{-1} = \cos(\theta) + \sin(\theta)\mathbf{e}$, $\vec{CD} = \mathbf{DC}^{-1} = \cos(\phi) + \sin(\phi)\mathbf{e}$.

If A and B are non-antipodal points, it then makes sense to define a directed great circle arc $\overrightarrow{AB}$ as a great circle arc $\overset{\frown}{AB}$ along with a designated initial and terminal point. (See Figure 8.10.) Let $\mathbf{A}$ and $\mathbf{B}$ denote the vectors pointing from the origin to A and B, respectively. Then two directed great circle arcs $\overrightarrow{AB}$ and $\overrightarrow{CD}$ are said to point in the same (respectively, opposite) direction(s) if A, B, C, and D all lie on a single great circle and $\mathbf{A} \times \mathbf{B}$ and $\mathbf{C} \times \mathbf{D}$ point in the same (respectively, opposite) direction(s).

The theorem creates an intriguing correspondence between quaternions of norm 1 and great circles on the sphere: it associates every directed great circle arc with a unit quaternion. The quaternion is the same for two directed arcs with the same axis, angle, and direction. Given a unit vector $\mathbf{e}$ and angle θ, the quaternion $\cos(\theta) + \sin(\theta)\mathbf{e}$ represents any directed great circle arc of measure θ on the polar circle of $\mathbf{e}$ which points counterclockwise when viewed from above $\mathbf{e}$. (See Figure 8.11 for examples.) Thus if we view directed arcs on the sphere as the "vectors" of spherical geometry, then they may be represented by quaternions. In space a directed line segment may be moved around without changing its length and direction via parallel displacement and be associated with the same vector. On the sphere, a directed arc may be rotated about the same great circle (keeping its length and direction the same) and be associated with the same quaternion.

Let us now suppose we are given a unit quaternion $\mathbf{q} = a + b\mathbf{i} + c\mathbf{j} + d\mathbf{k}$ and determine $\mathbf{e}$ and θ. We must have $\cos(\theta) = a$ and $\sin(\theta)\mathbf{e} = b\mathbf{i} + c\mathbf{j} + d\mathbf{k}$. Since $\mathbf{e}$ is a unit vector, $|\sin(\theta)| = \sqrt{b^2 + c^2 + d^2}$. We then have $\mathbf{e} = \pm(b\mathbf{i} + c\mathbf{j} + d\mathbf{k})/\sqrt{b^2 + c^2 + d^2}$. Either choice of sign is satisfactory: if we let $\mathbf{e} = (b\mathbf{i}+c\mathbf{j}+d\mathbf{k})/\sqrt{b^2 + c^2 + d^2}$ then $\sin(\theta) = \sqrt{b^2 + c^2 + d^2}$ and we may determine a unique θ in $[0, \pi]$ such that $(\cos(\theta), \sin(\theta)) = (a, \sqrt{b^2 + c^2 + d^2})$. So we have a directed arc of measure θ which points counterclockwise when viewed from above $\mathbf{e}$. At the same time we have $-\mathbf{e} = -(b\mathbf{i} + c\mathbf{j} + d\mathbf{k})/\sqrt{b^2 + c^2 + d^2}$ and $\sin(-\theta) = -\sqrt{b^2 + c^2 + d^2}$ so we may view the same arc as having measure θ but pointing clockwise when viewed from above $-\mathbf{e}$ since $-\theta$ is negative.

As an example, consider the unit quaternion $(2 - 3\mathbf{i} + 3\mathbf{j} + 5\mathbf{k})/\sqrt{47}$. The norm of $(-3\mathbf{i}+3\mathbf{j}+5\mathbf{k})/\sqrt{47}$ is $\sqrt{\frac{43}{47}}$, so we may take $\mathbf{e} = (-3\mathbf{i}+3\mathbf{j}+5\mathbf{k})/\sqrt{43}$. Then $(\cos(\theta), \sin(\theta)) = (\frac{2}{\sqrt{47}}, \sqrt{\frac{43}{47}})$ so θ is approximately 1.27 radians (73°).

But an algebraic representation of directed great circle arcs on a sphere as quaternions would not be terribly useful unless the algebraic properties of the quaternions had some relationship to the geometric properties of the directed arcs. In space, we see this when we add vectors: algebraic addition of vectors corresponds to the traditional head-to-tail addition of vectors in the plane. Does a similar correspondence exist on the sphere?

Suppose that A, B, and C are three points on the sphere not lying on a single great circle, and let $\mathbf{A}$, $\mathbf{B}$, and $\mathbf{C}$ be the three vectors (pure quaternions) in space pointing from the origin to A, B, and C, respectively. We let $\overrightarrow{AB}$ denote the directed great circle arc from A to B, $\overrightarrow{BC}$ the directed great circle

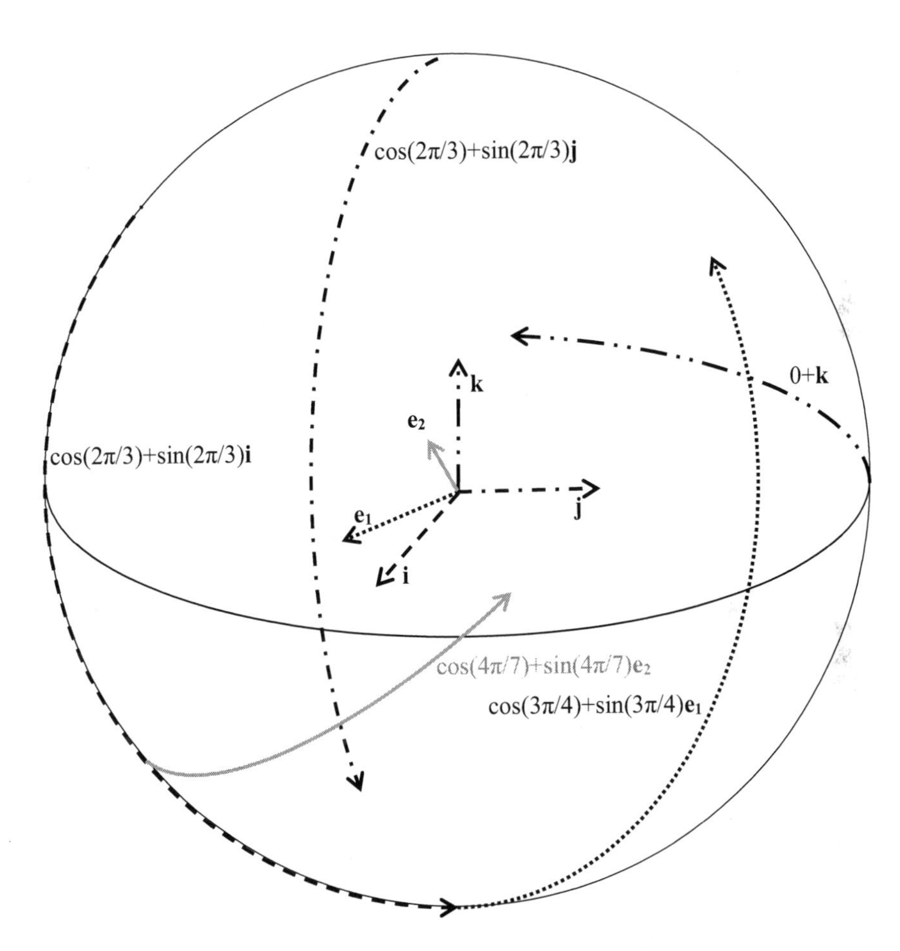

Figure 8.11: Examples of directed arcs associated with the quaternion $\cos(\theta)+\sin(\theta)\mathbf{e}$. Each vector $\mathbf{e}$ has the same dot pattern as its associated arc.

arc from B to C, and $\overrightarrow{AC}$ the directed great circle arc from A to C. (See Figure 8.12.) Then graphically, if we add "head to tail" $\overrightarrow{AB}$ to $\overrightarrow{BC}$, the resultant geometric arc by "combining" these two should take us from A to C. But note that the arcs $\overrightarrow{AB}$, $\overrightarrow{BC}$, and $\overrightarrow{AC}$ are associated with quaternions $\mathbf{BA}^{-1}$, $\mathbf{CB}^{-1}$, and $\mathbf{CA}^{-1}$ and $(\mathbf{CB}^{-1})(\mathbf{BA}^{-1})$ equals $\mathbf{CA}^{-1}$. Thus the graphical addition of the directed arcs corresponds exactly to multiplication of the associated quaternions!

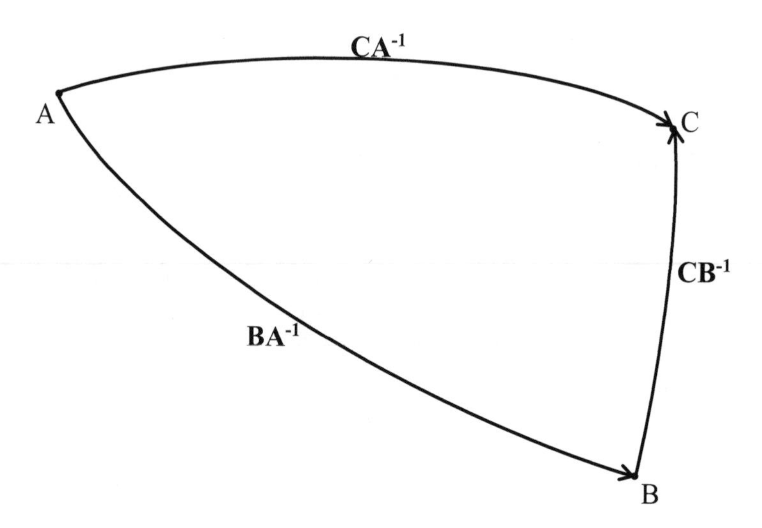

Figure 8.12: Graphical addition of directed arcs matches product of quaternions.

When vectors are added in space, they need not be situated so that the tail of the first vector is the same as the head of the second vector. The second vector may be moved first (in a manner that its direction and norm remain the same) so that its head coincides with the tail of the first; then they may be added graphically. Because the vector maintains the same algebraic representation as it is moved, the process of adding the algebraic representations of the vectors makes sense. The "moved" vector is viewed as being equivalent to the original.

Similarly, suppose that we are given two directed great circle arcs $\overrightarrow{AB}$ and $\overrightarrow{CD}$ on the sphere. (See Figure 8.13.) We may add them algebraically as follows. The great circles $\bigcirc AB$ and $\bigcirc CD$ meet in two antipodal points X and Y. We rotate $\overrightarrow{AB}$ on $\bigcirc AB$ until B coincides with X (and A becomes E.) Similarly we rotate $\overrightarrow{CD}$ on $\bigcirc CD$ until C coincides with X (and D becomes F). Then we may add the rotated directed arcs head to tail to obtain $\overrightarrow{EF}$. Furthermore, the arcs maintain the same algebraic representation as they are rotated, so the product of the quaternions representing them is the same as the product of the quaternions associated with the original vectors. A rotated great circle arc is viewed as the being equivalent to the original.

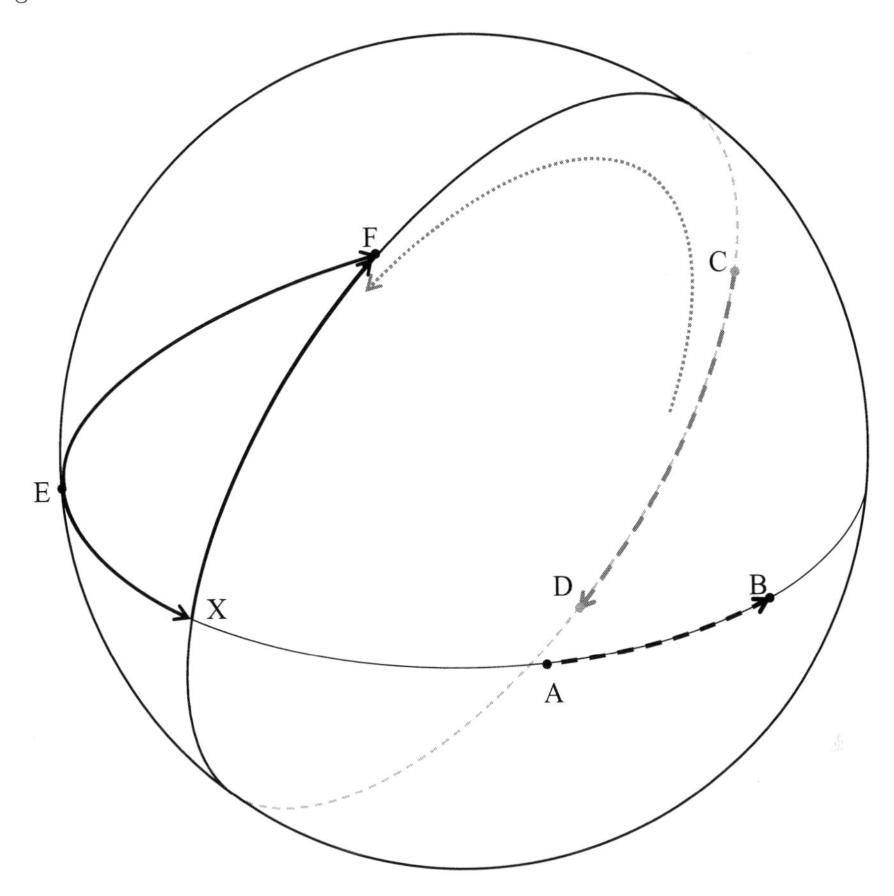

Figure 8.13: Adding $\overrightarrow{AB}$ and $\overrightarrow{CD}$ head to tail to obtain $\overrightarrow{EF}$.

We note that the quaternion 1 corresponds to any arc of measure 0 — that is, a point. The quaternion -1 corresponds to any arc of measure π — that is, a great semicircle. If the unit quaternion $\mathbf{p} = \cos(\theta) + \sin(\theta)\mathbf{e}$ corresponds to the directed arc $\overrightarrow{AB}$, then $\mathbf{p}^{-1} = \overline{\mathbf{p}} = \cos(\theta) + \sin(\theta)(-\mathbf{e})$ corresponds to an arc with the same measure θ but with the opposite orientation; this is the arc $\overrightarrow{BA}$. (See Exercise 7.)

We now use these ideas to prove a geometric theorem on the sphere.

Proposition 32.2 *Let $\triangle^s ABC$ be a spherical triangle on a sphere of radius r and let $\mathbf{A}$, $\mathbf{B}$, and $\mathbf{C}$ be the position vectors (from the center of the sphere) of its vertices. Then the three medians of the triangle meet at the point with position vector $r(\mathbf{A}+\mathbf{B}+\mathbf{C})/|\mathbf{A}+\mathbf{B}+\mathbf{C}|$.*

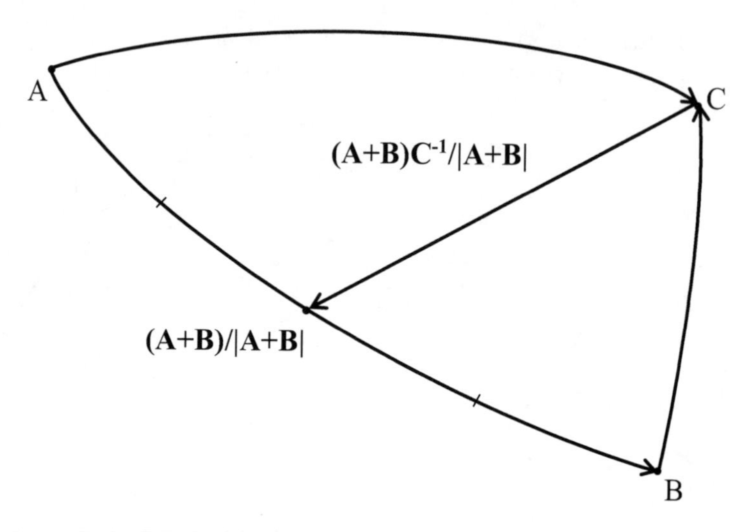

Figure 8.14: Directed median arc from C to $\widehat{AB}$.

Proof. It will suffice to prove this for $r = 1$; dilation will prove the cases when $r \neq 1$. So assume $r = 1$. We prove that the three medians of a spherical triangle all pass through a single point. First, we note that the midpoint of $\widehat{AB}$ is $(\mathbf{A}+\mathbf{B})/|\mathbf{A}+\mathbf{B}|$. (See Exercise 4.)

Next, the quaternion associated with the arc pointing from C to the midpoint of $\widehat{AB}$ is

$$M_C \equiv (\mathbf{A}+\mathbf{B})\overline{\mathbf{C}}/|\mathbf{A}+\mathbf{B}|,$$

recalling that $\mathbf{C}^{-1} = \overline{\mathbf{C}} = -\mathbf{C}$. (See Figure 8.14.) The other two medians have associated quaternions:

$$\begin{aligned} M_A &\equiv (\mathbf{B}+\mathbf{C})\overline{\mathbf{A}}/|\mathbf{B}+\mathbf{C}| \\ M_B &\equiv (\mathbf{A}+\mathbf{C})\overline{\mathbf{B}}/|\mathbf{A}+\mathbf{C}| \end{aligned}$$

The vector part of each median is a pole of that median. Next we calculate the quaternion product

$$\begin{aligned} &(\mathbf{A}+\mathbf{B}+\mathbf{C})(M_C) \\ = \;& \frac{(\mathbf{A}+\mathbf{B}+\mathbf{C})(\mathbf{A}+\mathbf{B})\overline{\mathbf{C}}}{|\mathbf{A}+\mathbf{B}|} \\ = \;& \frac{((\mathbf{A}+\mathbf{B})^2+\mathbf{C}(\mathbf{A}+\mathbf{B}))\overline{\mathbf{C}}}{|\mathbf{A}+\mathbf{B}|}. \end{aligned}$$

Now the square of a vector (pure quaternion) is the negative square of its norm, so is real, so $Re(\mathbf{A}+\mathbf{B})^2\overline{\mathbf{C}} = 0$. Thus

$$Re((\mathbf{A}+\mathbf{B}+\mathbf{C})(M_C)) \tag{8.14}$$

$$= \ Re\frac{\mathbf{C}(\mathbf{A}+\mathbf{B})\overline{\mathbf{C}}}{|\mathbf{A}+\mathbf{B}|} \tag{8.15}$$

$$= \ Re\frac{(\mathbf{A}+\mathbf{B})(\overline{\mathbf{C}})(\mathbf{C})}{|\mathbf{A}+\mathbf{B}|} \tag{8.16}$$

$$= \ Re\frac{(\mathbf{A}+\mathbf{B})|\mathbf{C}|^2}{|\mathbf{A}+\mathbf{B}|} \tag{8.17}$$

$$= \ 0, \tag{8.18}$$

where (8.16) holds by (8.4). Thus

$$\begin{aligned} 0 &= Re((\mathbf{A}+\mathbf{B}+\mathbf{C})(M_C)) \\ &= -(\mathbf{A}+\mathbf{B}+\mathbf{C})\cdot \mathrm{V}e(M_C). \end{aligned}$$

Thus $(\mathbf{A}+\mathbf{B}+\mathbf{C})$ is perpendicular to $\mathrm{V}e(M_C)$, the pole of the median from C to $\widehat{AB}$. So $(\mathbf{A}+\mathbf{B}+\mathbf{C})/|\mathbf{A}+\mathbf{B}+\mathbf{C}|$ lies on the great circle containing the median from C to $\widehat{AB}$. Similarly $(\mathbf{A}+\mathbf{B}+\mathbf{C})/|\mathbf{A}+\mathbf{B}+\mathbf{C}|$ lies on the great circles containing the other two medians. So the great circles containing the medians all pass through the same point. (In fact they pass through a pair of antipodal points.) One of these must be in the interior of the triangle where the median arcs are. ♢

The following proposition can be useful.

Proposition 32.3 *Three vectors* $\mathbf{A}$, $\mathbf{B}$, *and* $\mathbf{C}$ *on a sphere lie on a single great circle if and only if* $\mathrm{Re}(\mathbf{ABC}) = 0$.

The proof is Exercise 2. This allows us to write an "equation of a great circle" through two non-antipodal points A, B: it is $\mathrm{Re}(\mathbf{XAB}) = 0$. That is, X is on $\bigcirc AB$ if and only if $\mathrm{Re}(\mathbf{XAB}) = 0$.

We conclude with a quaternion proof of the existence of the *pentagramma mirificum* introduced in §19. Recall that to construct the pentagram, we begin with a $\triangle^s ABC$ which has acute sides and a right angle at C. (See Figure 8.15.) Let the polar triangle of $\triangle^s ABC$ be $\triangle^s A'B'C'$. We consider rays $\overrightarrow{AB}$ and choose point E on it a quarter circle from A. Similarly, on $\overrightarrow{CB}$ choose the point a quarter circle from C; this must be pole B' of $\bigcirc AC$. Now we can see $\triangle^s BEB'$ has a right angle at E: $\widehat{BE}$ has pole C' and $\widehat{B'E}$ has pole A. Since C' is perpendicular to A, $\widehat{BE}$ is perpendicular to $\widehat{B'E}$ at E. Then as seen earlier, $\triangle^s BB'E$ also has acute sides, and is called the complemental triangle of $\triangle^s ABC$. We then obtain the complemental triangle of $\triangle^s BB'E$ and repeat the process five times to obtain triangle $\triangle^s KML$. We successively obtain right angles at F, H, J, and L.

Proposition 32.4 *We have* $\triangle^s KML = \triangle^s ABC$.

Proof. We think of $\mathbf{A}$, $\mathbf{B}$, $\mathbf{C}$, $\mathbf{A}'$, $\mathbf{B}'$, and $\mathbf{C}'$ as being pure quaternions that are the position vectors of the points A, B, C, A', B', and C'. Then

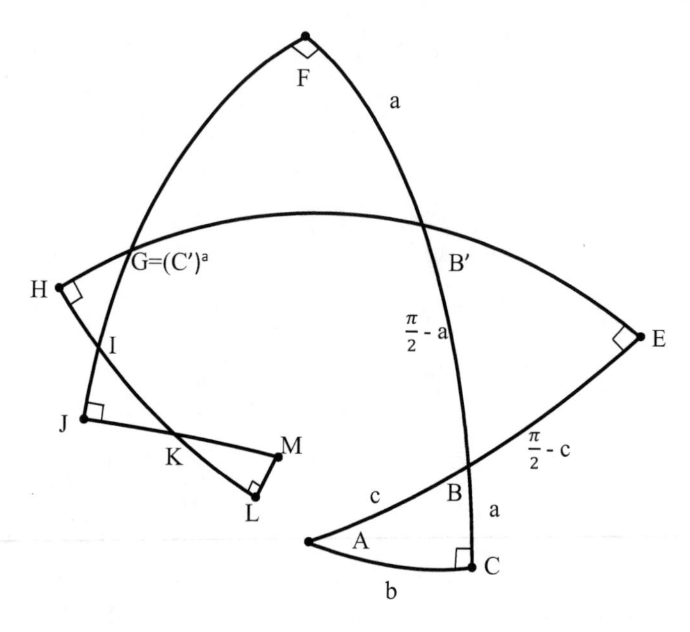

Figure 8.15: The *pentagramma mirificum*: $\triangle^s KML = \triangle^s ABC$.

$\mathbf{BC}^{-1} = \cos(\theta) + \sin(\theta)\mathbf{A}'$ where we assume (as the pictures indicates) that $\triangle^s ABC$ is oriented so that $\mathbf{A}' = \mathbf{C} \times \mathbf{B}/|\mathbf{C} \times \mathbf{B}|$ and thus $\theta > 0$. (The case where $\mathbf{A}' = -\mathbf{C} \times \mathbf{B}/|\mathbf{C} \times \mathbf{B}|$ so $\theta < 0$ is similar.) By Theorem 32.1,

$$\mathbf{B}'\mathbf{C}^{-1} = \cos(\frac{\pi}{2}) + \sin(\frac{\pi}{2})\mathbf{A}' = \mathbf{A}',$$

so $\mathbf{B}' = \mathbf{A}'\mathbf{C}$.

Similarly (see Exercise 3) C' is a pole of $\overset{\hookrightarrow}{BA}$ such that

$$\mathbf{C}' = \mathbf{B} \times \mathbf{A}/|\mathbf{B} \times \mathbf{A}| = \mathbf{E} \times \mathbf{A}/|\mathbf{E} \times \mathbf{A}|$$

so $\mathbf{C}' = \mathbf{AE}^{-1}$ and $\mathbf{E} = (\mathbf{C}')^{-1}\mathbf{A} = -\mathbf{C}'\mathbf{A}$ since the inverse of any pure unit quaternion is its negative.

The remainder of the proof is an application of Theorem 32.1 this way over and over again. Each time we identify a directed arc beginning at some point $\mathbf{a}$ and ending at $\mathbf{b}$ with pole $\mathbf{e}$ and with angle $\theta = \frac{\pi}{2}$ and conclude $\mathbf{ba}^{-1} = \mathbf{e}$. We list by table what these directed arcs are:

	a	b	e	$\mathbf{e} = \mathbf{b}\mathbf{a}^{-1}$	conclusion
1	$\mathbf{C}$	$\mathbf{B}'$	$\mathbf{A}'$	$\mathbf{A}' = \mathbf{B}'\mathbf{C}^{-1}$	$\mathbf{B}' = \mathbf{A}'\mathbf{C}$
2	$\mathbf{E}$	$\mathbf{A}$	$\mathbf{C}'$	$\mathbf{C}' = \mathbf{A}\mathbf{E}^{-1}$	$\mathbf{E} = -\mathbf{C}'\mathbf{A}$
3	$\mathbf{B}$	$\mathbf{F}$	$\mathbf{A}'$	$\mathbf{A}' = \mathbf{F}\mathbf{B}^{-1}$	$\mathbf{F} = \mathbf{A}'\mathbf{B}$
4	$\mathbf{E}$	$\mathbf{G}$	$\mathbf{A}$	$\mathbf{A} = \mathbf{G}\mathbf{E}^{-1}$	$\mathbf{G} = \mathbf{A}\mathbf{E}$
5	$\mathbf{B}'$	$\mathbf{H}$	$\mathbf{A}$	$\mathbf{A} = \mathbf{H}\mathbf{B}'^{-1}$	$\mathbf{H} = \mathbf{A}\mathbf{B}'$
6	$\mathbf{F}$	$\mathbf{I}$	$\mathbf{B}$	$\mathbf{B} = \mathbf{I}\mathbf{F}^{-1}$	$\mathbf{I} = \mathbf{B}\mathbf{F}$
7	$\mathbf{G}$	$\mathbf{J}$	$\mathbf{B}$	$\mathbf{B} = \mathbf{J}\mathbf{G}^{-1}$	$\mathbf{J} = \mathbf{B}\mathbf{G}$
8	$\mathbf{H}$	$\mathbf{K}$	$\mathbf{B}'$	$\mathbf{B}' = \mathbf{K}\mathbf{H}^{-1}$	$\mathbf{K} = \mathbf{B}'\mathbf{H}$
9	$\mathbf{I}$	$\mathbf{L}$	$\mathbf{B}'$	$\mathbf{B}' = \mathbf{L}\mathbf{I}^{-1}$	$\mathbf{L} = \mathbf{B}'\mathbf{I}$
10	$\mathbf{J}$	$\mathbf{M}$	$\mathbf{G}$	$\mathbf{G} = \mathbf{M}\mathbf{J}^{-1}$	$\mathbf{M} = \mathbf{G}\mathbf{J}$

Combining line 2 and line 4 we find that or $\mathbf{G} = -\mathbf{A}\mathbf{C}'\mathbf{A}$. Now $\mathbf{C}'$ is perpendicular to $\mathbf{A}$ by definition of pole and $\mathbf{A}$ and $\mathbf{C}'$ are both unit pure quaternions so $-\mathbf{A}\mathbf{C}'\mathbf{A}$ is $-\mathbf{C}'$ by §31, Exercise 4. So $\mathbf{G} = -\mathbf{C}'$, so points G and C' are antipodal.

Combining line 3 and line 6 we find

$$\mathbf{I} = \mathbf{B}\mathbf{A}'\mathbf{B} = \mathbf{A}' \tag{8.19}$$

again using §31, Exercise 4.

Combining line 5 and line 8 we find $\mathbf{K} = \mathbf{B}'\mathbf{A}\mathbf{B}' = \mathbf{A}$, again using §31, Exercise 4.

Combining line 7 and line 10 we find $\mathbf{M} = \mathbf{G}\mathbf{B}\mathbf{G} = \mathbf{B}$, again using §31, Exercise 4, as $\mathbf{G} = -\mathbf{C}'$.

Using line 9, (8.19) and line 1 we find $\mathbf{L} = \mathbf{B}'\mathbf{I} = \mathbf{B}'\mathbf{A}' = \mathbf{A}'\mathbf{C}\mathbf{A}' = \mathbf{C}$, again using §31, Exercise 4.

Thus the sixth triangle $\triangle^s KML$ is the same as $\triangle^s ABC$. ♢

Exercises §32

1. For each of the following unit quaternions, calculate the value of θ and $\mathbf{e}$ for which the quaternion represents a directed great circle arc of measure θ pointing counterclockwise about the origin of the plane perpendicular to $\mathbf{e}$ when viewed from above $\mathbf{e}$.

   (a) $(2 - 5\mathbf{i} + 3\mathbf{j} - \mathbf{k})/\sqrt{39}$
   (b) $(-4 - \mathbf{i} - 2\mathbf{j} + 3\mathbf{k})/\sqrt{30}$
   (c) $(1 + 2\mathbf{i} - 2\mathbf{j} + 4\mathbf{k})/5$
   (d) $(1 - 4\mathbf{k})/\sqrt{17}$

2. Prove Proposition 32.3. (Hint. See §31, Exercise 7.)

3. Suppose that $\mathbf{A}$, $\mathbf{B}$, and $\mathbf{C}$ are the vertices of a spherical triangle on the unit sphere and that the triangle is oriented so that $\mathbf{A}' = \mathbf{B}\times\mathbf{C}/|\mathbf{B}\times\mathbf{C}|$. Show that $\mathbf{B}' = \mathbf{C}\times\mathbf{A}/|\mathbf{C}\times\mathbf{A}|$ and $\mathbf{C}' = \mathbf{A}\times\mathbf{B}/|\mathbf{A}\times\mathbf{B}|$.

4. For two unit vectors **A** and **B**, show that the midpoint of the spherical arc on the unit sphere between their endpoints is $(\mathbf{A}+\mathbf{B})/|\mathbf{A}+\mathbf{B}|$.

5. Complete the proof of Theorem 32.1 by deriving the result for $|a| = |b|$ different from 1.

6. In defining the geometric sum of two directed spherical arcs, we could rotate them so that the tail of the first vector and the head of the second vector coincide at Y instead of X. Explain why the geometric sum of these directed arcs is the same.

7. What directed arc corresponds to the pure unit quaternion **e**? If the unit quaternion p corresponds to the directed arc $\overrightarrow{AB}$, what directed arc is $-p$?

8. Given a spherical $\triangle^s ABC$ on a sphere of radius r where **A**, **B**, and **C** are the position vectors to its vertices from the origin. Prove that the perpendicular bisectors of the sides of a spherical triangle meet at antipodal points with position vectors $\pm((\mathbf{BA}-\mathbf{AB})+(\mathbf{CB}-\mathbf{BC})+(\mathbf{AC}-\mathbf{CA}))r/|(\mathbf{BA}-\mathbf{AB})+(\mathbf{CB}-\mathbf{BC})+(\mathbf{AC}-\mathbf{CA})|$.

9. Let $\diamond ABCD$ be a spherical quadrilateral. Consider the two great circles passing through the midpoints of opposite sides, and the great circle passing through the midpoints of the diagonals. Prove that these three great circles all pass through the same pair of (antipodal) points.

10. Let A, B, C, and D be points on the sphere (pure quaternions) and let $\alpha = \overrightarrow{AB}$, $\beta = \overrightarrow{BA}$, $\gamma = \overrightarrow{CB}$, $\delta = \overrightarrow{DC}$, $\phi = \overrightarrow{CA}$, and $\psi = \overrightarrow{BD}$.

    (a) Prove that $\mathrm{Re}(\delta\beta) = \mathrm{Re}(\alpha\gamma)$. Conclude that similarly $\mathrm{Re}(\alpha\gamma^{-1}) = \mathrm{Re}(\phi\psi^{-1})$ and $\mathrm{Re}(\phi\psi) = \mathrm{Re}(\delta\beta^{-1})$.

    (b) Prove that $\mathrm{Re}(\phi\psi) + \mathrm{Re}(\alpha\gamma) = 2\mathrm{Re}(\beta)\mathrm{Re}(\delta)$. Similarly conclude that $\mathrm{Re}(\delta\beta) + \mathrm{Re}(\phi\psi^{-1}) = 2\mathrm{Re}(\alpha)\mathrm{Re}(\gamma)$ and $\mathrm{Re}(\alpha\gamma^{-1}) + \mathrm{Re}(\delta\beta^{-1}) = 2\mathrm{Re}(\phi)\mathrm{Re}(\psi)$.

    (c) Prove that

$$\begin{aligned}
\mathrm{Re}(\phi\psi) &= \mathrm{Re}(\phi)\mathrm{Re}(\psi) + \mathrm{Re}(\beta)\mathrm{Re}(\delta) - \mathrm{Re}(\alpha)\mathrm{Re}(\gamma) \\
\mathrm{Re}(\alpha\gamma) &= \mathrm{Re}(\alpha)\mathrm{Re}(\gamma) + \mathrm{Re}(\beta)\mathrm{Re}(\delta) - \mathrm{Re}(\phi)\mathrm{Re}(\psi) \\
\mathrm{Re}(\delta\beta) &= -\mathrm{Re}(\phi)\mathrm{Re}(\psi) + \mathrm{Re}(\beta)\mathrm{Re}(\delta) + \mathrm{Re}(\alpha)\mathrm{Re}(\gamma) \\
\mathrm{Re}(\phi\psi^{-1}) &= \mathrm{Re}(\phi)\mathrm{Re}(\psi) - \mathrm{Re}(\beta)\mathrm{Re}(\delta) + \mathrm{Re}(\alpha)\mathrm{Re}(\gamma) \\
\mathrm{Re}(\alpha\gamma^{-1}) &= \mathrm{Re}(\alpha)\mathrm{Re}(\gamma) - \mathrm{Re}(\beta)\mathrm{Re}(\delta) + \mathrm{Re}(\phi)\mathrm{Re}(\psi) \\
\mathrm{Re}(\delta\beta^{-1}) &= \mathrm{Re}(\phi)\mathrm{Re}(\psi) + \mathrm{Re}(\beta)\mathrm{Re}(\delta) - \mathrm{Re}(\alpha)\mathrm{Re}(\gamma)
\end{aligned}$$

(d) If $\diamond ABCD$ is a spherical quadrilateral, use Exercise 10c to show that the angle between $\bigcirc AC$ and $\bigcirc BD$ has cosine equal to

$$\pm(\cos(\alpha)\cos(\gamma) - \cos(\beta)\cos(\delta))\csc(\overset{\frown}{AC})\csc(\overset{\frown}{BD}),$$

the angle between $\bigcirc AB$ and $\bigcirc CD$ has cosine equal to

$$\pm(\cos(\alpha)\cos(\gamma) - \cos(\phi)\cos(\psi))\csc(\overset{\frown}{AB})\csc(\overset{\frown}{CD}),$$

and the angle between $\bigcirc AD$ and $\bigcirc BC$ has cosine equal to

$$\pm(\cos(\phi)\cos(\psi) - \cos(\beta)\cos(\delta))\csc(\overset{\frown}{AD})\csc(\overset{\frown}{BC}).$$

(See §16, Exercise 17.)

33 Triangles

In this section we show how to use the relationships between unit quaternions and directed spherical arcs to derive key formulas in spherical trigonometry.

A distinction regarding the orientation of triangles will be important. Let $\triangle^s ABC$ be a spherical triangle and $\triangle^s A'B'C'$ its polar triangle. Since the vector $\mathbf{A}'$ and $\mathbf{B}\times\mathbf{C}$ are both perpendicular to the plane containing $\bigcirc BC$, these two vectors point either in the same or opposite directions. It can be seen via the right-hand rule for cross product that the pair of vectors point in the same direction if the vertices A, B, C trace the perimeter of $\triangle^s ABC$ in the counterclockwise direction when viewed from above. This being the case, it will also be true that $\mathbf{B}'$ and $\mathbf{C}\times\mathbf{A}$ point in the same direction, and $\mathbf{C}'$ and $\mathbf{A}\times\mathbf{B}$ point in the same direction. The triangle $\triangle^s ABC$ will be said to have a *counterclockwise* (respectively, *clockwise*) labeling if A' and $\mathbf{B}\times\mathbf{C}$ point in the same (respectively, opposite) directions. (See Figures 8.16 and 8.17.) Equivalently, the triangle has a counterclockwise labeling if $\mathbf{A}$ and $\mathbf{B}\times\mathbf{C}$ point to the same side of the plane containing $\bigcirc BC$. This in turn occurs when $\mathbf{A}\cdot\mathbf{B}\times\mathbf{C}$ is positive because the angle between $\mathbf{A}$ and $\mathbf{B}\times\mathbf{C}$ is acute. Since $\mathbf{A}\cdot\mathbf{B}\times\mathbf{C} = \mathbf{B}\cdot\mathbf{C}\times\mathbf{A} = \mathbf{C}\cdot\mathbf{A}\times\mathbf{B}$, we see the same positivity condition to mean that the angle between $\mathbf{B}$ and $\mathbf{C}\times\mathbf{A}$ is acute, and that the angle between $\mathbf{C}$ and $\mathbf{A}\times\mathbf{B}$ is acute. For a triangle with a counterclockwise labeling,

$$\mathbf{A}' = \mathbf{B}\times\mathbf{C}/|\mathbf{B}\times\mathbf{C}|, \mathbf{B}' = \mathbf{C}\times\mathbf{A}/|\mathbf{C}\times\mathbf{A}|, \mathbf{C}' = \mathbf{A}\times\mathbf{B}/|\mathbf{A}\times\mathbf{B}| \quad (8.20)$$

Similarly, for a triangle with a clockwise labeling,

$$\mathbf{A}' = -\mathbf{B}\times\mathbf{C}/|\mathbf{B}\times\mathbf{C}|, \mathbf{B}' = -\mathbf{C}\times\mathbf{A}/|\mathbf{C}\times\mathbf{A}|, \mathbf{C}' = -\mathbf{A}\times\mathbf{B}/|\mathbf{A}\times\mathbf{B}|$$

Thus $\mathbf{A}\cdot\mathbf{A}'$, $\mathbf{B}\cdot\mathbf{B}'$ and $\mathbf{C}\cdot\mathbf{C}'$ all have the same sign; this sign is positive (negative) if $\triangle^s ABC$ has a counterclockwise (clockwise) labeling. Since

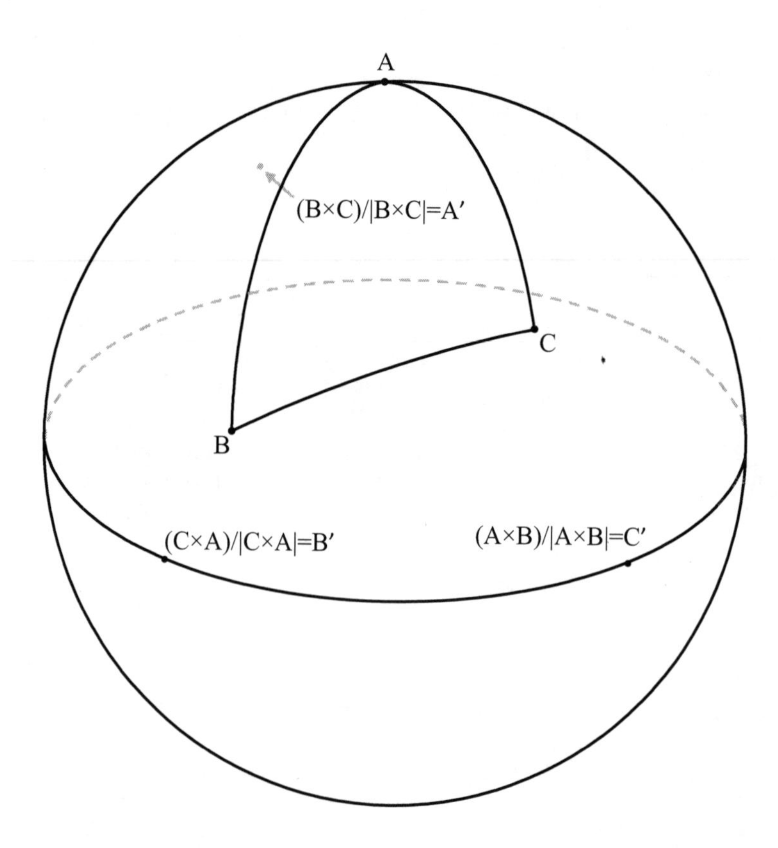

Figure 8.16: A triangle with counterclockwise labeling.

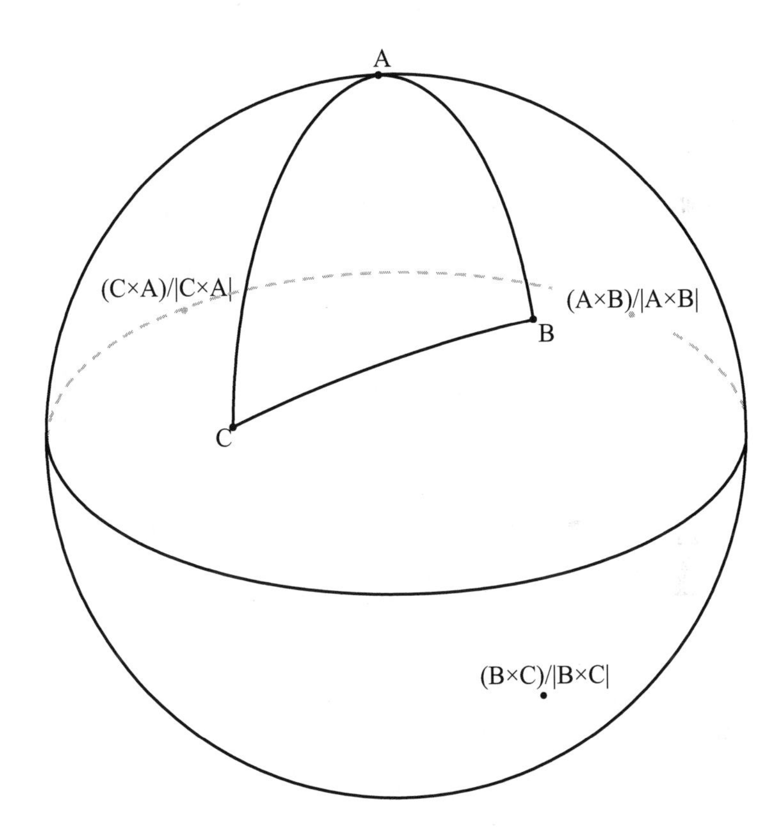

Figure 8.17: A triangle with clockwise labeling.

the polar triangle of $\triangle^s A'B'C'$ is $\triangle^s ABC$, we conclude that $\triangle^s ABC$ and $\triangle^s A'B'C'$ must have labelings with the same orientation (i.e., both are counterclockwise or both are clockwise.)

This situation makes our definition of polar triangle somewhat inconvenient with respect to vectors. Some authors (e.g., [Je1994]) define the polar triangle so that (8.20) always holds regardless of the labeling orientation of the triangle. In order to avoid this problem, we will instead generally assume for the rest of this chapter that $\triangle^s ABC$ comes with a counterclockwise labeling (so its polar triangle does also). This will mean that (8.20) holds. It will also mean that the following quantities are positive: $-\mathrm{Re}(\mathbf{ABC}) = \mathbf{A}\cdot\mathbf{B}\times\mathbf{C}$ and $-\mathrm{Re}(\mathbf{A'B'C'}) = \mathbf{A'}\cdot\mathbf{B'}\times\mathbf{C'}$. (For a clockwise triangle, they are both negative.) One other observation from (8.20): $|\mathbf{A}\times\mathbf{B}| = |\mathbf{A}||\mathbf{B}|\sin(\overset{\frown}{AB}) = \sin(c)$; similarly for other pairs. So we can write (8.20) as

$$\mathbf{A'} = \mathbf{B}\times\mathbf{C}/\sin(a), \mathbf{B'} = \mathbf{C}\times\mathbf{A}/\sin(b), \mathbf{C'} = \mathbf{A}\times\mathbf{B}/\sin(c) \tag{8.21}$$

Let $\triangle^s ABC$ be a spherical triangle with a counterclockwise labeling and let $\mathbf{A}$, $\mathbf{B}$, and $\mathbf{C}$ be the vectors pointing to the points A, B, and C, respectively, from the center of the sphere. Let $\triangle^s A'B'C'$ be the polar triangle of $\triangle^s ABC$ and let $\mathbf{A'}$, $\mathbf{B'}$, and $\mathbf{C'}$ be the vectors pointing from the center of the sphere to A', B', and C', respectively. (See Figure 8.16.) As observed earlier, $-\mathrm{Re}(\mathbf{ABC}) = \mathbf{A}\cdot\mathbf{B}\times\mathbf{C}$. Now

$$\begin{aligned} &-\mathrm{Re}(\mathbf{A'B'C'}) = \mathbf{A'}\cdot\mathbf{B'}\times\mathbf{C'} = \mathbf{A'}\cdot\mathbf{A}\sin(a') \\ = \;& \sin(\pi - A)\mathbf{A}\cdot\mathbf{A'} = \sin(A)\mathbf{A}\cdot\mathbf{B}\times\mathbf{C}/\sin(a) = -\frac{\sin(A)}{\sin(a)}\mathrm{Re}(\mathbf{ABC}). \end{aligned}$$

Thus $\frac{\sin(a)}{\sin(A)} = \frac{\mathbf{A}\cdot\mathbf{B}\times\mathbf{C}}{\mathbf{A'}\cdot\mathbf{B'}\times\mathbf{C'}}$; but the right side is the same if we permute A, B, C, so

$$\frac{\sin(a)}{\sin(A)} = \frac{\sin(b)}{\sin(B)} = \frac{\sin(c)}{\sin(C)} = \frac{\mathbf{A}\cdot\mathbf{B}\times\mathbf{C}}{\mathbf{A'}\cdot\mathbf{B'}\times\mathbf{C'}} = \frac{\mathrm{Re}(\mathbf{ABC})}{\mathrm{Re}(\mathbf{A'B'C'})} \tag{8.22}$$

and we obtain the spherical law of sines! (Note that if $\triangle^s ABC$ has a clockwise orientation both fractions on the right have negatives in the numerator and denominator; these cancel, and the statement holds.)

As observed earlier (see Figure 8.12), $\overrightarrow{AB}$ corresponds to the quaternion $\mathbf{BA}^{-1}$, $\overrightarrow{BC}$ corresponds to $\mathbf{CB}^{-1}$ and $\overrightarrow{AC}$ corresponds to $\mathbf{CA}^{-1}$. However, we have another way of writing these arcs. With $\overrightarrow{AB}$, note that

$$\mathbf{C'} = \mathbf{A}\times\mathbf{B}/|\mathbf{A}\times\mathbf{B}|$$

and the measure of $\overset{\frown}{AB}$ is c. Then

$$\overrightarrow{AB} = \cos(c) + \sin(c)\mathbf{C'}.$$

Continuing counterclockwise around the triangle:

$$\begin{aligned}\vec{BC} &= \cos(a) + \sin(a)\mathbf{A}' \\ \vec{CA} &= \cos(b) + \sin(b)\mathbf{B}'\end{aligned}$$

As noted earlier, $\vec{BA}$ is the conjugate of $\vec{AB}$, so $\vec{BA}= \cos(c) - \sin(c)\mathbf{C}'$. But $\vec{BA}= \mathbf{AB}^{-1}$, $\vec{CA}= \mathbf{AC}^{-1}$, and $\vec{BC}= \mathbf{CB}^{-1}$, so $\vec{BA}=\vec{CA}\vec{BC}$:

$$\begin{aligned}
& \cos(c) - \sin(c)\mathbf{C}' && (8.23)\\
= & (\cos(b) + \sin(b)\mathbf{B}')(\cos(a) + \sin(a)\mathbf{A}') && (8.24)\\
= & \cos(a)\cos(b) + \cos(b)\sin(a)\mathbf{A}' + \sin(b)\cos(a)\mathbf{B}' + \sin(a)\sin(b)\mathbf{B}'\mathbf{A}' \\
= & (\cos(a)\cos(b) - \sin(a)\sin(b)\mathbf{B}' \cdot \mathbf{A}') && (8.25)\\
+ & (\cos(b)\sin(a)\mathbf{A}' + \sin(b)\cos(a)\mathbf{B}' + \sin(a)\sin(b)\mathbf{B}' \times \mathbf{A}'),
\end{aligned}$$

where we used (8.1) and grouped the real terms first and vector terms second. Now $\mathbf{B}' \cdot \mathbf{A}' = |\mathbf{B}'||\mathbf{A}'|\cos(c')$, where $c' = m\ \overset{\frown}{A'B'}$. But since $\triangle^s A'B'C'$ is polar to $\triangle^s ABC$, $c' = \pi - C$, and so we obtain $\cos(c') = -\cos(C)$. Equating real parts of both sides of the equation (8.23)=(8.25) above, we obtain

$$\cos(c) = \cos(a)\cos(b) + \sin(a)\sin(b)\cos(C),$$

the spherical law of cosines (4.10)!

Next, we multiply both sides of the equation (8.23)=(8.24) on the right by $-\mathbf{A}'$ to obtain

$$\begin{aligned}
& -\cos(c)\mathbf{A}' + \sin(c)\mathbf{C}'\mathbf{A}' && (8.26)\\
= & (\cos(b) + \sin(b)\mathbf{B}')(-\cos(a)\mathbf{A}' + \sin(a)) && (8.27)\\
= & \cos(b)\sin(a) - \cos(b)\cos(a)\mathbf{A}' + \sin(b)\sin(a)\mathbf{B}' - \sin(b)\cos(a)\mathbf{B}'\mathbf{A}' \\
= & (\cos(b)\sin(a) + \sin(b)\cos(a)\mathbf{B}' \cdot \mathbf{A}') && (8.28)\\
+ & (-\cos(b)\cos(a)\mathbf{A}' + \sin(b)\sin(a)\mathbf{B}' - \sin(b)\cos(a)\mathbf{B}' \times \mathbf{A}'). && (8.29)
\end{aligned}$$

We follow the pattern above to find that

$$Re(\mathbf{C}'\mathbf{A}') = -\mathbf{C}' \cdot \mathbf{A}' = -\cos(\overset{\frown}{C'A'}) = -\cos(b') = -\cos(\pi - B) = \cos(B)$$

and similarly $\mathbf{B}' \cdot \mathbf{A}' = -\cos(C)$. Equating real parts of (8.26)=(8.28) we obtain $\sin(c)Re(\mathbf{C}'\mathbf{A}') = \cos(b)\sin(a) + \sin(b)\cos(a)\mathbf{B}' \cdot \mathbf{A}'$, or

$$\sin(c)\cos(B) = \cos(b)\sin(a) - \sin(b)\cos(a)\cos(C), \qquad (8.30)$$

which is (4.13), the analogue formula!

We obtain one other useful expression from (8.23)=(8.24). We multiply both sides on the right by $\mathbf{C}$ to obtain

$$\begin{aligned} & \cos(c)\mathbf{C} - \sin(c)\mathbf{C}'\mathbf{C} \\ = \; & (\cos(b) + \sin(b)\mathbf{B}')(\cos(a)\mathbf{C} + \sin(a)\mathbf{A}'\mathbf{C}) \\ = \; & \cos(a)\cos(b)\mathbf{C} + \cos(b)\sin(a)\mathbf{A}'\mathbf{C} \\ + \; & \sin(b)\cos(a)\mathbf{B}'\mathbf{C} + \sin(a)\sin(b)\mathbf{B}'\mathbf{A}'\mathbf{C}. \end{aligned} \tag{8.31}$$

We take real parts of the first and last lines. We have $Re(\mathbf{C}) = 0$ since $\mathbf{C}$ is a vector. We have $Re(\mathbf{A}'\mathbf{C}) = -\mathbf{A}' \cdot \mathbf{C} = 0$ since A' is a pole of $\bigcirc BC$. Similarly, $Re(\mathbf{B}'\mathbf{C}) = 0$. Next, $Re(\mathbf{C}'\mathbf{C}) = -\mathbf{C}' \cdot \mathbf{C} = -\mathbf{A} \times \mathbf{B} \cdot \mathbf{C}/\sin(c) = -\mathbf{B} \times \mathbf{C} \cdot \mathbf{A}/\sin(c) = -\mathbf{A} \cdot \mathbf{B} \times \mathbf{C}/\sin(c)$. Then $Re(\mathbf{B}'\mathbf{A}'\mathbf{C}) = Re(\mathbf{C}\mathbf{B}'\mathbf{A}')$ by §31, Exercise 6; this in turn is $-\mathbf{C} \cdot \mathbf{B}' \times \mathbf{A}' = \mathbf{C} \cdot \mathbf{A}' \times \mathbf{B}' = \mathbf{C} \cdot \mathbf{C}\sin(c') = \sin(\pi - C) = \sin(C)$. Substituting all these into the real parts of both sides of (8.31), we find that (for a triangle with counterclockwise orientation)

$$\mathbf{A} \cdot \mathbf{B} \times \mathbf{C} = \sin(a)\sin(b)\sin(C). \tag{8.32}$$

The letters may be permuted to obtain similar formulas. Note that orientation matters: if $\triangle^s ABC$ has a clockwise orientation, $\mathbf{A} \cdot \mathbf{B} \times \mathbf{C} = -\sin(a)\sin(b)\sin(C)$. The spherical law of sines may be extracted from (8.32); see Exercise 2.

The power of using quaternions on the sphere should now be evident. We have that the most important three identities involving components of a spherical triangle are, taken together, the real components of a quaternionic identity relating the three sides of the triangle.

Exercises §33

1. What happens if we take (8.23) and take the product of both sides with $\mathbf{B}'$? $\mathbf{A}$? $\mathbf{B}$? $\mathbf{C}'$?

2. Use (8.32) to prove the spherical law of sines.

34 Rotations and reflections

Key to the results of the previous sections was the fact that if we take the quaternion $\cos(\theta)+\sin(\theta)\mathbf{e}$ and multiply it on the left of a vector (pure quaternion) perpendicular to $\mathbf{e}$ then the latter is rotated an angle of θ. This might lead us to believe that multiplication by $\cos(\theta) + \sin(\theta)\mathbf{e}$ induces rotation by angle θ about the axis of $\mathbf{e}$ on the whole sphere. This would be a serious error — one made worse by the fact that it is close to being true. In fact it is best to identify this quaternion with a rotation in angle 2θ, although then the algebraic operation which causes this is not simply multiplication on the left. We prove two theorems which show how quaternions can be used to calculate reflections and rotations in space.

Theorem 34.1 *Suppose E is the plane in space perpendicular to the vector* $\mathbf{e}$. *Then if* $\mathbf{x}$ *is an arbitrary vector in space, its reflection in plane E is* $\mathbf{exe}$.

Proof. The plane E consists of all vectors $\mathbf{p}$ in space such that $\mathbf{p} \cdot \mathbf{e} = 0$. We first show that if $\mathbf{p}$ is in E, then $\mathbf{epe} = \mathbf{p}$. Since $\mathbf{e}$ and $\mathbf{p}$ are pure quaternions, $\mathbf{ep} = -\mathbf{e} \cdot \mathbf{p} + \mathbf{e} \times \mathbf{p}$ which is $\mathbf{e} \times \mathbf{p}$ since $\mathbf{p}$ and $\mathbf{e}$ are perpendicular. But this is $-\mathbf{p} \times \mathbf{e}$ which for similar reasons is $-\mathbf{pe}$. Thus $\mathbf{epe} = -\mathbf{pe}^2 = \mathbf{p}$, as desired.

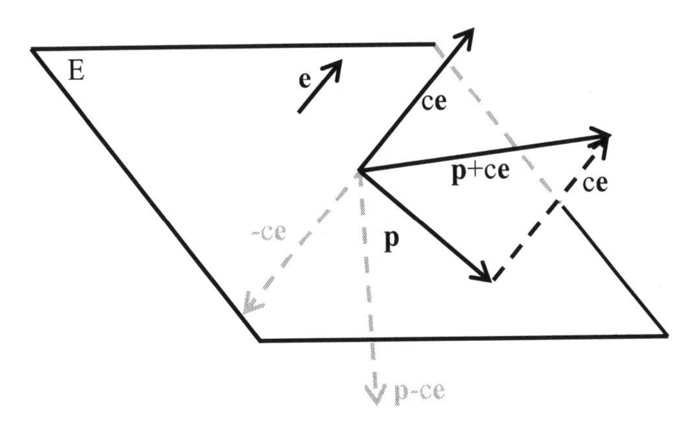

Figure 8.18: Reflection of $\mathbf{x} = \mathbf{p} + c\mathbf{e}$ in plane E.

We can write an arbitrary vector $\mathbf{x}$ in space as the sum of a component vector in E plus a component vector perpendicular to E: i.e., in the form $\mathbf{x} = \mathbf{p} + c\mathbf{e}$ where $\mathbf{p}$ is in E. Then $\mathbf{exe} = \mathbf{e}(\mathbf{p} + c\mathbf{e})\mathbf{e} = \mathbf{epe} + c\mathbf{eee}$. This is $\mathbf{p} - c\mathbf{e}$ since $\mathbf{p}$ is in E and $\mathbf{e}^2 = -1$. But $\mathbf{p} - c\mathbf{e}$ is the reflection of $\mathbf{x} = \mathbf{p} + c\mathbf{e}$ in plane E because the respective component vectors in E are the same but the respective components perpendicular to E are negatives of each other. $\diamondsuit$

Theorem 34.2 *Suppose that* $\mathbf{e}$ *is a unit vector in space and θ is a real number. Consider the unit quaternion* $\mathbf{p} = \cos(\frac{1}{2}\theta) + \sin(\frac{1}{2}\theta)\mathbf{e}$. *Then for any vector* $\mathbf{x}$ *in space the quaternion* $\mathbf{pxp}^{-1} = \mathbf{px\overline{p}}$ *is the rotation of the vector x about the axis of* $\mathbf{e}$ *by angle θ counterclockwise when viewed from above* $\mathbf{e}$.

Proof. (See Figure 8.19.) Let $\mathbf{a}$ and $\mathbf{b}$ be unit vectors perpendicular to $\mathbf{e}$ having angle $\frac{\theta}{2}$ between them and such that the angle from $\mathbf{a}$ to $\mathbf{b}$ winds counterclockwise around $\mathbf{e}$ when viewed from above $\mathbf{e}$. (That is, the cross product $\mathbf{a} \times \mathbf{b}$ points in the same direction as $\mathbf{e}$.) Then $\mathbf{ba} = -\mathbf{b} \cdot \mathbf{a} + \mathbf{b} \times \mathbf{a} = -\mathbf{a} \cdot \mathbf{b} - \mathbf{a} \times \mathbf{b} = -\cos(\frac{\theta}{2}) - \sin(\frac{\theta}{2})\mathbf{e} = -\mathbf{p}$. We now consider the reflection in the plane perpendicular to $\mathbf{a}$ and the plane perpendicular to $\mathbf{b}$. As observed in §30, the composition of the first reflection followed by the second is a rotation in an angle θ counterclockwise about $\mathbf{e}$ when viewed from above $\mathbf{e}$. The image of vector $\mathbf{x}$ under the composition of these reflections is $\mathbf{b}(\mathbf{axa})\mathbf{b} = (\mathbf{ba})\mathbf{x}(-\mathbf{a})(-\mathbf{b}) = (\mathbf{ba})\mathbf{x}(\mathbf{a}^{-1})(\mathbf{b}^{-1}) = (\mathbf{ba})\mathbf{x}(\mathbf{ba})^{-1} = \mathbf{pxp}^{-1}$. $\diamondsuit$

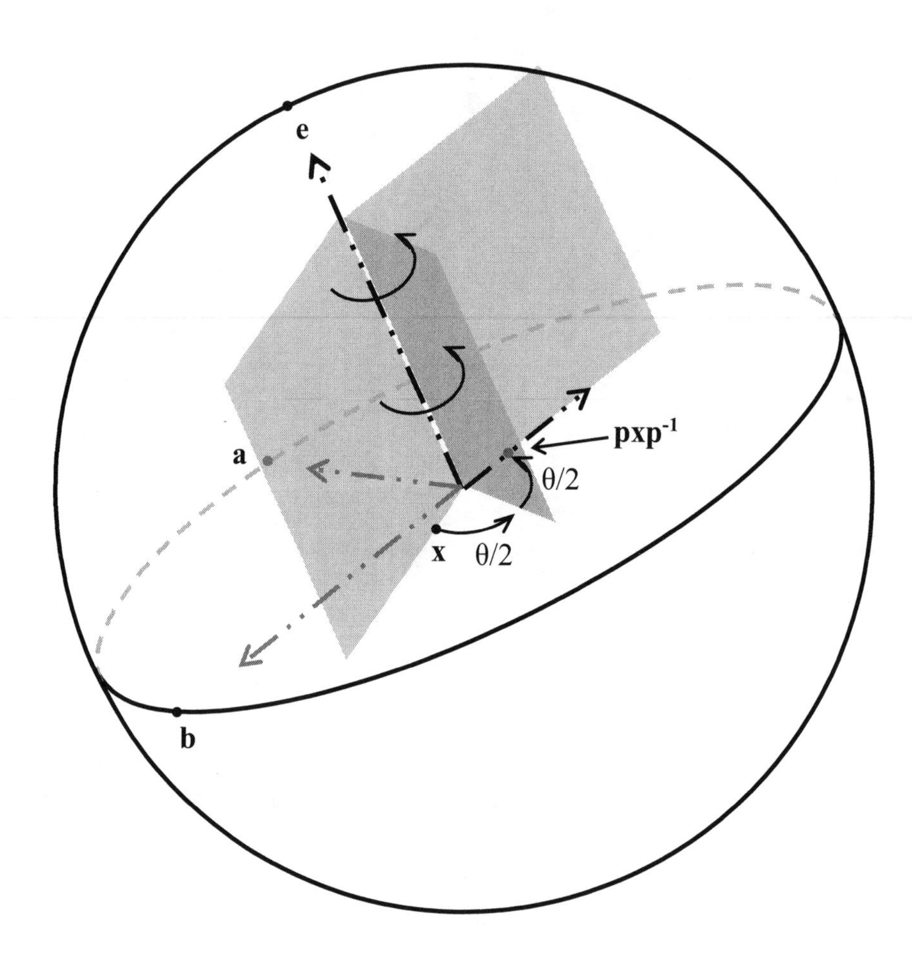

Figure 8.19: Rotation about axis of $\mathbf{e}$ by angle θ.

Theorem 34.2 allows us to calculate the composition of two rotations in space fairly easily. Suppose that we have two unit vectors $\mathbf{e}_1$ and $\mathbf{e}_2$ about which we rotate space (counterclockwise from above each) by angles θ_1 and θ_2. Then we form the unit quaternions $\mathbf{p}_1 = \cos(\frac{\theta_1}{2}) + \sin(\frac{\theta_1}{2})\mathbf{e}_1$ and $\mathbf{p}_2 = \cos(\frac{\theta_2}{2}) + \sin(\frac{\theta_2}{2})\mathbf{e}_2$. Rotation of vector $\mathbf{x}$ counterclockwise about $\mathbf{e}_1$ by θ_1 is given by $\mathbf{p}_1\mathbf{x}\mathbf{p}_1^{-1}$; rotation of vector $\mathbf{x}$ counterclockwise about $\mathbf{e}_2$ by θ_2 is given by $\mathbf{p}_2\mathbf{x}\mathbf{p}_2^{-1}$. By rotating $\mathbf{x}$ first around $\mathbf{e}_1$ and then around $\mathbf{e}_2$ we obtain $\mathbf{p}_2(\mathbf{p}_1\mathbf{x}\overline{\mathbf{p}_1})\overline{\mathbf{p}_2} = \mathbf{p}_2\mathbf{p}_1\mathbf{x}\overline{(\mathbf{p}_2\mathbf{p}_1)} = \mathbf{p}_2\mathbf{p}_1\mathbf{x}(\mathbf{p}_2\mathbf{p}_1)^{-1}$. The quaternion $\mathbf{p}_2\mathbf{p}_1$ is a unit quaternion and can be written in the form $\cos(\frac{\phi}{2}) + \sin(\frac{\phi}{2})\mathbf{e}$. Thus the product of the two rotations is a rotation by angle ϕ counterclockwise about vector $\mathbf{e}$.

Here is a geometric method of determining the axis of rotation of the composition of these two rotations. (See Figure 8.20.) Let E_1 be the terminal point of $\mathbf{e}_1$ and E_2 the terminal point of $\mathbf{e}_2$. Let $\overset{\hookrightarrow}{r_1}$ be the spherical ray found by rotating $\overset{\hookrightarrow}{E_1E_2}$ clockwise about E_1 by angle $\frac{\theta_1}{2}$. Let $\overset{\hookrightarrow}{r_2}$ be the spherical ray found by rotating $\overset{\hookrightarrow}{E_2E_1}$ counterclockwise about E_2 by angle $\frac{\theta_2}{2}$. By §9, Exercise 25, $\overset{\hookrightarrow}{r_1}$ intersects $\overset{\hookrightarrow}{r_2}$ in a single point. We call this point E_3 and let $\mathbf{e}_3$ denote the vector pointing to E_3. Then we claim that E_3 is the center of the rotation found by taking the composition of the rotations associated with $\mathbf{p}_1$ and $\mathbf{p}_2$ above. If we rotate about the axis of $\mathbf{e}_1$ by angle θ_1 counterclockwise, the point E_3 is moved to a point called E_4 on the opposite side of $\bigcirc E_1E_2$ such that $m \prec E_4E_1E_2 = \frac{\theta_1}{2} = m \prec E_3E_1E_2$. Since $\overset{\frown}{E_4E_1} \cong \overset{\frown}{E_3E_1}$ and $\overset{\frown}{E_1E_2} \cong \overset{\frown}{E_1E_2}$, $\triangle^s E_4E_1E_2 \cong \triangle^s E_3E_1E_2$ by spherical SAS congruence (Theorem 12.3). Thus $\overset{\frown}{E_2E_4} \cong \overset{\frown}{E_2E_3}$ and $m \prec E_4E_2E_1 = \prec E_3E_2E_1 = \frac{\theta_2}{2}$. Then rotation about E_2 by angle θ_2 counterclockwise maps E_4 back to E_3. So the composition of the rotations about the axes $\mathbf{e}_1$ and $\mathbf{e}_2$ has E_3 as a fixed point, and this means the composition must be rotation about E_3 by some angle.

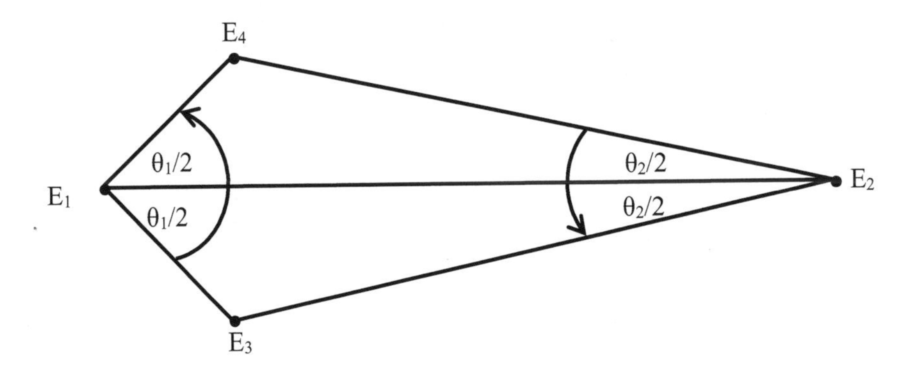

Figure 8.20: E_3 is the axis of rotation of the composition rotations at E_1, E_2.

We illustrate with an example. Suppose our first rotation is counterclockwise about the **i** direction by 45° and the second rotation is counterclockwise about $(\mathbf{i}+\mathbf{j})/\sqrt{2}$ by 30°. Then $\mathbf{p}_1 = \cos(45°) + \sin(45°)\mathbf{i}$ and $\mathbf{p}_2 = \cos(30°) + (\sin(30°))(\frac{\mathbf{i}+\mathbf{j}}{\sqrt{2}})$. So

$$\begin{aligned}
\mathbf{p}_2\mathbf{p}_1 &= (\frac{\sqrt{3}}{2} + \frac{1}{2}(\frac{\mathbf{i}+\mathbf{j}}{\sqrt{2}}))(\frac{1}{\sqrt{2}} + \frac{1}{\sqrt{2}}\mathbf{i}) \\
&= \frac{\sqrt{6}-1}{4} + \frac{1+\sqrt{6}}{4}\mathbf{i} + \frac{1}{4}\mathbf{j} - \frac{1}{4}\mathbf{k} \\
&\approx \cos(68.75°) + \sin(68.75°)(0.925\mathbf{i} + 0.268\mathbf{j} - 0.268\mathbf{k})
\end{aligned}$$

So the composition of these two rotations (illustrated in Figure 8.21) is a rotation about the axis of $0.925\mathbf{i}+0.268\mathbf{j}-0.268\mathbf{k}$ of 68.75° counterclockwise.

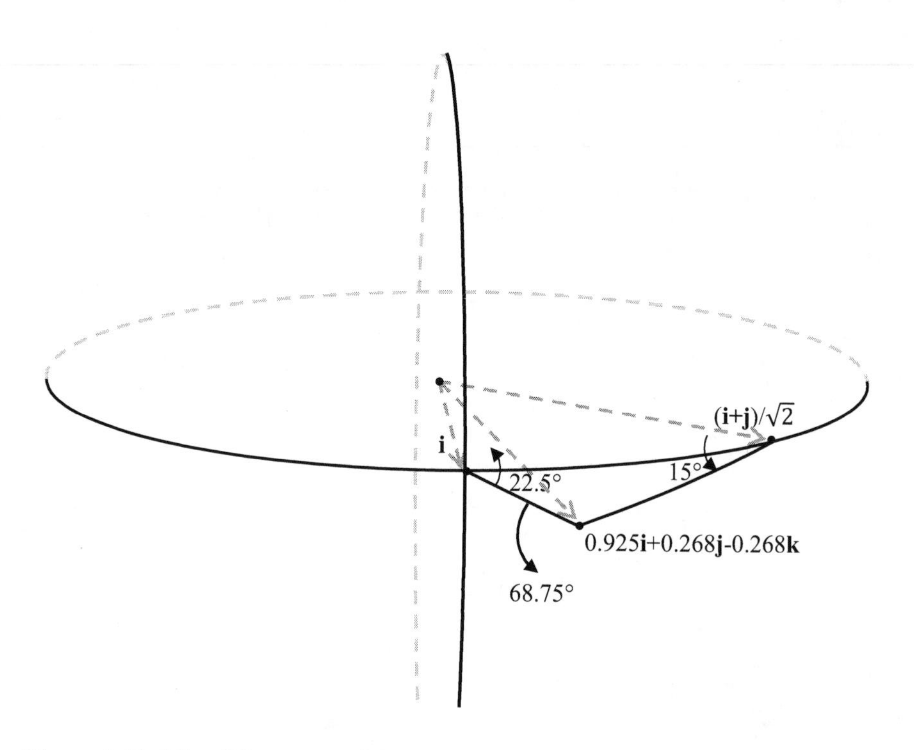

Figure 8.21: The 45° rotation followed by the 30° rotation is the 68.75° rotation.

We now demonstrate the power of Theorems 34.1 and 34.2 in proving formulas for spherical triangles.

Theorem 34.3 (Delambre's analogies) *In any spherical $\triangle^s ABC$:*

$$\frac{\sin(\frac{1}{2}(A+B))}{\cos(\frac{1}{2}C)} = \frac{\cos\frac{1}{2}(a-b)}{\cos(\frac{1}{2}c)} \tag{8.33}$$

$$\frac{\cos(\frac{1}{2}(A-B))}{\sin(\frac{1}{2}C)} = \frac{\sin\frac{1}{2}(a+b)}{\sin(\frac{1}{2}c)} \tag{8.34}$$

$$\frac{\sin(\frac{1}{2}(A-B))}{\cos(\frac{1}{2}C)} = \frac{\sin\frac{1}{2}(a-b)}{\sin(\frac{1}{2}c)} \tag{8.35}$$

$$\frac{\cos(\frac{1}{2}(A+B))}{\sin(\frac{1}{2}C)} = \frac{\cos\frac{1}{2}(a+b)}{\cos(\frac{1}{2}c)} \tag{8.36}$$

Proof. We assume without loss of generality that $\triangle^s ABC$ is presented with a counterclockwise orientation. We consider the composition of two rotations. (See Figure 8.22.) The first (which we call R_1) fixes C and rotates side $\overset{\hookrightarrow}{CA}$ into $\overset{\hookrightarrow}{CB}$. The second rotation R_2 fixes A' and takes $R_1(A)$ to B; this has measure $a-b$ clockwise (which is a counterclockwise rotation of measure $b-a$ if $a-b<0$.) The composition $R_2 \circ R_1$ is then also a rotation which maps $\overset{\frown}{CA}$ to $\overset{\frown}{CB}$ and A to B. Similarly consider rotations R_3 and R_4. R_3 fixes C' and rotates counterclockwise by measure c: this takes A to B. Then R_4 rotates counterclockwise by angle $\pi - A - B$ about B (again, this is a clockwise rotation if $\pi - A - B < 0$.) Because $R_2 \circ R_1$ and $R_4 \circ R_3$ are both rotations which coincide at two distinct points (A and C), they must be identical.

Now we calculate the quaternions associated with these four rotations. R_1 is represented by $\cos(\frac{C}{2}) + \sin(\frac{C}{2})\mathbf{C}$, R_2 by $\cos(\frac{b-a}{2}) + \sin(\frac{b-a}{2})\mathbf{A}'$, R_3 by $\cos(\frac{c}{2})+\sin(\frac{c}{2})\mathbf{C}'$, and R_4 by $\cos(\frac{\pi-A-B}{2})+\sin(\frac{\pi-A-B}{2})\mathbf{B}$. Then the statement that $R_2 \circ R_1 = R_4 \circ R_3$ gives the quaternion equation

$$(\cos(\tfrac{b-a}{2}) + \sin(\tfrac{b-a}{2})\mathbf{A}')(\cos(\tfrac{C}{2}) + \sin(\tfrac{C}{2})\mathbf{C}) \tag{8.37}$$

$$= (\cos(\tfrac{\pi-A-B}{2}) + \sin(\tfrac{\pi-A-B}{2})\mathbf{B})(\cos(\tfrac{c}{2}) + \sin(\tfrac{c}{2})\mathbf{C}'). \tag{8.38}$$

Equating real parts,

$$\begin{aligned} &\cos(\tfrac{b-a}{2})\cos(\tfrac{C}{2}) - \sin(\tfrac{b-a}{2})\sin(\tfrac{C}{2})\mathbf{A}'\cdot\mathbf{C} \\ = \;&\cos(\tfrac{\pi-A-B}{2})\cos(\tfrac{c}{2}) - \sin(\tfrac{\pi-A-B}{2})\sin(\tfrac{c}{2})\mathbf{B}\cdot\mathbf{C}'. \end{aligned}$$

Now $\mathbf{A}'\cdot\mathbf{C} = 0$ since A is a pole of $\bigcirc BC$; similarly $\mathbf{B}\cdot\mathbf{C}' = 0$. Then using the fact that $\sin(\frac{\pi-A-B}{2}) = \cos(\frac{A+B}{2})$ and $\cos(\frac{\pi-A-B}{2}) = \sin(\frac{A+B}{2})$, we have (8.33). Equation (8.34) follows by dualizing (8.33) via the polar triangle: replace $A = \pi - a$, $B = \pi - b$, $C = \pi - c$.

We next prove (8.36). Multiply (8.37) by $\mathbf{B}$ to obtain

$$\begin{aligned} &(\cos(\tfrac{b-a}{2}) + \sin(\tfrac{b-a}{2})\mathbf{A}')(\cos(\tfrac{C}{2}) + \sin(\tfrac{C}{2})\mathbf{C})\mathbf{B} \\ = \;&(\cos(\tfrac{\pi-A-B}{2}) + \sin(\tfrac{\pi-A-B}{2})\mathbf{B})(\cos(\tfrac{c}{2}) + \sin(\tfrac{c}{2})\mathbf{C}')\mathbf{B}. \end{aligned}$$

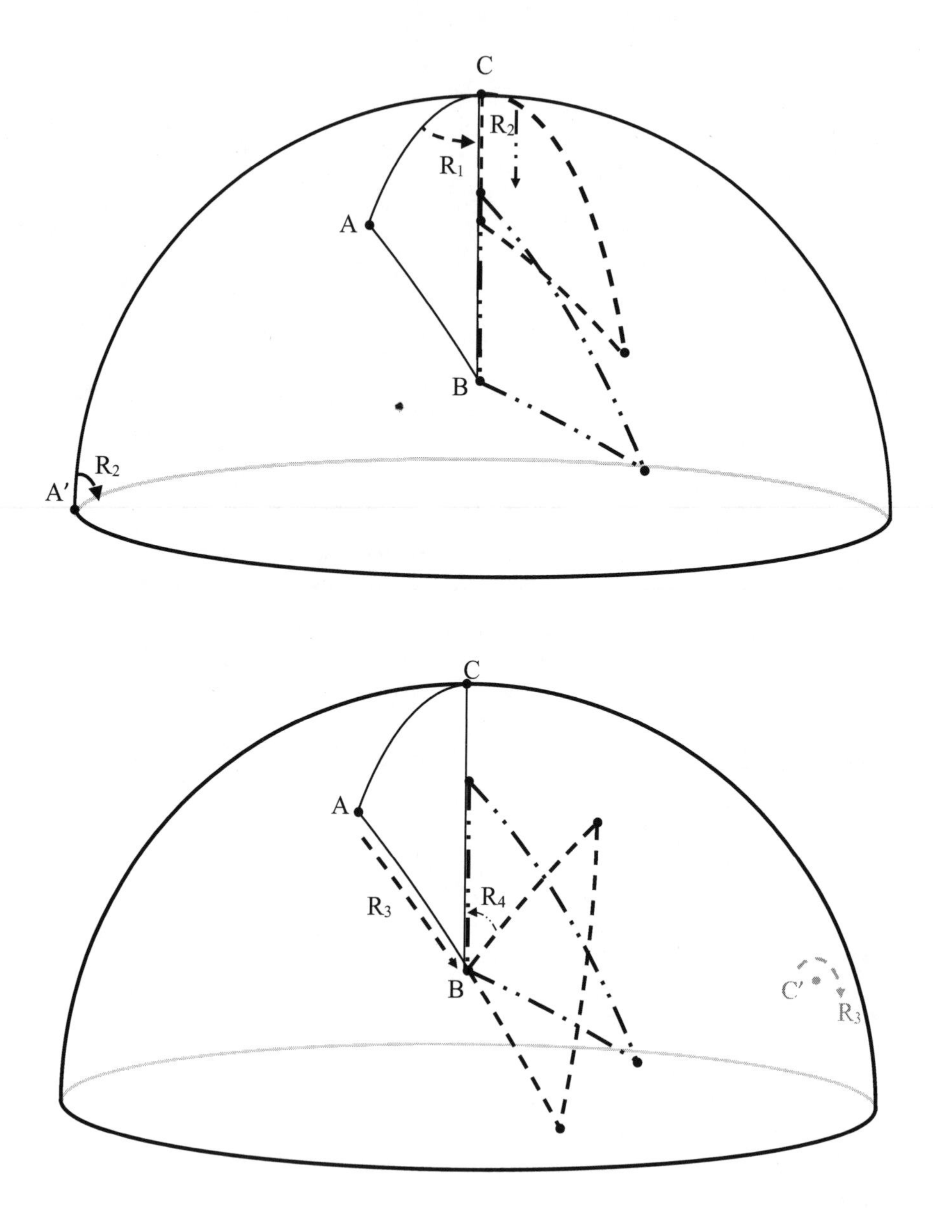

Figure 8.22: The rotation R_1 moves $\triangle^s ABC$ to the triangle with dashed sides, and R_2 moves the triangle with dashed sides to the triangle with dotted and dashed sides. R_3 moves $\triangle^s ABC$ to the triangle with dashed sides and R_4 moves the triangle with dashed sides to the triangle with dotted and dashed sides. So $R_2 \circ R_1 = R_4 \circ R_3$.

Taking real parts of both sides and using (8.4),

$$\begin{aligned} & Re\left((\cos(\tfrac{b-a}{2}) + \sin(\tfrac{b-a}{2})\mathbf{A}')(\cos(\tfrac{C}{2}) + \sin(\tfrac{C}{2})\mathbf{C})\mathbf{B}\right) \\ = \; & Re\left(\mathbf{B}(\cos(\tfrac{\pi-A-B}{2}) + \sin(\tfrac{\pi-A-B}{2})\mathbf{B})(\cos(\tfrac{c}{2}) + \sin(\tfrac{c}{2})\mathbf{C}')\right). \end{aligned}$$

Multiplying out, note that $Re(\mathbf{A}'\mathbf{B}) = -\mathbf{A}' \cdot \mathbf{B} = 0$ since $\mathbf{A}'$ is perpendicular to $\mathbf{B}$. Similarly, $Re(\mathbf{B}\mathbf{C}') = 0$. We find that

$$\begin{aligned} & Re\left(\cos(\frac{b-a}{2})\sin(\frac{C}{2})\mathbf{C}\mathbf{B} + \sin(\frac{b-a}{2})\sin(\frac{C}{2})\mathbf{A}'\mathbf{C}\mathbf{B}\right) \\ = \; & -\sin(\frac{\pi - A - B}{2})\cos(\frac{c}{2}). \end{aligned}$$

Now $Re(\mathbf{C}\mathbf{B}) = -\mathbf{C} \cdot \mathbf{B} = -\cos(\widehat{BC}) = -\cos(a)$ and

$$Re(\mathbf{A}'\mathbf{C}\mathbf{B}) = -\mathbf{A}' \cdot \mathbf{C} \times \mathbf{B} = \mathbf{A}' \cdot \mathbf{B} \times \mathbf{C} = \mathbf{A}' \cdot \mathbf{A}' \sin(a) = \sin(a).$$

This gives us

$$\begin{aligned} & -\cos(\frac{b-a}{2})\sin(\frac{C}{2})\cos(a) + \sin(\frac{b-a}{2})\sin(\frac{C}{2})\sin(a) \\ = \; & -\sin(\frac{\pi - A - B}{2})\cos(\frac{c}{2}), \end{aligned}$$

so

$$\begin{aligned} \sin(\frac{C}{2})\left(\cos(\frac{b-a}{2})\cos(a) - \sin(\frac{b-a}{2})\sin(a)\right) &= \cos(\frac{A+B}{2})\cos(\frac{c}{2}) \\ \sin(\frac{C}{2})\cos(\frac{b-a}{2} + a) &= \cos(\frac{A+B}{2})\cos(\frac{c}{2}) \\ \sin(\frac{C}{2})\cos(\frac{a+b}{2}) &= \cos(\frac{A+B}{2})\cos(\frac{c}{2}), \end{aligned}$$

which produces (8.36). Equation (8.35) follows by dualizing (8.36) via a colunar triangle: replace A by $\pi - A$, a by $\pi - a$, C by $\pi - C$, c by $\pi - c$, and leave B and b alone.♢

We are using the word "analogy" in a usage that is now archaic. For us an analogy will simply be an equality of proportions. The reader may be used to analogies of the form $A{:}B{::}C{:}D$ among objects A, B , C, and D, which may be loosely translated as "A is to B as C is to D." If A, B, C, and D are numbers, this is understood to mean that A/B is equal to C/D. We retain the usage of the word "analogy" here because it is traditional in spherical geometry.

Corollary 34.4 (Napier's Analogies) *In any spherical triangle $\triangle^s ABC$,*

$$\frac{\sin \frac{1}{2}(A - B)}{\sin \frac{1}{2}(A + B)} = \frac{\tan \frac{1}{2}(a - b)}{\tan \frac{1}{2}c} \tag{8.39}$$

$$\frac{\cos\frac{1}{2}(A-B)}{\cos\frac{1}{2}(A+B)} = \frac{\tan\frac{1}{2}(a+b)}{\tan\frac{1}{2}c} \tag{8.40}$$

$$\frac{\sin\frac{1}{2}(a-b)}{\sin\frac{1}{2}(a+b)} = \frac{\tan\frac{1}{2}(A-B)}{\cot\frac{1}{2}C} \tag{8.41}$$

$$\frac{\cos\frac{1}{2}(a-b)}{\cos\frac{1}{2}(a+b)} = \frac{\tan\frac{1}{2}(A+B)}{\cot\frac{1}{2}C} \tag{8.42}$$

where the fractions are defined.

Proof. Divide (8.35) by (8.33) to obtain (8.39) and divide (8.34) by (8.36) to obtain (8.40). Dualization via the polar triangle of (8.39) and (8.40) produces (8.41) and (8.42). ◇

We next use quaternions to prove three key theorems which relate the spherical excess of a spherical triangle to the measures of its sides and angles.

Theorem 34.5 (Euler's formula) *Let E be one-half of the spherical excess of a spherical triangle: $E = \frac{1}{2}(A+B+C-\pi)$. Then we have*

$$\cos(E) = \frac{1+\cos(a)+\cos(b)+\cos(c)}{4\cos(\frac{a}{2})\cos(\frac{b}{2})\cos(\frac{c}{2})}. \tag{8.43}$$

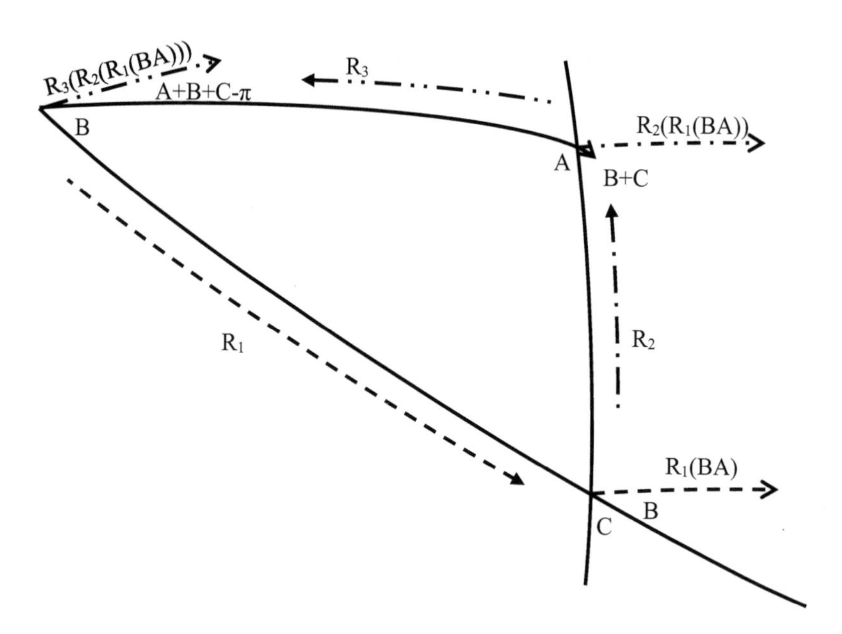

Figure 8.23: Successive rotations about A', B', and C'.

Proof. We again assume without loss of generality that the triangle is presented with a counterclockwise orientation. (See Figure 8.23.) Let A', B' and

C' be the vertices of the polar triangle of $\triangle^s ABC$. Then we consider the result of composition of the following three rotations: (R_1) rotation counterclockwise by angle a about A', (R_2) rotation counterclockwise by angle b about B' and (R_3) rotation counterclockwise by angle c about C'. We claim that the composition is rotation counterclockwise by $A+B+C-\pi$ about B. To see this, observe the position of $\overset{\hookrightarrow}{BA}$ after each successive rotation. After rotation about A', $R_1(\overset{\hookrightarrow}{BA})$ forms an angle of $B+C$ with $\overset{\hookrightarrow}{AC}$. Then $R_2 \circ R_1(\overset{\hookrightarrow}{BA})$ forms an angle of $B+C$ with $\overset{\hookrightarrow}{AC}$, so an angle of $A+B+C-\pi$ with $\overset{\hookrightarrow}{BA}$. Hence $R_3 \circ R_2 \circ R_1(\overset{\hookrightarrow}{BA})$ forms an angle of $A+B+C-\pi$ with $\overset{\hookrightarrow}{BA}$. Thus the composition is a counterclockwise rotation of $A+B+C-\pi$ about B. We let $E = \frac{1}{2}(A+B+C-\pi)$ be the spherical excess.

In terms of quaternions, R_1 is represented by $\cos(\frac{a}{2}) + \sin(\frac{a}{2})\mathbf{A}'$, R_2 is represented by $\cos(\frac{b}{2}) + \sin(\frac{b}{2})\mathbf{B}'$, R_3 is represented by $\cos(\frac{c}{2}) + \sin(\frac{c}{2})\mathbf{C}'$, and counterclockwise rotation of $A+B+C-\pi$ about B is represented by $\cos(E) + \sin(E)\mathbf{B}$. Thus

$$\begin{aligned}&\cos(E) + \sin(E)\mathbf{B} \\ &= \left(\cos\left(\tfrac{c}{2}\right) + \sin\left(\tfrac{c}{2}\right)\mathbf{C}'\right)\left(\cos\left(\tfrac{b}{2}\right) + \sin\left(\tfrac{b}{2}\right)\mathbf{B}'\right)\left(\cos\left(\tfrac{a}{2}\right) + \sin\left(\tfrac{a}{2}\right)\mathbf{A}'\right)\end{aligned}$$

or

$$\begin{aligned}&\left(\cos\left(\frac{c}{2}\right) - \sin\left(\frac{c}{2}\right)\mathbf{C}'\right)(\cos(E) + \sin(E)\mathbf{B}) \\ =\ &\left(\cos\left(\frac{b}{2}\right) + \sin\left(\frac{b}{2}\right)\mathbf{B}'\right)\left(\cos\left(\frac{a}{2}\right) + \sin\left(\frac{a}{2}\right)\mathbf{A}'\right).\end{aligned} \tag{8.44}$$

Distributing,

$$\begin{aligned}&\cos(\tfrac{c}{2})\cos(E) - \sin(\tfrac{c}{2})\cos(E)\mathbf{C}' + \cos(\tfrac{c}{2})\sin(E)\mathbf{B} - \sin(E)\sin(\tfrac{c}{2})\mathbf{C}'\mathbf{B} \\ =\ &\cos(\tfrac{b}{2})\cos(\tfrac{a}{2}) + \cos(\tfrac{b}{2})\sin(\tfrac{a}{2})\mathbf{A}' + \sin(\tfrac{b}{2})\cos(\tfrac{a}{2})\mathbf{B}' + \sin(\tfrac{b}{2})\sin(\tfrac{a}{2})\mathbf{B}'\mathbf{A}'.\end{aligned}$$

On each side of this equation the first terms are real, and the second and third terms are pure quaternion. In the last terms, $Re(\mathbf{C}'\mathbf{B}) = -\mathbf{C}' \cdot \mathbf{B}$ which is zero since $\mathbf{C}'$ is the pole of $\overset{\frown}{AB}$. Also, $Re(\mathbf{B}'\mathbf{A}') = -\mathbf{B}' \cdot \mathbf{A}' = -\cos(c') = -\cos(\pi - C) = \cos(C)$. Thus equating real parts of the above equation,

$$\cos\left(\frac{c}{2}\right)\cos(E) = \cos\left(\frac{b}{2}\right)\cos\left(\frac{a}{2}\right) + \sin\left(\frac{b}{2}\right)\sin\left(\frac{a}{2}\right)\cos(C). \tag{8.45}$$

Solving for $\cos(E)$ and multiplying the fraction's numerator and denominator by $4\cos(\frac{b}{2})\cos(\frac{a}{2})$ we get

$$\cos(E) = \frac{2\cos^2(\frac{b}{2})2\cos^2(\frac{a}{2}) + 2\sin(\frac{b}{2})\cos(\frac{b}{2})2\sin(\frac{a}{2})\cos(\frac{a}{2})\cos(C)}{4\cos(\frac{c}{2})\cos(\frac{b}{2})\cos(\frac{a}{2})}.$$

Using the double angle formulas for sine and cosine,

$$\begin{aligned}\cos(E) &= \frac{(1+\cos(b))(1+\cos(a))+\sin(a)\sin(b)\cos(C)}{4\cos(\frac{c}{2})\cos(\frac{b}{2})\cos(\frac{a}{2})} \\ &= \frac{1+\cos(a)+\cos(b)+\cos(a)\cos(b)+\sin(a)\sin(b)\cos(C)}{4\cos(\frac{c}{2})\cos(\frac{b}{2})\cos(\frac{a}{2})} \\ &= \frac{1+\cos(a)+\cos(b)+\cos(c)}{4\cos(\frac{c}{2})\cos(\frac{b}{2})\cos(\frac{a}{2})},\end{aligned}$$

as desired (where the last equality holds by the spherical law of cosines). ♢

Theorem 34.6 (Cagnoli's formula) *Let E be one-half of the spherical excess of a spherical triangle: $E = \frac{1}{2}(A+B+C-\pi)$. Let $s = (a+b+c)/2$. Then we have*

$$\sin(E) = \frac{\sqrt{\sin(s)\sin(s-a)\sin(s-b)\sin(s-c)}}{2\cos(\frac{a}{2})\cos(\frac{b}{2})\cos(\frac{c}{2})}. \tag{8.46}$$

Proof. By writing $\sin(E) = \sqrt{1-\cos^2(E)}$ we may prove this from Euler's formula. This is left as Exercise 4. ♢

Exercises §34

1. For each of the following pairs (R_1, R_2) of rotations with center and counterclockwise angle of rotation given, determine the center and counterclockwise angle of rotation for $R_2 \circ R_1$. For $i = 1, 2$, the center of R_i is $\mathbf{e}_i$ and angle is θ_i.

   (a) $\mathbf{e}_1 = \mathbf{i}, \theta_1 = 60°, \mathbf{e}_2 = (-\mathbf{i}+\mathbf{j})/\sqrt{2}, \theta_2 = 45°$

   (b) $\mathbf{e}_1 = \mathbf{i}, \theta_1 = 30°, \mathbf{e}_2 = \mathbf{j}, \theta_2 = 45°$

   (c) $\mathbf{e}_1 = \mathbf{k}, \theta_1 = 60°, \mathbf{e}_2 = (\mathbf{i}+\mathbf{j})/\sqrt{2}, \theta_2 = 60°$

   (d) $\mathbf{e}_1 = (\mathbf{i}+\mathbf{j}+\mathbf{k})/\sqrt{3}, \theta_1 = 60°, \mathbf{e}_2 = (-\mathbf{i}+\mathbf{j})/\sqrt{2}, \theta_2 = 45°$

2. On a sphere with $r = 5$ find the area of a spherical triangle whose sides have measures

   (a) $a = 80°$, $b = 70°$, $c = 60°$.

   (b) $a = 110°$, $b = 90°$, $c = 75°$.

   (c) $a = 120°$, $b = 110°$, $c = 100°$.

   (d) $a = 145°$, $b = 125°$, $c = 30°$.

   (e) $a = 140°$, $b = 135°$, $c = 10°$.

3. On a sphere with $r = 10$ find the area of a spherical triangle whose sides have measures

(a) $a = \frac{\pi}{2}$, $b = \frac{\pi}{3}$, $c = \frac{\pi}{3}$.
(b) $a = \frac{2\pi}{3}$, $b = \frac{2\pi}{3}$, $c = \frac{\pi}{2}$.
(c) $a = \frac{4\pi}{5}$, $b = \frac{3\pi}{5}$, $c = \frac{2\pi}{5}$.
(d) $a = \frac{5\pi}{7}$, $b = \frac{4\pi}{7}$, $c = \frac{3\pi}{7}$.
(e) $a = \frac{7\pi}{9}$, $b = \frac{2\pi}{3}$, $c = \frac{2\pi}{9}$.

4. Complete the proof of Cagnoli's formula.

5. What happens when we dualize Delambre's analogies via the colunar triangle? Compare to dualization by the polar triangle.

6. Breitschneider's analogies are (from [Ca1889]):

$$\begin{aligned}
\frac{\sin(\frac{1}{2}E)\cos(\frac{1}{2}(C-E))}{\sin(\frac{1}{2}C)} &= \frac{\sin(\frac{1}{2}s)\sin(\frac{1}{2}(s-c))}{\cos(\frac{1}{2}c)}\\
\frac{\cos(\frac{1}{2}E)\sin(\frac{1}{2}(C-E))}{\sin(\frac{1}{2}C)} &= \frac{\cos(\frac{1}{2}s)\cos(\frac{1}{2}(s-c))}{\cos(\frac{1}{2}c)}\\
\frac{\sin(\frac{1}{2}E)\sin(\frac{1}{2}(C-E))}{\cos(\frac{1}{2}C)} &= \frac{\sin(\frac{1}{2}(s-b))\sin(\frac{1}{2}(s-a))}{\cos(\frac{1}{2}c)}\\
\frac{\cos(\frac{1}{2}E)\cos(\frac{1}{2}(C-E))}{\cos(\frac{1}{2}C)} &= \frac{\cos(\frac{1}{2}(s-b))\cos(\frac{1}{2}(s-a))}{\cos(\frac{1}{2}c)}\\
\frac{\cos(\frac{1}{2}(B-E))\cos(\frac{1}{2}(A-E))}{\sin(\frac{1}{2}C)} &= \frac{\sin(\frac{1}{2}s))\cos(\frac{1}{2}(s-c))}{\sin(\frac{1}{2}c)}\\
\frac{\sin(\frac{1}{2}(B-E))\sin(\frac{1}{2}(A-E))}{\sin(\frac{1}{2}C)} &= \frac{\cos(\frac{1}{2}s))\sin(\frac{1}{2}(s-c))}{\sin(\frac{1}{2}c)}\\
\frac{\cos(\frac{1}{2}(B-E))\sin(\frac{1}{2}(A-E))}{\cos(\frac{1}{2}C)} &= \frac{\sin(\frac{1}{2}(s-b)))\cos(\frac{1}{2}(s-a))}{\sin(\frac{1}{2}c)}\\
\frac{\sin(\frac{1}{2}(B-E))\cos(\frac{1}{2}(A-E))}{\cos(\frac{1}{2}C)} &= \frac{\cos(\frac{1}{2}(s-b)))\sin(\frac{1}{2}(s-a))}{\sin(\frac{1}{2}c)}
\end{aligned}$$

The role of C and c in the denominators is not special; we may permute A, B, and C (and a, b, and c) to obtain sixteen other analogies.

Prove some of Breitschneider's analogies by subtracting 1 from each of Delambre's analogies. Prove the rest of Breitschneider's analogies by adding 1 to each of Delambre's analogies.

7. Use Breitschneider's analogies to prove Lhuilier's formula:

$$\tan(\frac{E}{2}) = \sqrt{\tan(\frac{s}{2})\tan(\frac{s-a}{2})\tan(\frac{s-b}{2})\tan(\frac{s-c}{2})}. \tag{8.47}$$

8. Use Breitschneider's analogies to prove

$$\tan(\frac{s}{2}) = \sqrt{\tan(\frac{E}{2})\cot(\frac{A-E}{2})\cot(\frac{B-E}{2})\cot(\frac{C-E}{2})}. \quad (8.48)$$

9. Use Breitschneider's analogies to prove

$$\tan(\frac{s-c}{2}) = \sqrt{\tan(\frac{E}{2})\tan(\frac{A-E}{2})\tan(\frac{B-E}{2})\cot(\frac{C-E}{2})}. \quad (8.49)$$

10. Explore multiplying (8.37) by other quaternions and take real parts. What other formulas can be found?

11. Instead of the rotations used in the proof of Delambre's analogies, try these: R_1 fixes C and rotates side $\overleftrightarrow{CA}$ into $\overleftrightarrow{CB}^a$. (Recall B^a is the antipode of B.) R_2 fixes A' and takes $R_1(A)$ to B; this rotation has angle $a + b$ clockwise. The composition $R_2 \circ R_1$ is then also a rotation which maps $\overset{\frown}{CA}$ to $\bigcirc CB$ and A to B. Next, R_3 fixes C' and rotates counterclockwise by angle c: this takes A to B. Then R_4 rotates clockwise by angle $A + B$ about B. Prove (8.36) from an associated quaternion equation.

12. Suppose that m is the measure of the spherical arc connecting the midpoints of $\overset{\frown}{CA}$ and $\overset{\frown}{CB}$. Prove that $\cos(E) = \cos(m)/\cos(\frac{c}{2})$.

13. Use (8.44) to obtain the following expression:

$$\sin(E) = \sin(\frac{a}{2})\sin(\frac{b}{2})\sin(C)/\cos(\frac{c}{2}). \quad (8.50)$$

14. Use the result of Exercise 13 to produce another way of proving Cagnoli's formula.

15. Prove that

$$\cot(E) = \frac{\cot(\frac{a}{2})\cot(\frac{b}{2}) + \cos(C)}{\sin(C)}. \quad (8.51)$$

16. On a sphere with $r = 10$ find the area of a spherical triangle given the measures of two sides and an included angle:

    (a) $a = \frac{\pi}{2}$, $b = \frac{\pi}{3}$, $C = \frac{\pi}{3}$.
    (b) $a = \frac{2\pi}{3}$, $b = \frac{2\pi}{3}$, $C = \frac{\pi}{2}$.
    (c) $a = \frac{4\pi}{5}$, $b = \frac{3\pi}{5}$, $C = \frac{3\pi}{5}$.
    (d) $a = \frac{5\pi}{7}$, $b = \frac{4\pi}{7}$, $C = \frac{3\pi}{7}$.
    (e) $a = \frac{7\pi}{9}$, $b = \frac{5\pi}{6}$, $C = \frac{2\pi}{3}$.

17. Suppose that we are given the values of real numbers a and b. We consider the problem of determining the value of C which maximizes the area of a spherical triangle $\Delta^s ABC$ with sides a and b and included angle C. (Recall from Theorem 14.4 that the area of $\Delta^s ABC$ is $2Er^2$.)

    (a) If $a + b > \pi$ then show that although the area must be less than $2\pi r^2$, no triangle attains area $2\pi r^2$.

    (b) If $a + b = \pi$ then show although the area must be less than πr^2, no triangle attains area πr^2.

    (c) Suppose that $a + b < \pi$. Use Exercise 15 to show that a maximum area occurs when $\cos(C) = -\tan(\frac{a}{2})\tan(\frac{b}{2})$. Explain why the angle at C must be obtuse. (Hint. The area is maximized when the quantity $\tan(E)$ is maximized. Interpret the expression for $\tan(E)$ one obtains from Exercise 15 as a slope between the point $(\cos(C), \sin(C))$ and the point $(-\cot(\frac{a}{2})\cot(\frac{b}{2}), 0)$ and determine when the slope is largest.)

    (d) Use §16, Exercise 18a to show that this maximum occurs when the median from A to $\overset{\frown}{BC}$ is half as long as $\overset{\frown}{BC}$. (That is, the triangle is diametrical.)

Appendix A

Selected solutions to exercises

§1

1. Let D be the projection of A to $\overleftrightarrow{BC}$. We claim D cannot be on line $\overleftrightarrow{AB}$. If it were, then (since D is on $\overleftrightarrow{BC}$) $D = B$. But then $\overleftrightarrow{AD}$ would be perpendicular to $\overleftrightarrow{BC}$ at $B = D$ — and then $\angle ABC$ would be a right angle, a contradiction. Then A, B, and D form a triangle. Since $\angle ADB$ is a right angle, $\angle ABD$ must be acute. Now rays $\overrightarrow{BC}$ and $\overrightarrow{BD}$ lie on the same line, so are either the same ray or are opposite — and then angles $\angle ABC$ and $\angle ABD$ are respectively either the same or are in a linear pair, so have the same measure or are supplementary. If $\angle ABC$ is acute then $\angle ABC = \angle ABD$, so $\overrightarrow{BC}$ and $\overrightarrow{BD}$ are the same, so C and D are on the same side of B. If $\angle ABC$ is obtuse, then $\angle ABC$ and $\angle ABD$ are a linear pair, so C and D are on opposite sides of B.

5. The angle formed by the two radii bounding the circular sector has measure θ, by the definition of arc measure. Then the proportion of the area inside the circle which is inside the sector is $\frac{\theta}{2\pi}$. Since the area inside the whole circle is πr^2, the proportion inside the sector is $\frac{\theta}{2\pi}\pi r^2$, or $\frac{1}{2}r^2\theta$.

9. Given $\triangle ABC$ and $\triangle DEF$ with right angles at B and E, respectively, $\overline{AB} \cong \overline{DE}$ and $\overline{AC} \cong \overline{DF}$. Then $AB = DE$ and $AC = DF$, and by the Pythagorean theorem, $BC^2 = AC^2 - AB^2 = DF^2 - DE^2 = EF^2$, so $BC = EF$ and $\overline{BC} \cong \overline{EF}$. Thus we have an SSS correspondence between $\triangle ABC$ and $\triangle DEF$, so they are congruent.

§2

1. Suppose the two lines are $\overleftrightarrow{AB}$ and $\overleftrightarrow{CD}$ and both are perpendicular to the plane p at B and D, respectively. Then A does not lie in p; otherwise, so would $\overleftrightarrow{AB}$ (by Proposition 2.1), but it does not since $\overleftrightarrow{AB}$ is perpendicular to p. Thus

A, B, and D determine a unique plane q. By Proposition 2.1, q contains $\overleftrightarrow{AB}$, so q is perpendicular to p by Theorem 2.18. Similarly, B, C, and D determine a unique plane r perpendicular to p. Suppose q and r are distinct. Then by Proposition 2.4, their intersection must be a line containing B and D. The points B and D are distinct by Proposition 2.7. Thus B and D determine a unique line which must be the intersection of q and r. Since B and D lie in p, so does $\overleftrightarrow{BD}$, by Proposition 2.1. Since q and r are both perpendicular to p, their intersection $\overleftrightarrow{BD}$ must be perpendicular to p (Proposition 2.19.) But $\overleftrightarrow{BD}$ cannot be both perpendicular to and contained in p. Thus the assumption that q and r are distinct is false: in fact, they are the same plane, which contains all of A, B, C, and D. Since $\overleftrightarrow{AB}$ and $\overleftrightarrow{CD}$ lie in the same plane perpendicular to the same line ($\overleftrightarrow{BD}$), they must be parallel as desired.

§3

6a. We have $\sin(B) = 5\sin(40°) = 3.2\ldots > 1$, so there is no solution.

6b. We have $\sin(B) = 5\sin(40°)/4 = 0.803\ldots$, so $B = 53.464\ldots°$ or $B = 126.5\ldots°$. Both these values satisfy $A+B < 180°$, so we get two solutions: first one where $C = 180° - 93.464° = 86.5\ldots°$ and $c = 4\frac{\sin(86.5°)}{\sin(40°)} = 6.2\ldots$; and then one where $C = 180° - 166.5\ldots° = 13.4°$ and $c = 4\frac{\sin(13.4°)}{\sin(40°)} = 1.44\ldots$.

6c. We have $\sin(B) = 5\sin(40°)/7 = 0.459\ldots$, so $B = 27.331°$ or $B = 152.668°$. According to the algorithm we keep only $B = 27.331°$, which leads to the solution $C = 112.668\ldots°$ and $c = 7\sin(112.668\ldots°)/\sin(40°) = 10.048\ldots$.

9. We have $\frac{a+b}{c} = \frac{a/b+1}{c/b} = \frac{\sin(A)/\sin(B)+1}{\sin(C)/\sin(B)} = \frac{\sin(A)+\sin(B)}{\sin(C)}$. Using the sum-to-product formulas (1.15) on the numerator and a half-angle formula (1.14) on the denominator, we obtain $\frac{2\sin(\frac{A+B}{2})\cos(\frac{A-B}{2})}{2\sin(\frac{C}{2})\cos(\frac{C}{2})}$. Since $A+B+C$ is $180°$, $\cos(\frac{C}{2}) = \sin(\frac{A+B}{2})$; cancellation of these factors yields the desired formula.

13. Given a, b, and c, we define C by $\tan(\frac{C}{2}) = \sqrt{\frac{(s-b)(s-a)}{s(s-c)}}$. (By assumption, the radicand is always positive, so always has a positive square root.) Then there exists an angle $C/2$ between 0 and $\frac{\pi}{2}$ whose tangent is this quantity, which gives us a C between 0 and π. Now construct a triangle with adjacent sides of length a and b and angle C in between. We claim this is the desired triangle. Let $\tilde{c}$ be the length of the side opposite angle C. Then $\tilde{c}^2 = a^2 + b^2 - 2ab\cos(C)$. We also have $\tan^2(\frac{C}{2}) = \frac{(s-b)(s-a)}{s(s-c)}$. Replace the half-angle tangent with $\frac{1-\cos(C)}{1+\cos(C)}$. Cross-multiplying, and gathering like terms yields $c^2 = a^2 + b^2 - 2ab\cos(C)$. Thus $\tilde{c}^2 = c^2$, so $\tilde{c} = c$. Thus the triangle desired exists.

Alternately, using the previous problem: define C by $\cos(C) = \frac{c^2-a^2-b^2}{2ab}$. This determines a unique C between 0 and π provided $-1 < \frac{c^2-a^2-b^2}{2ab} < 1$. To see this, note that $\frac{c^2-a^2-b^2}{2ab} - 1 = \frac{c^2-(a+b)^2}{2ab}$, which is $\frac{(a+b+c)(c-a-b)}{2ab}$. Since a, b and c are positive and $c > a+b$, this quantity is negative, so

$\frac{c^2-a^2-b^2}{2ab} < 1$. A similar argument shows $-1 < \frac{c^2-a^2-b^2}{2ab}$. So C is well-defined. We form a triangle with elements a, b and C. If the third side is $\tilde{c}$, then $\tilde{c}^2 = a^2 + b^2 - 2ab\cos(C)$. But we already have $c^2 = a^2 + b^2 - 2ab\cos(C)$ by the definition of C. So $c^2 = \tilde{c}^2$, and $c = \tilde{c}$. So the desired triangle exists.

§4

1. Let A', B', and M' be the projections of A, B, and M to ℓ. If the segment is parallel to the line then $a_1 = a_2$ and $b_1 = b_2$ and the statement follows. Let $\vec{A_1} = \vec{AB}$ and $\vec{A_2} = \vec{A'B'}$ be the vector projection of $\vec{A_1}$ to the line ℓ. Let $\vec{B_1} = \vec{MM'}$ and let $\vec{B_2} = \vec{MN}$. Then $(\vec{A_1} - \vec{A_2}) \times \vec{B_1} = 0$ since $\vec{A_1} - \vec{A_2}$ is parallel to $\vec{B_1}$. Similarly, $\vec{A_2} \times (\vec{B_2} - \vec{B_1}) = 0$. Thus $\vec{A_1} \times \vec{B_1} = \vec{A_2} \times \vec{B_1}$ and $\vec{A_2} \times \vec{B_2} = \vec{A_2} \times \vec{B_1}$. From this we get $\vec{A_1} \times \vec{B_1} = \vec{A_2} \times \vec{B_2}$, so $\|\vec{A_1} \times \vec{B_1}\| = \|\vec{A_2} \times \vec{B_2}\|$. Now the angle $\theta \neq 0$ between $\vec{A_1}$ and $\vec{B_1}$ is the same as the angle between $\vec{A_2}$ and $\vec{B_2}$, so $\|\vec{A_1}\|\|\vec{B_1}\|\sin(\theta) = \|\vec{A_2}\|\|\vec{B_2}\|\sin(\theta)$, from which we get $a_1b_1 = a_2b_2$.

§5

1a. $\frac{25\pi}{2}$ 1b. $\frac{5000\pi}{9}$ 1c. $\frac{7000\pi}{9}$

5. Let Γ be any great circle passing through P (which also passes through P^a). Then Q satisfies $m \stackrel{\frown}{PQ} = \rho$ if and only if $m \stackrel{\frown}{P^aQ} = \pi - \rho$ by Exercise 3. Thus Q is on the circle centered at P if and only if it is also on the circle centered at P^a with supplementary radius.

9. Let C be the spherical center of the small circle. By definition of spherical distance, $m\angle COA = m\angle COB = \rho$. Thus the radius in space of the small circle is $r\sin(\rho)$. Since the measure of the arc in question is ϕ radians, its length is $r\phi\sin(\rho)$. The full circle is the case $\phi = 2\pi$.

§6

1a. Plane DEF is perpendicular to $\overleftrightarrow{OD}$ because $\overleftrightarrow{DE}$ and $\overleftrightarrow{DF}$ are perpendicular to $\overleftrightarrow{OD}$ (using Proposition 2.8). Thus plane DEF is perpendicular to plane OAC (by Proposition 2.18). Since both planes OBC and DEF are perpendicular to plane OAC, their intersection line $\overleftrightarrow{EF}$ is perpendicular to plane OAC (by Proposition 2.19). Then $\overleftrightarrow{EF}$ is perpendicular to both $\overleftrightarrow{OF}$ and $\overleftrightarrow{DF}$ (by Definition 2.6). So $\angle EFO$ and $\angle DFO$ are right angles.

1b. The measure of the spherical angle $\sphericalangle BAC$ is the same as the measure of the dihedral angle $\angle B - OA - C$, which in turn has the same measure as plane angle $\angle EDF$.

§7

1. The proportion of the area of the zone that is inside the lune is $\frac{\theta}{2\pi}$. So, using Theorem 7.3, the area is $\frac{\theta}{2\pi}2\pi rd = rd\theta$.

5. This is a zone of height $d = r\sqrt{2}$, so the proportion is $d/2r = 1/\sqrt{2}$.

§9

1. This is plain from the definition of antipode. B is the antipode of A if a great circle with one-one correspondence ℓ exists such that $\ell(B) - \ell(A) = \pi(\text{mod } 2\pi)$. But then $\ell(A) - \ell(B) = -\pi = \pi(\text{mod } 2\pi)$, which means that A is

the antipode of B.

5. The reduction of $(\ell(A) - \ell(B)) \bmod 2\pi$ to a number in $(-\pi, \pi]$ is the value $\ell(A) - \ell(B) - 2\pi n$ where n is the unique integer such that $-\pi < \ell(A) - \ell(B) - 2\pi n \leq \pi$. But then $-\pi \leq \ell(B) - \ell(A) + 2\pi n < \pi$, so the reduction of $(\ell(B) - \ell(A)) \bmod 2\pi$ is the negative of the reduction of $\ell(A) - \ell(B)$ (unless $\ell(B) - \ell(A) + 2\pi n = -\pi$), so their absolute values are the same, so $d(B, A) = d(A, B)$. If $\ell(B) - \ell(A) + 2\pi n = -\pi$, then $d(A, B) = d(B, A) = \pi$.

9. Let P be a point satisfying the properties of the axiom. Let P^a be its antipode. Let A be any point on the given great circle Γ. Then since A is at distance from P other than π radians, A is not the antipode of P, and so there exists a unique great circle $\bigcirc PA$ — which also passes through P^a. Let ℓ be the one-one correspondence for $\bigcirc PA$. Then $\ell(P) - \ell(P^a) = \pi (\bmod 2\pi)$ and $\ell(A) - \ell(P) = \pm\pi/2 (\bmod 2\pi)$. Adding these, $\ell(A) - \ell(P^a) = \pi \pm \pi/2 = \pm\pi/2 (\bmod 2\pi)$. Since A was chosen arbitrarily on Γ, we have shown that P^a satisfies the same property as P. As for there being no other such point: suppose Q is a point at spherical distance $\pi/2$ from each point on Γ. If Q is different from P and P^a then there exists a unique great circle passing through P and Q called $\bigcirc PQ$. Let ℓ be the one-one correspondence to $\mathbf{R}/2\pi$. Since $\bigcirc PQ$ contains P (which is not on Γ) $\bigcirc PQ$ is different from Γ. So $\bigcirc PQ$ and Γ meet in distinct antipodal points B and B^a. Also, P^a must be on $\bigcirc PQ$. Then consider the points $\ell(P), \ell(B), \ell(Q)$ and $\ell(P^a)$. Then $\ell(P), \ell(Q)$ and $\ell(P^a)$ must satisfy $\ell(P) - \ell(B) = \pm\pi/2 (\bmod 2\pi)$, $\ell(Q) - \ell(B) = \pm\pi/2 (\bmod 2\pi)$, $\ell(P^a) - \ell(B) = \pi/2 (\bmod 2\pi)$, and $\ell(P) - \ell(P^a) = \pi (\bmod 2\pi)$. Subtracting the first two, $\ell(P) - \ell(Q) = 0$ or π modulo 2π. In the first case, since ℓ is one-one, $P = Q$. In the second case, we subtract $\ell(P) - \ell(P^a) = \pi$ and $\ell(P) - \ell(Q) = \pi$ to find $\ell(P^a) - \ell(Q) = 0$ modulo 2π, so $P^a = Q$.

13. Ray $\overleftrightarrow{AB}$ consists of those C on $\bigcirc AB$ such that $C = A$, $C = B$, C is between A and B, or B is between A and C. Ray $\overleftrightarrow{BA}$ consists of those C on $\bigcirc AB$ such that $C = A$, $C = B$, C is between A and B, or A is between B and C. Point C clearly satisfies both of these conditions if it satisfies either $C = A$, $C = B$, or C is between A and B. (So the points of $\overset{\frown}{AB}$ are on both rays.) Whether there are any other possibilities for C requires us to settle the question of whether we could have A between B and C and also B between A and C. If this were true, then applying Proposition 9.17, we would conclude that $d(A, C) < d(B, C)$ if A is between B and C and $d(B, C) < d(A, C)$ if B is between A and C. But we cannot have both of these. Thus C is on both $\overleftrightarrow{AB}$ and $\overleftrightarrow{BA}$ if and only if $C = A$, $C = B$ or C is between A and B; that is, C is on $\overset{\frown}{AB}$.

17. Suppose sets S and T are both spherically convex. Let A and B be non-antipodal points in their intersection. Since S and T are each spherically convex, $\overset{\frown}{AB}$ lies in S and lies in T, so it lies in the intersection of S and T. Thus the intersection of S and T is spherically convex.

21. Let the points be A, B, and C. Consider the rays $\overleftrightarrow{AC}$ and $\overleftrightarrow{CA}$. If B is

on either of these, it leads to one of three cases: B is between A and C, C is between A and B, or A is between B and C. If B is not on either of $\overset{\hookrightarrow}{AC}$ or $\overset{\hookrightarrow}{CA}$, then B is on the intersection of the complement sets $\overset{\hookrightarrow}{A^aC^a}$ and $\overset{\hookrightarrow}{C^aA^a}$, which means B is between A^a and C^a.

25. Suppose that the two rays are $\overset{\hookrightarrow}{AB}$ and $\overset{\hookrightarrow}{CD}$. Assume the rays are on the same side of $\bigcirc AC$ except for A and C. Ray $\overset{\hookrightarrow}{AB}$ is not contained in $\bigcirc CD$ since A is neither C nor C^a (as the only place where $\bigcirc AC$ meets $\bigcirc CD$ is at C and C^a). By Proposition 9.37 $\overset{\hookrightarrow}{AB}$ meets $\bigcirc CD$ in a single point we call E. This point lies on the same side of $\bigcirc AC$ as B, which is the same side of $\bigcirc AC$ as D. E must lie on either $\overset{\hookrightarrow}{CD}$ or $\overset{\hookrightarrow}{CD^a}$. Since the latter consists of points on the opposite side of $\bigcirc AC$ from D (except for C), E must lie on $\overset{\hookrightarrow}{CD}$. Thus the rays meet at E only.

§10

1. If A^a, B^a, and C^a lie on a single great circle, then that circle would contain their antipodes A, B, and C, a contradiction. So $\prec A^aB^aC^a$ is a well-defined spherical angle. Then $\overset{\hookrightarrow}{B^aA^a}, \overset{\hookrightarrow}{B^aC^a}$ are rays antipodal to $\overset{\hookrightarrow}{BA}, \overset{\hookrightarrow}{BC}$, respectively. Furthermore, suppose D, E are chosen on $\overset{\hookrightarrow}{BA}, \overset{\hookrightarrow}{BC}$, respectively, at a quarter circle from B. Then D^a, E^a are on $\overset{\hookrightarrow}{B^aA^a}, \overset{\hookrightarrow}{B^aC^a}$, respectively, at a quarter circle from B^a. (Note that $m\ \overset{\frown}{B^aD^a} = m\ \overset{\frown}{BD} = m\ \overset{\frown}{B^aE^a} = m\ \overset{\frown}{BE} = \frac{\pi}{2}$ by §9, Exercise 6.) Then $m \prec A^aB^aC^a = m\ \overset{\frown}{D^aE^a} = m\ \overset{\frown}{DE} = m \prec ABC$, as desired.

5. Let A' be the pole of $\bigcirc BC$ on the same side of $\bigcirc BC$ as A. Then $\overset{\hookrightarrow}{BA}$ and $\overset{\hookrightarrow}{BA'}$ are both perpendicular to $\bigcirc BC$ at B, and A, A' are on the same side of $\bigcirc BC$. By Proposition 10.8, $\overset{\hookrightarrow}{BA} = \overset{\hookrightarrow}{BA'}$.

Since side $\overset{\frown}{AB}$ is right, then $A = A'$ by Proposition 9.30. Then by Proposition 10.12, $\bigcirc AC$ is perpendicular to $\bigcirc BC$, so $\prec C$ is a right angle, as desired. Furthermore, since A is the pole of $\bigcirc BC$, $\overset{\frown}{AC}$ is a quarter circle.

9. Let the two distinct great circles Γ_1, Γ_2 meet at P, P^a; then the desired great circle is the polar of P. It is perpendicular to both Γ_1, Γ_2 by Proposition 10.12. Any great circle perpendicular to both Γ_1, Γ_2 must have poles on both Γ_1, Γ_2 by the same proposition, so the polar of P, P^a is the only possibility.

§11

1. If A^a, B, and C lay on a single great circle, this circle would also contain the antipode of A^a (i.e., A), so A, B, and C would lie on a single great circle, a contradiction. We have $m\ \overset{\frown}{A^aB} = \pi - m\ \overset{\frown}{AB}$ and $m\ \overset{\frown}{A^aC} = \pi - m\ \overset{\frown}{AC}$ by §9, Exercise 14. We have $m \prec A^aBC = \pi - m \prec ABC$ and $m \prec A^aCB = \pi - m \prec ACB$ by Proposition 10.4.

5. Let D and E be the foot of the shorter and longer perpendiculars, respectively, from A to $\bigcirc BC$. If D is B or the antipode of B, then $\prec ABC$

is a right angle, a contradiction. Since D and E are a semicircle apart, both lie on $\bigcirc BC$ and are different from B and its antipode, exactly one of D or E lies on $\overrightarrow{BC}$.

If D lies on $\overrightarrow{BC}$, then $\prec ABC = \prec ABD$ and $\overset{\frown}{AD}$ is acute. Applying Proposition 11.12 to $\triangle^s ABD$, $\prec ABD$ is acute, so $\prec ABC$ is acute. Similarly, if E lies on $\overrightarrow{BC}$, $\overset{\frown}{AE}$ is obtuse, and by applying Proposition 11.12 to $\triangle^s ABE$, $\prec ABE$ is obtuse, so $\prec ABC$ is obtuse. The previous two implications put together imply that if $\prec ABC$ is acute, D lies on $\overrightarrow{BC}$ and if $\prec ABC$ is obtuse, E lies on $\overrightarrow{BC}$.

9. This problem can be done much like Exercise 8. However, we can shorten our work by making use of Exercise 8. Let B^a be the antipode of B and consider $\triangle^s AB^aC$, which has a right angle at C and $\overset{\frown}{AB^a}$ — whose measure is supplementary to that of $\overset{\frown}{AB}$ — must be acute. Applying Exercise 8 to $\triangle^s AB^aC$, we conclude that $\overset{\frown}{AC}$ and $\overset{\frown}{B^aC}$ are both acute or both obtuse. Since the measure of $\overset{\frown}{B^aC}$ is supplementary to that of $\overset{\frown}{BC}$, we conclude that of $\overset{\frown}{AC}$ and $\overset{\frown}{BC}$, one is acute and the other is obtuse.

13. Let the triangle be $\triangle^s ABC$ and X and Y in its interior. Then X and Y are in the interior of each angle, so in the interior of $\prec ABC$. Thus X and Y are on the same side of $\bigcirc AB$ as C, so the arc between them is also. Also, X and Y are on the same side of $\bigcirc BC$ as A, so the arc $\overset{\frown}{XY}$ is also. Thus $\overset{\frown}{XY}$ is in the interior of $\prec ABC$. Similarly, $\overset{\frown}{XY}$ is in the interiors of $\prec ACB$ and $\prec BAC$. Thus $\overset{\frown}{XY}$ is in the interior of $\triangle^s ABC$.

17. Let the triangle be $\triangle^s ABC$. Let Γ be the great circle with a pole at A. Since the measures of the sides of $\triangle^s ABC$ are less than $\frac{\pi}{2}$, both B and C are within $\frac{\pi}{2}$ of A, so the points of $\overset{\frown}{AB}$ and $\overset{\frown}{AC}$ are on the same side of Γ as A. By Proposition 9.42, $\overset{\frown}{BC}$ is in the same hemisphere. But $\triangle^s ABC$ is the polar triangle of $\triangle^s A'B'C'$. So Γ contains the side $\overset{\frown}{B'C'}$ of the polar triangle of $\triangle^s ABC$. Thus $\triangle^s ABC$ and $\overset{\frown}{B'C'}$ have no point in common. A similar argument about B and C shows that the other sides of $\triangle^s A'B'C'$ also have no point in common with $\triangle^s ABC$, as desired.

21. The polar triangle $\triangle^s A'B'C'$ has a right angle at B. We apply Proposition 11.12. Reverting to the original triangle using Theorem 11.19, we obtain the desired conclusion.

25. If two of the A'_i were the same or antipodal then at least three of the A_i would lie on the same great circle, a contradiction. It must be shown that no three of the A'_i lie on a great circle. This is left to the reader. Since for all i, A'_{i-1} is a pole of $\overset{\frown}{A_{i-1}A_i}$ and A'_i is a pole of $\overset{\frown}{A_iA_{i+1}}$, A_i is a quarter circle from both A'_{i-1} and A'_i. Since the latter two are not the same or antipodal, A_i is a pole of $\overset{\frown}{A'_{i-1}A'_i}$. Furthermore, for $i = 1, 2, \ldots n-2$, A_n and A'_i are on the same

side of $\widehat{A_i A_{i+1}}$, so A'_i is within a quarter circle of A_n. From above we have that A_n is a pole of $\bigcirc A'_{n-1}A'_n$. Thus we may say that $A'_1, A'_2, \ldots, A'_{n-2}$ are all on the same side of $\bigcirc A'_{n-1}A'_n$ as A_n. A similar argument shows that for any other side $\widehat{A'_{i-1}A'_i}$ the other points A'_j are all on the same side of $\bigcirc A'_{i-1}A'_i$ as A_i. This proves that $A'_1 A'_2 \ldots A'_n$ forms a spherically convex n-gon.

§12

1. Assume without loss of generality that $\prec B \cong \prec C$. Then $\widehat{BC} \cong \widehat{BC}$, so $\triangle^s ABC \cong \triangle^s ACB$ by ASA. Then by the definition of congruent triangles, $\widehat{AB} \cong \widehat{AC}$; these are the sides opposite B and C.

5. We compare $\triangle^s ADB$ and $\triangle^s ADC$. We have $AD = AD$, $AB = AC$ and $\prec ADB, \prec ADC$ are both right. If $\widehat{AD}$ were right, A would be a pole of $\bigcirc BC$, so $\widehat{AB}, \widehat{AC}$ would also be right. Since $\widehat{AD}$ is not right, $\triangle^s ADB \cong \triangle^s ADC$ by the hypotenuse-leg theorem. So $\widehat{DB} \cong \widehat{DC}$.

9. Suppose that in $\triangle^s ABC$, $\widehat{AB}$ is supplementary to $\widehat{BC}$. Then in the colunar $\triangle^s A^a BC$, $\widehat{A^a B}$ is supplementary to $\widehat{AB}$, so is congruent to $\widehat{BC}$. By Theorem 11.10, $\prec BA^aC \cong \prec BCA^a$. Since $\prec BCA$ and $\prec BCA^a$ are supplementary and $\prec BAC \cong \prec BA^aC$, we conclude that $\prec BAC$ and $\prec BCA$ are supplementary, as desired.

Alternately, consider $\triangle^s AB^aC$: then since $\widehat{AB}$ and $\widehat{BC}$ are supplementary to $\widehat{AB^a}$ and $\widehat{B^aC}$, respectively, $\widehat{AB} \cong \widehat{B^aC}$ and $\widehat{BC} \cong \widehat{AB^a}$. Then $\triangle^s ABC \cong \triangle^s CB^aA$ by SAS. But then $\prec BAC \cong \prec B^aCA$. Since $\prec BCA$ is supplementary to $\prec B^aCA$, we conclude that $\prec BAC$ and $\prec BCA$ are supplementary.

The converse is similar with either approach.

17. We consider the colunar $\triangle^s A^aCB$. We have $\prec BAD \cong \prec CA^aD$, $\widehat{BD} \cong \widehat{CD}$, $\prec ADB \cong \prec A^aDC$ and $m\ \widehat{A^aC} = \pi - m\ \widehat{AC}$ is not supplementary to $m\ \widehat{AB}$ by our assumption. By $SSAA$ congruence, $\triangle^s ABD \cong \triangle^s A^aCD$. We conclude that $m\ \widehat{AB} + m\ \widehat{AC} = \pi$. If $m\ \widehat{AB} = m\ \widehat{AC}$, then by Exercise 4, we would not necessarily have $m\ \widehat{AB} + m\ \widehat{AC} = \pi$; $\triangle^s ABC$ could be any isosceles triangle with vertex A.

21. Let A^a be the antipode of A. Then consider $\triangle^s ABD$ and $\triangle^s A^aCE$. We have $\widehat{AB} \cong \widehat{A^aC}$, $\widehat{AD} \cong \widehat{A^aE}$. Applying Exercise 9, $\prec ABC \cong \prec A^aCB$ and $\prec ADE$ and $\prec AED$ are supplementary, so $\prec ADB \cong \prec A^aEC$. Since $m\ \widehat{AB} \neq m\ \widehat{AC}$, $\widehat{AB}$ and $\widehat{A^aC}$ are not supplementary. Then $\triangle^s ABD \cong \triangle^s A^aCE$ by $SSAA$ congruence. This implies that $\widehat{BD} \cong \widehat{CE}$ and $\prec BAD \cong \prec CA^aE$; but $\prec CA^aE \cong \prec CAE$, so $\prec BAD \cong \prec CAE$. We cannot remove that condition that $m\ \widehat{AB} \neq m\ \widehat{AC}$ because then both would be right sides. Then any choice of D and E on $\widehat{BC}$ would satisfy the conditions required without necessarily having $\widehat{BD} \cong \widehat{CE}$ or $\prec BAD \cong \prec CAE$.

25. We suppose that the congruent legs are $\overset{\frown}{BC}$ and $\overset{\frown}{EF}$. Then by Proposition 11.12, $\prec A$ and $\prec D$ are both acute, both right, or both obtuse. They cannot both be right since then $\overset{\frown}{BC}$ and $\overset{\frown}{EF}$ would be hypotenuses by Proposition 11.11. So $\prec A$ and $\prec D$ are both acute or both obtuse, which means they are not supplementary. By Theorem 12.6, $\triangle^s ABC \cong \triangle^s DEF$.

29. Suppose the triangle is $\triangle^s ABC$ and the right angle is at B. We show that $m\ \overset{\frown}{BC} < m \prec A$. Since $\prec A$ is acute, so is $\overset{\frown}{BC}$ by Proposition 11.12.

We first assume that the hypotenuse $\overset{\frown}{AC}$ is acute. Then so are both legs by §11, Exercise 8 (and the fact that we already know that leg $\overset{\frown}{BC}$ is acute.) Consider $\overset{\hookrightarrow}{AB}$ and $\overset{\hookrightarrow}{AC}$, and choose D and E on each (respectively) at a quarter circle from A. Then A is a pole of $\bigcirc DE$ by Proposition 5.7. Since the sides of $\triangle^s ABC$ are all acute, B is between A and D and C is between A and E. Now $\bigcirc BC$ meets $\bigcirc DE$ in two points. Since $\bigcirc AB$ is perpendicular to both of these great circles (at B and D), the poles of $\bigcirc AB$ must be the intersection points of $\bigcirc BC$ and $\bigcirc DE$ (see Proposition 10.12.) Those poles must be a quarter circle from B and D; we let F be the pole which is on $\overset{\hookrightarrow}{BC}$. Since $\overset{\frown}{BC}$ is acute, C is between B and F, and so $\overset{\frown}{CF}$ is acute. Thus $\overset{\hookrightarrow}{AC}$ lies in the hemisphere centered at F (except for A) by Proposition 9.44. Thus E is within a quarter circle of F, which means F is on $\overset{\hookrightarrow}{DE}$. Since A is a pole of $\bigcirc DE$, $m \prec A = m\ \overset{\frown}{DE}$ by Definition 10.2. Thus $\overset{\frown}{DE}$ is acute. Since $\overset{\frown}{DE}$ is acute, $\overset{\frown}{DF}$ is right, and F is on $\overset{\hookrightarrow}{DE}$, we must have that $\overset{\frown}{EF}$ is acute also. We now obtain a $\triangle^s FCE$ which has a right angle at E. Since $\overset{\frown}{CE}$ is acute, the opposite angle $\prec CFE$ is acute by Proposition 11.12. Similarly, side $\overset{\frown}{FE}$ is acute, so opposite angle $\prec FCE$ is acute. This $\triangle^s FCE$ satisfies the assumptions in Exercise 28, so we may conclude that $m\ \overset{\frown}{FE} < m\ \overset{\frown}{FC}$. Since $\overset{\frown}{DE}$ is complementary to $\overset{\frown}{EF}$ and $\overset{\frown}{BC}$ is complementary to $\overset{\frown}{CF}$, we conclude that $m\ \overset{\frown}{BC} < m\ \overset{\frown}{DE}$. But as noted above, $m\ \overset{\frown}{DE} = m \prec BAC = m \prec A$, so $m\ \overset{\frown}{BC} < m \prec A$, as desired.

If the hypotenuse $\overset{\frown}{AC}$ is obtuse, consider the colunar $\triangle^s A^a BC$; its hypotenuse $\overset{\frown}{A^a C}$ is acute. Applying the above argument to $\triangle^s A^a BC$, we conclude that $m\ \overset{\frown}{BC} < m \prec BA^aC$, but $m \prec BA^aC = m \prec A$, so we get $m\ \overset{\frown}{BC} < m \prec A$.

If the hypotenuse is right, then by §11, Exercise 7, $\triangle^s ABC$ has a right angle at C. Then by Proposition 11.11, $\overset{\frown}{AC}$ and $\overset{\frown}{AB}$ are right, so $m\angle A = m\ \overset{\frown}{BC}$ by Definition 10.2.

33. Let the quadrilateral be $\diamond ABCD$. Then $\triangle^s ABC \cong \triangle^s CDA$ by SSS. So $\prec CBA \cong \prec CDA$. Similarly, $\prec BAD \cong \prec BCD$. Furthermore, $\prec CAB \cong \prec ACD$, $\prec CAD \cong \prec ACB$, $\prec DBA \cong \prec BDC$, and $\prec CBD \cong \prec BDA$. The diagonals meet at some point X by §11, Exercise 23. Then $\triangle^s XBA \cong$

$\triangle^s XDC$ by ASA, so $\widehat{BX} \cong \widehat{XD}$, so X is the midpoint of $\widehat{BD}$. Similarly, it is the midpoint of $\widehat{AC}$.

§13

1. By Theorem 13.9 applied at C, $|m \prec A - m \prec B| < \frac{\pi}{2} < m \prec A + m \prec B$. By Theorem 13.9 applied at A, $\pi - m \prec A > |m \prec B - \frac{\pi}{2}| \geq m \prec B - \frac{\pi}{2}$ so $m \prec A + m \prec B < \frac{3\pi}{2}$.

5. Let $\overleftrightarrow{AD}$ meet $\widehat{BC}$ at E. Suppose that $m\ \widehat{CA} \leq m\ \widehat{CD}$. Then by Theorem 13.10, $m \prec CDA \leq m \prec CAD$, so $\pi - m \prec CDE \leq m \prec CAD$ and $m \prec CAD + m \prec CDE \geq \pi$. Then $m \prec BAC + m \prec BDC > m \prec CAD + m \prec CDE \geq \pi$, a contradiction. Thus $m\ \widehat{CD} < m\ \widehat{AC}$ and a similar argument shows that $m\ \widehat{BD} < m\ \widehat{BA}$.

9. Let the triangle be $\triangle^s ABC$. Suppose that $m\ \widehat{AB} + m\ \widehat{AC} < \pi$. Then $m\ \widehat{AC} - (\pi - m\ \widehat{AB}) < 0$. Construct the colunar $\triangle^s B^a AC$. Then $m\ \widehat{B^a A} = \pi - m\ \widehat{AB}$ and so $m\ \widehat{AC} < m\ \widehat{B^a A}$. Since larger angles are opposite larger sides in $\triangle^s AB^aC$, $m \prec AB^aC < m \prec ACB^a$. But $m \prec AB^aC = m \prec ABC$ and $m \prec ACB^a = \pi - m \prec ACB$. So $m \prec ABC < \pi - m \prec ACB$, or $m \prec ABC + m \prec ACB < \pi$, as desired. Since we also get $m \prec ABC < m \prec ACB^a$, the exterior angle at C is larger than the interior angle at B, as desired. A similar argument holds if $m\ \widehat{AB} + m\ \widehat{AC}$ is equal to or greater than π.

13. By §11, Exercise 16, $\overleftrightarrow{DE}$ meets $\triangle^s ABC$ in some point F other than B or C. Suppose this point is on $\widehat{AB}$. Since $m \prec ABC = m \prec FDC$, $m \prec ABC + m \prec FDB = \pi$. By Exercise 9, $m\ \widehat{FB} + m\ \widehat{FD} = \pi$. But $m\ \widehat{FB} < m\ \widehat{AB} \leq \frac{\pi}{2}$. So $m\ \widehat{FD} > \frac{\pi}{2}$. Since $m\ \widehat{FD} > m\ \widehat{FB}$, $m \prec ABC > m \prec FDB$. So $\prec ABC$ is obtuse and $\prec FDB$ is acute (since they add to π.) Then $\prec ABC^a$ is acute. Since $m\ \widehat{AB} + m\ \widehat{AC} < \pi$, $m\ \widehat{AB} < \pi - m\ \widehat{AC} = m\ \widehat{AC^a}$. Thus $m \prec AC^aB < m \prec ABC^a$, so $\prec AC^aB$ is acute, so $\prec ACB$ is also. Since both $\prec ABC^a$ and $\prec AC^aB$ are acute, the foot of the shorter perpendicular from A to $\bigcirc BC$ lies between B and C^a. By Exercise 7, $m\ \widehat{AB} < m\ \widehat{AD} < m\ \widehat{AC}$. Then $\pi = m\ \widehat{FB} + m\ \widehat{FD} < m\ \widehat{FB} + m\ \widehat{FA} + m\ \widehat{AD} = m\ \widehat{AB} + m\ \widehat{AD} < m\ \widehat{AB} + m\ \widehat{AC} < \pi$, a contradiction. Thus the assumption is false, and F must lie between A and C.

17. If $m \prec BAD / m \prec CAD$ is rational and equal to the ratio of positive integers r/s, it is possible to divide each of $\angle BAD$ and $\angle CAD$ into angles of equal measure u where $ru = m \prec BAD$ and $su = m \prec CAD$. These sub-angles are formed by rays which meet $\widehat{BC}$ in points $D_0 = B, D_1, D_2, D_3, \ldots, D_{r+s} = C$ so that $m \prec BAD_k = ku$, $k = 0, 1, 2, 3, \ldots$. Then $m\ \widehat{BD_k} < m\ \widehat{BD_{k+1}}$ for $k = 0, 1, 2, 3, \ldots$. Because the foot of the shorter perpendicular from A to $\bigcirc BC$ is not on $\widehat{BC}$, $m\ \widehat{AD_k} < m\ \widehat{AD_{k+1}}$ for $k = 0, 1, 2, 3, \ldots$. By applying

Exercise 15 to $\triangle^s AD_iD_{i+2}$ for $i = 0, 1, 2, \ldots$, we conclude that $m\ \overset{\frown}{D_0D_1} < m\ \overset{\frown}{D_1D_2} < m\ \overset{\frown}{D_2D_3} < \ldots < m\ \overset{\frown}{D_{r+s-1}D_{r+s}}$. Then

$$\frac{m\ \overset{\frown}{BD}}{m\ \overset{\frown}{CD}} = \frac{m\ \overset{\frown}{D_0D_1} + m\ \overset{\frown}{D_1D_2} + \ldots m\ \overset{\frown}{D_{r-1}D_r}}{m\ \overset{\frown}{D_rD_{r+1}} + m\ \overset{\frown}{D_{r+1}D_{r+2}} + \ldots + \overset{\frown}{D_{r+s-1}D_{r+s}}} \quad \text{(A.1)}$$

$$< \frac{(r)m\ \overset{\frown}{D_{r-1}D_r}}{(s)m\ \overset{\frown}{D_rD_{r+1}}} \quad \text{(A.2)}$$

$$< \frac{r}{s} \quad \text{(A.3)}$$

as desired, applying Exercise 15 once more to $\triangle^s D_{r-1}AD_{r+1}$.

27. Let P be the center for the circumscribed circle. Suppose that P lies on none of $\bigcirc AB$, $\bigcirc BC$ or $\bigcirc AC$. Then we have three isosceles triangles $\triangle^s PAB$, $\triangle^s PBC$ and $\triangle^s PAC$. The circle's spherical radius is acute. So by §12, Exercise 30, $m\ \overset{\frown}{AB} < m \prec APB$, $m\ \overset{\frown}{AC} < m \prec APC$ and $m\ \overset{\frown}{BC} < m \prec BPC$. Then $m\ \overset{\frown}{AB} + m\ \overset{\frown}{BC} + m\ \overset{\frown}{AC} < m \prec APB + m \prec APC + m \prec BPC$. If A is in the interior of $\prec BPC$, then $m \prec APB + m \prec APC = m \prec BPC$, and so $m \prec APB + m \prec APC + m \prec BPC = 2m \prec BPC < 2\pi$ since any angle has measure less than π, as desired. A similar conclusion holds if B is in the interior of $\prec APC$ or C is in the interior of $\prec APB$. If none of these three holds, then the sum $m \prec APB + m \prec APC + m \prec BPC$ is exactly 2π, so the measures of the sides again is less than 2π.

Lastly, suppose that P lies on one of the great circles containing a side of $\triangle^s ABC$ — say $\bigcirc AB$. Then since P is within a spherical distance of $\frac{\pi}{2}$ of both A and B, P must be the midpoint of $\overset{\frown}{AB}$. Then $m\ \overset{\frown}{AC} + m\ \overset{\frown}{BC} < m \prec APC + m \prec BPC$ via the above isosceles triangle method. But the last sum is π since P is the midpoint of $\overset{\frown}{AB}$. Then $m\ \overset{\frown}{AB} + m\ \overset{\frown}{AC} + m\ \overset{\frown}{BC} < m\ \overset{\frown}{AB} + \pi$, which is less than 2π, since $m\ \overset{\frown}{AB}$ must be less than π.

31. Sketch: choose G on $\overset{\frown}{EF}$ so that $m\ \overset{\frown}{BC} = m\ \overset{\frown}{EG}$. Construct ray $\overset{\hookrightarrow}{GH}$ on the same side of $\bigcirc EF$ as D so that $m \prec EGH = m \prec BCA$. Let H be the point where this ray meets $\overset{\hookrightarrow}{ED}$ and let I be the point where this ray meets $\overset{\hookrightarrow}{FD}$. We have $\triangle^s ABC \cong \triangle^s HEG$ by ASA, so $\overset{\frown}{AB} \cong \overset{\frown}{HE}$ and $\overset{\frown}{AC} \cong \overset{\frown}{HG}$. Also, $\prec IFG, \prec IGF$ are supplementary so $m\ \overset{\frown}{IG} + m\ \overset{\frown}{IF} = \pi$. Ray $\overset{\hookrightarrow}{GH}$ meets $\triangle^s DEF$ in one point in addition to G. Either this point is $D = H = I$ or it is not D, in which case it is (mutually exclusively) H or I.

If $D = H = I$, $m\ \overset{\frown}{DF} + \overset{\frown}{DG} = \pi$, $\overset{\frown}{AB} \cong \overset{\frown}{DE}$ and $\overset{\frown}{AC} \cong \overset{\frown}{DG}$. Thus $m\ \overset{\frown}{AC} + m\ \overset{\frown}{DF} = \pi$, as desired.

If $\overset{\hookrightarrow}{GH}$ meets $\overset{\frown}{DE}$ at $H \neq D$, then $m\ \overset{\frown}{AB} < m\ \overset{\frown}{DE}$ and I is between D and F^a. So $m\ \overset{\frown}{AC} + m\ \overset{\frown}{DF} = m\ \overset{\frown}{HG} + m\ \overset{\frown}{DF} < m\ \overset{\frown}{IG} + m\ \overset{\frown}{IF} = \pi$.

If $\overleftrightarrow{GH}$ meets $\widehat{DF}$ at $I \neq D$, then H is between D and E^a, so $m\ \widehat{AB}>$ $m\ \widehat{DE}$. Also, $m\ \widehat{AC} + m\ \widehat{DF} = m\ \widehat{HG} + m\ \widehat{DF} > m\ \widehat{IG} + m\ \widehat{IF} = \pi$.

35. Consider $\triangle^s ABC$ and let D be the midpoint of $\widehat{AB}$ and E be the midpoint of $\widehat{AC}$. If $m\ \widehat{DE} \geq \frac{\pi}{2}$, we are done since $m\ \widehat{BC} < \pi$. So we may assume $m\ \widehat{DE} < \frac{\pi}{2}$. Let F be the point on $\overleftrightarrow{DE}$ such that $m\ \widehat{DF} = 2m\ \widehat{DE} < \pi$. Now $\triangle^s AED \cong \triangle^s CEF$ by SAS. So $\prec ADE \cong \prec EFC$ and $\widehat{CF} \cong \widehat{AD}$ $(\cong \widehat{BD})$. Since C is on $\overleftrightarrow{AE}$, C and E are on the same side of $\bigcirc AD$. Since F is on $\overleftrightarrow{DE}$, F and E are on the same side of $\bigcirc AD$. So C and F are on the same side of $\bigcirc AD$, so $\widehat{CF}$ does not meet $\bigcirc AD$. Let $\overleftrightarrow{CF}$ meet $\bigcirc DA$ at point G; since G is not on $\widehat{CF}$, F is between C and G. So C and G are on opposite sides of $\bigcirc DF$. Now A and C are on opposite sides of $\bigcirc DF$ (since $\widehat{AC}$ meets $\bigcirc DF$ at E.) So A and G are on the same side of $\bigcirc DF$. Since we already have that G is on $\bigcirc AD$, G must be on $\overleftrightarrow{DA}$. Now C and G are on opposite sides of $\bigcirc DF$, so $\prec EFC$ is exterior to $\prec EDG$ of $\triangle^s DFG$. Since $\prec EFC \cong \prec EDA$, $m\ \widehat{GD}$ $+m\ \widehat{GF} = \pi$. Since C and G are on opposite sides of $\bigcirc DF$, $m\ \widehat{GC} > m\ \widehat{GF}$, so $m\ \widehat{GD} + m\ \widehat{GC} > \pi$ by Exercise 9. So $m \prec GDC + m \prec GCD > \pi$. Note that A and B are on opposite sides of $\bigcirc DF$, so B and G are on opposite sides of $\bigcirc DF$. Thus in $\triangle^s GDC$, $\prec BDC$ is exterior, and so $m \prec BDC < m \prec GCD$ by Exercise 9. Comparing $\triangle^s BDC$ and $\triangle^s FCD$, we see that $\widehat{BD} \cong \widehat{FC}$, $\widehat{DC}$ is common to both, and the angles between these two pairs of sides are unequal: $m \prec BDC < m \prec FCD$. By the Spherical Hinge Theorem, $m\ \widehat{DF} > m\ \widehat{BC}$; that is $2m\ \widehat{DE} > m\ \widehat{BC}$.

38. We prove the desired property for the vertex angle. Let the triangle be $\triangle^s ABC$, the vertex A, D the foot of the angle bisector from A to $\widehat{BC}$, and E the point opposite A on $\widehat{BC}$ such that $m\ \widehat{AE}$ is half the sum of the measures of $\widehat{AB}$ and $\widehat{AC}$. Then $\widehat{AE}$ does not bisect the angle at A by Exercise 37. Suppose that $m\ \widehat{AB} < m\ \widehat{AC}$ without loss of generality; then $m\ \widehat{AB} < (b+c)/2$. By Exercise 37, $m\ \widehat{AD} < (b+c)/2$. Thus by convexity of the small circle centered at A (Exercise 12), every point on $\widehat{BD}$ has distance to A less than $(b+c)/2$. Thus E must be between C and D. Thus $m \prec EAC < m \prec EAB$, as desired.

42. We construct the polar triangles of $\triangle^s ABC$, $\triangle^s DBC$ which we call $\triangle^s A'B'C'$ and $\triangle^s D'EF$. We have $A' = D'$, B' is between A' and E (because $\widehat{AC}, \widehat{BC}$ and $\widehat{DC}$ meet at C) and C' is between A' and F (because $\widehat{BA}, \widehat{BD}$ and $\widehat{BC}$ meet at B). We also construct the polar triangle of $\triangle^s DBM$ which is $\triangle^s A'GF$ for some G. G is between B' and $(C')^a$ (because $\bigcirc AB, \bigcirc AC$ and $\bigcirc DM$ meet at A). Also, G is between E and F^a (because $\bigcirc DM, \bigcirc CD$ and $\bigcirc BD$ meet at D). The measure of $\widehat{BC}$ is twice the measure of $\widehat{BM}$, so

$\pi - m \prec B'A'C' = 2(\pi - m \prec GA'F)$, i.e., $m \prec B'A'F^a = 2(m \prec GA'F^a)$. Thus $\overleftrightarrow{A'G}$ bisects $\prec B'A'F^a$. Note that F^a is between A' and $(C')^a$. Since $m\ \widehat{AB} + m\ \widehat{AC} < \pi$, by properties of polar triangles we have $m \prec A'C'B' + m \prec A'B'C' > \pi$. Thus $m \prec A'(C')^a B' > \pi - m \prec A'B'C' = m \prec A'B'(C')^a$. So in $\triangle^s A'B'(C')^a$ the opposite sides are also unequal: $m\ \widehat{A'B'} > m\ \widehat{A'(C')^a}$. By Exercise 41, $m\ \widehat{EF^a} > m\ \widehat{B'(C')^a}$, so $\pi - m\ \widehat{EF} > \pi - m\ \widehat{B'C'}$, so $m\ \widehat{EF} < m\ \widehat{B'C'}$. By polar triangles again, $\pi - m \prec BDC < \pi - m \prec BAC$, so $m \prec BDC > m \prec BAC$, as desired.

49. Let $\triangle^s DEF$ be the polar triangle $\triangle^s A'B'C'$ of $\triangle^s ABC$. Then since both triangles are equilateral and equiangular, we find that $\triangle^s DEF$ satisfies the desired properties. To wit: $m \prec A' + m \prec B' + m \prec C' = (\pi - m\ \widehat{BC}) + (\pi - m\ \widehat{AC}) + (\pi - m\ \widehat{AB}) \leq 3\pi - (2\pi - 6\delta) = \pi + 6\delta$. Similarly, $m\ \widehat{A'B'} + m\ \widehat{A'C'} + m\ \widehat{B'C'} = (\pi - m \prec C) + (\pi - m \prec B) + (\pi - m \prec A) < 3\pi - (3\pi - 12\delta) = 12\delta$.

53. Given a spherically convex n-gon, its polar n-gon is also spherically convex, and by Exercise 52, the sum of the measures of its angles is at least $(n-2)\pi$. But since every side of the original n-gon is supplementary to a corresponding angle in the polar n-gon, the sum of the measures of the sides of the n-gon is $n\pi$ minus the sum of the measures of the angles in the polar n-gon. But this last quantity is less than $n\pi - (n-2)\pi = 2\pi$, as desired.

§14

4. The key is to show that a convex n-gon can be subdivided into n overlapping spherical triangles. This is easily done: simply connect any vertex of the n-gon to all the others via diagonal. These diagonals are all in the region enclosed by the n-gon because it is the intersection of hemispheres (so spherically convex). These triples of vertices all form spherical triangles because we assumed that no three lay on a single great circle. Then apply the area of triangle formula to all of these and add them together.

If the polygon is not convex, the formula can still work if the region can be triangulated. One issue that arises is how to define the region bounded by the polygon (interior or exterior?), how to define the angles we add together (they must be the interior angles, which sometimes have measures larger than π). Another issue to consider is what happens if diagonals or sides do not all have measures less than π.

§15

1. If two opposite angles (say $\prec A$ and $\prec B$) are supplementary, then $\sin(A)/\sin(B) = 1$, so $\sin(a)/\sin(b) = 1$. Thus $a = b$ or $a + b = \pi$. If $a \neq b$ then $a + b = \pi$ and we are done. If $a = b$ then by Theorem 11.10 we also have $A = B$, so $A = B = \frac{\pi}{2}$. By Theorem 11.11, $a = b = \frac{\pi}{2}$, so the sides opposite $\prec A$ and $\prec B$ are supplementary, as desired. The converse is similar.

5. Using the spherical law of tangents (4.5), we find that $\tan(\frac{A-B}{2})$ and $\tan(\frac{a-b}{2})$ are both positive, both negative, or both zero since the arguments are between $-\frac{\pi}{2}$ and $\frac{\pi}{2}$ and by Theorem 13.10. If either is zero, the other is

zero and then $A = B$ and $a = b$. In this case, the conclusion of the problem is identical to that of Theorem 11.14. Otherwise $\tan(\frac{A-B}{2})$ and $\tan(\frac{a-b}{2})$ have the same sign, so $\cot(\frac{A+B}{2})$ and $\cot(\frac{a+b}{2})$ are both positive, both negative, or both zero, which implies that $\frac{A+B}{2}$ and $\frac{a+b}{2}$ are both acute, both obtuse, or both right, respectively.

9. Let a, b, c, A, B, and C have their usual meaning. Then $b < c$ and $b + c < \pi$, so by Exercise 7, $\sin(b) < \sin(c)$. By Exercise 6,

$$\sin(\overset{\frown}{CD})/\sin(\overset{\frown}{BD}) < 1,$$

so $\sin(\overset{\frown}{CD}) < \sin(\overset{\frown}{BD})$. Since $m\ \overset{\frown}{BD} + m\ \overset{\frown}{CD} = m\ \overset{\frown}{BC} < \pi$, we conclude that $m\ \overset{\frown}{CD} < m\ \overset{\frown}{BD}$ by Exercise 7.

13. We make use of Exercise 12. Suppose the circles have spherical radii r_1, r_2, and r_3. A point X at distance d_1, d_2, and d_3 from the centers which is on the great circle passing through the intersection of the first two circles must satisfy $\cos(d_1)\cos(r_2) - \cos(d_2)\cos(r_1) = 0$. If X is on the great circle passing through the intersection of the second pair of small circles then $\cos(d_2)\cos(r_3) - \cos(d_3)\cos(r_2) = 0$. If we eliminate $\cos(d_2)$ from these two equations, we obtain $\cos(d_1)\cos(r_3) - \cos(d_3)\cos(r_1) = 0$, using the fact that we can divide by $\cos(r_2) \neq 0$. Thus X is on the great circle passing through the intersection of the third pair of small circles.

17. Suppose (4.8) holds. Then consider the ray with vertex A which passes through the intersection of the two cevians based at B and C. This ray splits the opposite side into arcs of length $x, a - x$, whereas the other cevian splits it into arcs of length $y, a - y$. Then $\sin(x)/\sin(a - x) = \sin(y)/\sin(a - y)$. We must show this implies that $x = y$. Clearing fractions, $\sin(x)\sin(a - y) = \sin(y)\sin(a-x)$. Using sum-to-product formulas, $\cos(x-a+y)-\cos(x+a-y) = \cos(y - a + x) - \cos(y + a - x)$, so $\cos(a + x - y) - \cos(a + y - x) = 0$, so $\sin(a)\sin(x - y) = 0$, so $\sin(x - y) = 0$, which means $x = y$, since $|x - y| < \pi$.

21. Simply apply Exercises 6 and 17.

§16

3a. (i) The coordinates of Boston (point A) are $42.36°N, 71.06°W$, or $(+42.36, -71.06)$. For Paris (point B), we have $(+48.87, +2.35)$. Then the polar distance for Boston is $90° - 42.36° = 47.64°$ and for Paris is $90 - 48.87° = 41.13°$. Let C be the north pole. The difference in longitudes is $73.41°$. Then if c is the spherical distance between them, $\cos(c) = \cos(47.64°)\cos(41.13°) + \sin(47.64°)\sin(41.13°)\cos(73.41°) = 0.646$, so $c = 49.74°$, or 0.87 radians. Multiplying by the radius of the earth (about 3964 miles) we find a distance of about 3449 miles.

(ii) Using the analogue formula,

$$\begin{aligned} & \sin(49.74°)\cos(A) \\ = \ & \cos(41.13°)\sin(47.64°) - \sin(41.13°)\cos(47.64°)\cos(73.41°), \end{aligned}$$

so $\cos(A) = 0.5635$ and $A = 55.7°$. So the bearing from A is $N55.7°E$. Similarly,

$$\begin{aligned} & \sin(49.74°)\cos(B) \\ = & \sin(41.13°)\cos(47.64°) - \cos(41.13°)\sin(47.64°)\cos(73.41°), \end{aligned}$$

so $\cos(B) = 0.373$, $B = 68.13°$. So the bearing from Paris is $N68.13°W$.

(iii) Since both of the angles at A and B are acute, the foot of the shorter perpendicular from C to $\bigcirc AB$ lies between A and B. Call this point D. Then $\sin(A) = \sin(\overset{\frown}{CD})/\sin(\overset{\frown}{AC})$, so

$$\sin(\overset{\frown}{CD}) = \sin(A)\sin(\overset{\frown}{AC}) = \sin(55.7°)\sin(47.64°) = 0.61,$$

so $m\ \overset{\frown}{CD}= 37.62°$, which is about $2,602$ miles.

(iv) We have to find $m\ \overset{\frown}{AD}$ and $m\ \overset{\frown}{BD}$. Since $\overset{\frown}{AD}$ is the third side to a right triangle $\Delta^s ACD$ where the other two sides are known, we find

$$\cos(\overset{\frown}{AD}) = \cos(\overset{\frown}{AC})/\cos(\overset{\frown}{CD}) = \cos(47.64°)/\cos(37.62°) = 0.85,$$

so $m\ \overset{\frown}{AD}= 31.71°$, or 0.55 radians, which is about $2,180$ miles. We can find $m\ \overset{\frown}{BD}$ in a similar manner — since $\Delta^s BCD$ is right — or just subtract the answer just found from $m\ \overset{\frown}{AB}$ (the latter was found in part (i)) to obtain $1,269$ miles.

(v) The latitude of D is the complement of the measure of $\overset{\frown}{CD}$, which is $90°-37.62° = 52.38°$. For the longitude, we should find $m \prec ACD$ and add this to the longitude for Boston. By (4.26), $\cos(\prec ACD) = \cos(\overset{\frown}{AD})\sin(\prec CAD) = (0.85)\sin(55.7°) = 0.70$, so $m \prec ACD = 45.39°$. So the longitude of D is $-71.06° + 45.39° = -25.67°$.

5. Let E be the halfway point. For Boston/Paris, let A be Boston, B be Paris and C be the north pole. Then $m\ \overset{\frown}{AE}= \frac{1}{2}m\ \overset{\frown}{AB}= 24.87°$. Also $m\ \overset{\frown}{AC}= 47.64°$, and $m \prec C = 55.7°$. Then using the law of cosines, $\cos(\overset{\frown}{CE}) = \cos(24.87°)\cos(47.64°) + \sin(24.87°)\sin(47.64°)\cos(55.7°) = 0.78$, so $m\ \overset{\frown}{CE}= 38.14°$. Thus the latitude of E is $90° - 38.14° = 51.96°$. Next, the analogue formula applied to $\Delta^s ACD$ gives

$$\begin{aligned} & \sin(38.14°)\cos(\prec ACE) \\ = & \cos(\overset{\frown}{AE})\sin(\overset{\frown}{AC}) - \cos(\overset{\frown}{AC})\sin(\overset{\frown}{AE})\cos(\prec A) \\ = & \cos(24.87°)\sin(47.64°) - \sin(24.87°)\cos(47.64°)\cos(55.7°) \\ = & 0.511, \end{aligned}$$

so $\cos(\prec ACE) = 0.83$, and $m \prec ACE = 34.21°$. Adding this to the longitude of Boston, the longitude of E is $-71.06° + 34.21° = -36.84°$, or $36.84°W$.

6. The route is due east/west at some point if the foot of the altitude from the pole sits on the route at some point. This point has already been studied in Exercise 3a.

9. Let N be the north pole, A the initial point and B the point reached after 500 miles of travel. Then $m\ \overset{\frown}{NB}= 45°$, $m\ \prec NBA = 150°$ and $m\ \overset{\frown}{AB}=$ $(500/3964)(180°/\pi) \approx 7.23°$. We first find $m\ \overset{\frown}{NA}$ which is the complement of the initial latitude. We have

$$\cos(\overset{\frown}{NA}) = \cos(7.23°)\cos(45°) + \sin(7.23°)\sin(45°)\cos(150°) \approx 0.62,$$

so $m\ \overset{\frown}{NA}\approx 51.36°$. So the initial latitude is about 38.64°. We next use the analogue formula on $\triangle^s ANB$ to find $m\ \prec ANB$. We have

$$\sin(51.36°)\cos(\prec ANB) = \cos(7.23°)\sin(45°) - \sin(7.23°)\cos(45°)\cos(150°)$$

so $\cos(\prec ANB) \approx 0.996$ and $m\ \prec ANB \approx 4.61°$. So the initial longitude was $44.61°W$. For the initial bearing we use the analogue formula to find $m\ \prec NAB$:

$$\sin(51.36°)\cos(\prec NAB) = \sin(7.23°)\cos(45°) - \cos(7.23°)\sin(45°)\cos(150°).$$

So $\cos(\prec NAB) = 0.89$ and $m\ \prec NAB = 26.9°$. Thus the initial course was $N26.9°E$.

13. From the law of cosines equation (4.10), note that $\cos(C) > -1$, so $\cos(c) > \cos(a)\cos(b) - \sin(a)\sin(b) = \cos(a+b)$. But since the cosine is decreasing in the angle, $c < a + b$.

17. Using the law of cosines four times,

$$\begin{aligned}
\cos(\overset{\frown}{AD}) &= \cos(\overset{\frown}{AE})\cos(\overset{\frown}{ED}) + \sin(\overset{\frown}{AE})\sin(\overset{\frown}{ED})\cos(\prec AED)\\
\cos(\overset{\frown}{CD}) &= \cos(\overset{\frown}{CE})\cos(\overset{\frown}{ED}) - \sin(\overset{\frown}{CE})\sin(\overset{\frown}{ED})\cos(\prec AED)\\
\cos(\overset{\frown}{AB}) &= \cos(\overset{\frown}{AE})\cos(\overset{\frown}{EB}) - \sin(\overset{\frown}{AE})\sin(\overset{\frown}{EB})\cos(\prec AED)\\
\cos(\overset{\frown}{BC}) &= \cos(\overset{\frown}{BE})\cos(\overset{\frown}{EC}) + \sin(\overset{\frown}{BE})\sin(\overset{\frown}{EC})\cos(\prec AED)
\end{aligned}$$

Then $\cos(\overset{\frown}{AD})\cos(\overset{\frown}{BC}) - \cos(\overset{\frown}{AB})\cos(\overset{\frown}{CD})$ equals

$$\begin{aligned}
&\cos(\overset{\frown}{AE})\cos(\overset{\frown}{ED})\cos(\overset{\frown}{BE})\cos(\overset{\frown}{EC})\\
&+\cos(\overset{\frown}{AE})\cos(\overset{\frown}{ED})\sin(\overset{\frown}{BE})\sin(\overset{\frown}{EC})\cos(\prec AED)\\
&+\cos(\overset{\frown}{BE})\cos(\overset{\frown}{EC})\sin(\overset{\frown}{AE})\sin(\overset{\frown}{ED})\cos(\prec AED)\\
&+\sin(\overset{\frown}{AE})\sin(\overset{\frown}{ED})\sin(\overset{\frown}{BE})\sin(\overset{\frown}{EC})\cos^2(\prec AED)\\
&-\\
&(\cos(\overset{\frown}{AE})\cos(\overset{\frown}{ED})\cos(\overset{\frown}{BE})\cos(\overset{\frown}{EC})
\end{aligned}$$

$$
\begin{aligned}
&-\cos(\overset{\frown}{AE})\cos(\overset{\frown}{EB})\sin(\overset{\frown}{DE})\sin(\overset{\frown}{EC})\cos(\prec AED)\\
&-\cos(\overset{\frown}{DE})\cos(\overset{\frown}{EC})\sin(\overset{\frown}{AE})\sin(\overset{\frown}{EB})\cos(\prec AED)\\
&+\sin(\overset{\frown}{AE})\sin(\overset{\frown}{ED})\sin(\overset{\frown}{BE})\sin(\overset{\frown}{EC})\cos^2(\prec AED))
\end{aligned}
$$

which is

$$
\begin{aligned}
&\cos(\overset{\frown}{AE})\cos(\overset{\frown}{ED})\sin(\overset{\frown}{BE})\sin(\overset{\frown}{EC})\cos(\prec AED)\\
&+\cos(\overset{\frown}{BE})\cos(\overset{\frown}{EC})\sin(\overset{\frown}{AE})\sin(\overset{\frown}{ED})\cos(\prec AED)\\
&+\cos(\overset{\frown}{AE})\cos(\overset{\frown}{EB})\sin(\overset{\frown}{DE})\sin(\overset{\frown}{EC})\cos(\prec AED)\\
&+\cos(\overset{\frown}{DE})\cos(\overset{\frown}{EC})\sin(\overset{\frown}{AE})\sin(\overset{\frown}{EB})\cos(\prec AED),
\end{aligned}
$$

i.e.,

$$
\begin{aligned}
&\cos(\prec AED)\\
\cdot\ &\Big(\cos(\overset{\frown}{AE})\sin(\overset{\frown}{EC})\big(\cos(\overset{\frown}{ED})\sin(\overset{\frown}{BE})+\cos(\overset{\frown}{EB})\sin(\overset{\frown}{DE})\big)\\
+\ &\cos(\overset{\frown}{EC})\sin(\overset{\frown}{AE})\big(\cos(\overset{\frown}{ED})\sin(\overset{\frown}{BE})+\cos(\overset{\frown}{EB})\sin(\overset{\frown}{DE})\big)\Big),
\end{aligned}
$$

or

$$
\begin{aligned}
&\cos(\prec AED)\\
\cdot\ &\big(\cos(\overset{\frown}{ED})\sin(\overset{\frown}{BE})+\cos(\overset{\frown}{EB})\sin(\overset{\frown}{DE})\big)\\
\cdot\ &\big(\cos(\overset{\frown}{AE})\sin(\overset{\frown}{EC})+\cos(\overset{\frown}{EC})\sin(\overset{\frown}{AE})\big)
\end{aligned}
$$

i.e., $\cos(\overset{\frown}{AD})\cos(\overset{\frown}{BC})-\cos(\overset{\frown}{AB})\cos(\overset{\frown}{CD}) = \cos(\prec AED)\sin(\overset{\frown}{AC})\sin(\overset{\frown}{BD})$, so $\cos(\prec AED) = \big(\cos(\overset{\frown}{AD})\cos(\overset{\frown}{BC}) - \cos(\overset{\frown}{AB})\cos(\overset{\frown}{CD})\big)\csc(\overset{\frown}{AC})\csc(\overset{\frown}{BD})$, as desired.

21. Let the triangle be $\triangle^s ABC$. Let D be a point of $\overset{\frown}{BC}$, let $a = m\ \overset{\frown}{BC}$, $b = m\ \overset{\frown}{AC}$, $c = m\ \overset{\frown}{AB}$, and $d = m\ \overset{\frown}{AD}$.

We assume $b + c < \pi$ and D divides $\overset{\frown}{BC}$ so that $a_1 = m\ \overset{\frown}{BD}$ and $a_2 = m\ \overset{\frown}{CD}$. Using Exercise 18b,

$$\sin(A/2)(\cot(b) + \cot(c)) = \cot(d)\sin(A) = 2\sin(A/2)\cos(A/2)\cot(d).$$

Thus $\cot(b) + \cot(c) = 2\cos(A/2)\cot(d)$. The left side is positive since $b + c < \pi$. So the right side is also, so $\cot(d) > 0$, hence $d < \pi/2$. Now by §3, Exercise 16,

$$\cot(\frac{b+c}{2}) \le \frac{\cot(b)+\cot(c)}{2}.$$

Then we find that $\cot(\frac{b+c}{2}) \leq \cos(A/2)\cot(d) < \cot(d)$, which will imply that $d < \frac{b+c}{2}$.

28. Suppose a spherical ray emanating from B intersects $\overset{\frown}{AC}$ between A and C. Let point D in the interior of $\triangle^s ABC$ vary on this ray, and assume that always $m \prec ADB > m \prec ACB$. If the ray meets $\overset{\frown}{AC}$ at point E, then by continuity, $m \prec AEB \geq m \prec ACB$ for any E between A and C. Then $m\,\overset{\frown}{BC} + m\,\overset{\frown}{BE} \leq \pi$ for any E between A and C by §13, Exercise 9. Since the measure of $\overset{\frown}{BE}$ varies continuously with E on $\overset{\frown}{AC}$, the same inequality holds when $E = A$ (so $a + c \leq \pi$) and when $E = C$ (so $2a \leq \pi$, or $a \leq \frac{\pi}{2}$.) A similar argument shows that $b + c \leq \pi$ and $b \leq \frac{\pi}{2}$.)

§17

1a. We have $c = 70°$ between $25°$ and $155°$ so a unique solution exists. We get $\cos(b) = \cos(70°)/\cos(25°) \approx 0.38$, so $b \approx 67.8°$. Since a and b are acute, A and B are acute also. Then $\sin(A) = \sin(a)/\sin(c) \approx 0.45$ so $A \approx 26.7°$. Then $\cos(B) = \tan(a)/\tan(c) \approx 0.17$ so $B \approx 80.2°$.

1b. We have $c = 125°$ between $25°$ and $155°$ so a unique solution exists. We get $\cos(b) = \cos(125°)/\cos(25°) \approx -0.63$, so $b \approx 129°$. Since a is acute and b is obtuse, A is acute and B is obtuse. Then $\sin(A) = \sin(a)/\sin(c) \approx 0.52$ so $A \approx 31.1°$. Then $\cos(B) = \tan(a)/\tan(c) \approx -0.32$. so $B \approx 109.1°$.

1c. We have $c = 70°$ between $60°$ and $120°$ so a unique solution exists. We get $\cos(b) = \cos(70°)/\cos(120°) \approx -0.68$, so $b \approx 133.1°$. Since a is obtuse and b is obtuse, A is obtuse and B is obtuse. Then $\sin(A) = \sin(a)/\sin(c) \approx 0.92$ so $A \approx 112.8°$. Then $\cos(B) = \tan(a)/\tan(c) \approx -0.63$. so $B \approx 129°$.

1d. Since $c = 90°$, $0 = \cos(90°) = \cos(40°)\cos(b)$, so $b = 90°$. Then $B = 90°$ and $A = 40°$.

1e. We need $\cos(40°) = \cos(90°)\cos(b)$, so $\cos(40°) = 0$, a contradiction. So there is no solution.

1f. Since $40°$ is not between $60°$ and $120°$, there is no solution.

1g. Since $40°$ is not between $150°$ and $30°$, there is no solution.

5a. We have $\cos(c) = \cot(80°)\cot(20°) \approx 0.48$, so $c \approx 61°$. Next, $\cos(a) = \cos(A)/\sin(B) = \cos(80°)/\sin(20°) \approx 0.51$. So $a \approx 59.5°$. Then $\cos(b) = \cos(B)/\sin(A) = \cos(20°)/\sin(80°) \approx 0.95$. So $b \approx 17.4°$.

5b. We have $\cos(c) = \cot(80°)\cot(70°) \approx 0.06$, so $c \approx 86.3°$. Next, $\cos(a) = \cos(A)/\sin(B) = \cos(80°)/\sin(70°) \approx 0.18$. So $a \approx 79.4°$. Then $\cos(b) = \cos(B)/\sin(A) = \cos(70°)/\sin(80°) \approx 0.34$. So $b \approx 69.6°$.

5c. We have $\cos(c) = \cot(47°)\cot(46°) \approx 0.9$, so $c \approx 25.7°$. Next, $\cos(a) = \cos(A)/\sin(B) = \cos(47°)/\sin(46°) \approx 0.94$. So $a \approx 18.5°$. Then $\cos(b) = \cos(B)/\sin(A) = \cos(46°)/\sin(47°) \approx 0.95$. So $b \approx 18.2°$.

5d. We have $\cos(c) = \cot(135°)\cot(50°) \approx -0.83$, so $c \approx 147°$. Next, $\cos(a) = \cos(A)/\sin(B) = \cos(135°)/\sin(50°) \approx -0.92$. So $a \approx 157°$. Then $\cos(b) = \cos(B)/\sin(A) = \cos(50°)/\sin(135°) \approx 0.91$. So $b \approx 24.6°$.

5e. We have $\cos(c) = \cot(135°)\cot(133°) \approx 0.93$, so $c \approx 21.2°$. Next, $\cos(a) = \cos(A)/\sin(B) = \cos(135°)/\sin(133°) \approx -0.97$. So $a \approx 165.2°$. Then

$\cos(b) = \cos(B)/\sin(A) = \cos(133°)/\sin(135°) \approx -0.96$. So $b \approx 165.7°$.

5f. We have $\cos(c) = \cot(45.1°)\cot(45.1°) \approx 0.99$, so $c \approx 6.8°$. Next, $\cos(a) = \cos(A)/\sin(B) = \cos(45.1°)/\sin(45.1°) \approx 0.99$. So $a \approx 4.78°$. Then $\cos(b) = \cos(B)/\sin(A) = \cos(45.1°)/\sin(45.1°) \approx 0.99$. So $b \approx 4.78°$.

5g. We have $\cos(c) = \cot(65.2°)\cot(154.9°) \approx -0.99$, so $c \approx 171°$. Next, $\cos(a) = \cos(A)/\sin(B) = \cos(65.2°)/\sin(154.9°) \approx 0.98$. So $a \approx 8.8°$. Then $\cos(b) = \cos(B)/\sin(A) = \cos(154.9°)/\sin(65.2°) \approx -0.99$. So $b \approx 176°$.

5h. Since $A = C = 90°$, $a = c = 90°$. Then $b = B = 40°$.

5i. We have $\cos(c) = \cot(100°)\cot(70°) \approx -.06$, so $c \approx 93.6°$. Next, $\cos(a) = \cos(A)/\sin(B) = \cos(100°)/\sin(70°) \approx -0.18$. So $a \approx 100.6°$. Then $\cos(b) = \cos(B)/\sin(A) = \cos(70°)/\sin(100°) \approx 0.34$. So $b \approx 69.7°$.

5j. Because $A + B = 80° < 90°$ there is no solution.

5k. Because $|A - B| = 100° > 90°$ there is no solution.

5l. Because $|A + B| = 280° > 270°$ there is no solution.

9a. Using (4.22), we find that $\cos(c) = \cot(A)\cot(B)$. If $c = \frac{\pi}{2}$, $\cos(c) = 0$, so either $\cos(A) = 0$ or $\cos(B) = 0$, so either A or B is $\frac{\pi}{2}$.

9b. Using (4.21), $\cos(c) = \cos(a)\cos(b)$. If the hypotenuse is acute, $\cos(c) > 0$, so $\cos(a)$ and $\cos(b)$ have the same sign — so these sides are both acute or both obtuse (and are not right.)

9c. Using (4.21), $\cos(c) = \cos(a)\cos(b)$. If the hypotenuse is obtuse, $\cos(c) < 0$, so $\cos(a)$ and $\cos(b)$ have differerent (nonzero) signs, so the associated sides are not right, one is acute and the other obtuse. (One could also use (4.22) in connection with Proposition 11.12.)

9d. Suppose that $\triangle^s ABC$ has a right angle at C. Then by (4.30), since $\sin(B) > 0$, $\cos(A)$ and $\cos(a)$ have the same sign, so a and A are both right, both acute, or both obtuse.

9e. Suppose we are given $\triangle^s ABC$ and $\triangle^s DEF$ with right angles at C and F, $c = f$, and $A = D$ are both non-right. So $\cot(A) = \cot(D)$. By (4.22), $\cot(A)\cot(E) = \cot(D)\cot(E) = \cot(f) = \cot(c) = \cot(A)\cot(B)$. Since A is non-right, $\cot(A) \neq 0$, so $\cot(E) = \cot(B)$. Since B and E are both between $0°$ and $180°$, we find $B = E$. Thus there is an ASA correspondence between $\triangle^s ABC$ and $\triangle^s DEF$, so by Theorem 12.4, $\triangle^s ABC \cong \triangle^s DEF$.

9f. Suppose that the triangle is $\triangle^s ABC$, with $\frown{AB} \cong \frown{AC}$. Let D be the midpoint of $\frown{BC}$. Then by §12, Exercise 3, $\prec ADB$ is a right angle. Applying (4.24), we find that $\cot(b) = \cot(a/2)\cos(B)$, using the fact that $B = C$ and $b = c$. Since $a/2$ is acute, its cotangent is positive. So $\cot(b)$ and $\cos(B)$ are both positive, negative, or zero, which means that b and B are both acute, right, or obtuse.

13. Suppose $a = A$. Then from (4.30), $\cos(A) = \sin(B)\cos(a)$. Either $\cos(A) = 0$ (so $A = a = 90°$) or $\cos(A) \neq 0$ so $\sin(B) = 1$ and $B = 90°$. Then by Proposition 11.12, $b = c = 90°$.

17. Any two triangles having corresponding hypotenuses with measure c and angles with measure $B \neq 90°$ must be congruent by the hypotenuse-angle theorem (Theorem 12.10.) This proves uniqueness. To prove existence, we have

different options. Option 1: Construct spherical ray $\overset{\hookrightarrow}{BA}$ so that $m\ \widehat{BA}= c$. Then choose ray $\overset{\hookrightarrow}{r}$ with endpoint B which forms an angle of measure B with $\overset{\hookrightarrow}{BA}$. By §11 Exercise 5 there is a point C on $\overset{\hookrightarrow}{r}$ such that $\prec ACB$ is a right angle. Then $\triangle^s ABC$ satisfies the desired properties. Option 2: if $c = 90°$, a solution would be a triangle where $c = C = a = A = 90°$ and $b = B$ are as given. If $c \neq 90°$, $\tan(c)$ is defined, so let a be the unique solution to the equation $\tan(a) = \tan(c)\cos(B)$ for $0 < a < 180°$, $a \neq 90°$. Then we may construct an angle $\prec ABC$ of measure B where $m\ \widehat{BC}= a$ and $m\ \widehat{BA}= c$. We only have to check that $\triangle^s ABC$ has $C = 90°$ and this will prove existence. For this note that the analogue formula (4.16) gives $\sin(b)\cos(C) = \cos(c)\sin(a) - \sin(c)\cos(a)\cos(B) = \cos(c)\cos(a)(\tan(a) - \tan(c)\cos(B)) = 0$, so $\cos(C) = 0$ and the angle at C is right, as desired.

21. Let D be the midpoint of side $\widehat{AB}$. Then $\triangle^s CDB$ is a right triangle with a right angle at D. (See §12, Exercise 3.) Applying (4.22), we obtain $\cos(a) = \cot(B)\cot(\frac{C}{2})$ and applying (4.24) we obtain $\cos(B)\cot(\frac{c}{2}) = \cot(a)$. If B is not a right angle, then $\cos(B) \neq 0$ and $\cot(B) \neq 0$ so that in the above equations we may divide by $\cos(B)$ and $\cot(B)$ to obtain values for $\cot(\frac{C}{2})$ and $\cot(\frac{c}{2})$. Since C and c are between 0 and π, $\frac{C}{2}$ and $\frac{c}{2}$ are between 0 and $\frac{\pi}{2}$, so $\cot(\frac{C}{2})$ and $\cot(\frac{c}{2})$ are positive. Then we take an inverse cotangent (providing a unique value for the argument between 0 and $\frac{\pi}{2}$) and solve for c and C (obtaining a unique answer for C and c between 0 and π).

If A and B are right angles, $\cot(B) = \cos(B) = 0$, so $\cos(a) = 0$, and we find that $a = b = \frac{\pi}{2}$. Thus $\widehat{CA}$ and $\widehat{CB}$ are right sides, and by Definition 10.2, $C = c$. But C and c could otherwise take any value between 0 and π and are not determined by A, B, a, and b.

25. We apply the law of sines to $\triangle^s PCE$:

$$\frac{\sin(\widehat{CE})}{\sin \prec CPE} = \frac{\sin(\widehat{PE})}{\sin \prec PCE}.$$

Since P is the pole of $\bigcirc AB$, $m \prec CPE = m \prec BPD = m\ \widehat{BD}$, $\prec CBA$ and $\prec EDA$ are right angles, and $m\ \widehat{BC}$, $m\ \widehat{DE}$, are complementary to $m\ \widehat{PC}$, $m\ \widehat{PE}$, respectively. Since $\prec PCE$ and $\prec ACB$ are vertical angles, $\prec PCE \cong \prec ACB$. By (4.36), $\sin(\prec ACB) = \cos(\prec BAC)/\cos(\widehat{BC})$. Thus

$$\begin{aligned}\sin(\widehat{CE})/\sin(\widehat{BD}) &= \cos(\widehat{DE})/(\cos(\prec BAC)/\cos(\widehat{BC})) \\ &= \cos(\widehat{DE})\cos(\widehat{BC})/\cos(\prec BAC).\end{aligned}$$

Inverting both sides gives us the desired result.

§18

2. We have

$$\sin(A) = 2\sin(\frac{A}{2})\cos(\frac{A}{2}) = \frac{2\sqrt{\sin(s)\sin(s-a)\sin(s-b)\sin(s-c)}}{\sin(b)\sin(c)}.$$

Then $\sin(a)/\sin(A)$ is as given in (4.55).

5. For equations (4.25) and (4.35), we begin with the four-parts formula (4.42) and let $C = \frac{\pi}{2}$; then $\sin(C) = 1$ and $\cos(C) = 0$, so $\cot(a)\sin(b) = \cot(A)(1) + (0)\cos(b)$, which gives $\cot(a)\sin(b) = \cot(A)$, or (4.25). We obtain (4.35) when we solve for $\tan(A)$. For equation (4.34), we take the four-parts formula (4.42) and reverse the roles of A, a and C, c to get

$$\cot(c)\sin(b) = \cot(C)\sin(A) + \cos(A)\cos(b)$$

and let $C = \frac{\pi}{2}$; then $\cot(C) = 0$. Then $\cot(c)\sin(b) = \cos(A)\cos(b)$, so solving for $\cot(c)$, we get (4.24). We get (4.34) when we can solve for $\cos(A)$.

9. Assuming a, b, and c given, note that the assumption $a + b > c$ implies $s-c > 0$ and $a+b-c < a+b+c < 2\pi$. Thus $0 < s-c < \pi$ and so $\sin(s-c) > 0$. Similarly from our assumptions we obtain $\sin(s-a) > 0$ and $\sin(s-b) > 0$. Lastly, we assume that $0 < a + b + c < 2\pi$, so $0 < s < \pi$ and $\sin(s) > 0$. Now we let C be defined by (4.51); because the factors of the radicand are positive, there exists a $C/2$ satisfying (4.51) such that $0 < C/2 < \pi/2$. Hence $0 < C < \pi$. By (1.14),

$$\frac{1-\cos(C)}{1+\cos(C)} = \frac{\sin(s-a)\sin(s-b)}{\sin(s)\sin(s-c)}. \tag{A.4}$$

By sum-to-product formulas,

$$2\sin(s-a)\sin(s-b) = \cos(a-b) - \cos(c). \tag{A.5}$$

An analogous use of sum-to-product formulas yields

$$2\sin(s)\sin(s-c) = \cos(c) - \cos(a+b). \tag{A.6}$$

Dividing these, and combining with (A.4) we find that

$$\frac{1-\cos(C)}{1+\cos(C)} = \frac{\cos(a-b)-\cos(c)}{\cos(c)-\cos(a+b)}, \tag{A.7}$$

so $(1 - \cos(C))(\cos(c) - \cos(a + b)) = (1 + \cos(C))(\cos(a - b) - \cos(c))$. Multiplying out and gathering like terms, we get $\cos(c) = \cos(a)\cos(b) + \sin(a)\sin(b)\cos(C)$. Now suppose we construct a spherical triangle with adjacent sides of spherical measure a and b with an angle C between them. If the side opposite the angle C has length c', it must also satisfy $\cos(c') = \cos(a)\cos(b) + \sin(a)\sin(b)\cos(C)$, which means $\cos(c) = \cos(c')$. Since both c

and c' are between 0 and π, we have $c = c'$, and the desired triangle has been constructed.

§19

1. The dual is

$$\sin(\pi - C)\cos(\pi - b) = \cos(\pi - B)\sin(\pi - A) - \sin(\pi - B)\cos(\pi - A)\cos(\pi - c),$$

or

$$-\sin(C)\cos(b) = -\cos(B)\sin(A) - \sin(B)\cos(A)\cos(c),$$

i.e.,

$$\sin(C)\cos(b) = \cos(B)\sin(A) + \sin(B)\cos(A)\cos(c).$$

5. We divide (4.63) by (4.62) to obtain (4.64).

9. For equation (4.22), we refer to (4.60) and let $C = \frac{\pi}{2}$:

$$0 = -\cos(A)\cos(B) + \sin(A)\sin(B)\cos(c).$$

Solving for $\cos(c)$, we obtain $\cos(c) = \cot(A)\cot(B)$. (Note that since A and B are both between 0 and π, $\sin(A)$ and $\sin(B)$ are both nonzero, so the cotangent is always defined for these values of A and B.)

For equation (4.36), we start with (4.60) and reverse the roles of B, b and C, c to get

$$\cos(B) = -\cos(C)\cos(A) + \sin(C)\sin(A)\cos(b).$$

We let $C = \frac{\pi}{2}$; then $\sin(C) = 1$ and $\cos(C) = 0$ again, so $\cos(B) = 0 + \sin(A)\cos(b)$, which is (4.26). If we can solve for $\sin(A)$, we get (4.36).

13. Use formulas (4.63) and (4.62) to substitute for $\sin(\frac{a}{2})$, $\sin(\frac{b}{2})\cos(\frac{c}{2})$ and obtain the first formula for $\sin(E)$. Then multiply top and bottom by $4\cos(\frac{a}{2})\cos(\frac{b}{2})$ and use the double angle formula for the sine on the top to get

$$\sin(E) = \frac{\sin(a)\sin(b)\sin(C)}{4\cos(\frac{a}{2})\cos(\frac{b}{2})\cos(\frac{c}{2})}. \tag{A.8}$$

Use the last equality in (4.55) to get another expression for $\sin(a)\sin(b)\sin(C)$. Substitute into (A.8) to get (4.68).

17. In the polar triangle $\triangle^s A'B'C'$, we have that the value of $a' = \pi - A$ is no closer to $\frac{\pi}{2}$ than the value of $b' = \pi - B$. By §16, Exercise 14, $\prec A'$ is acute, right, or obtuse if and only if the opposite side is acute, right, or obtuse, respectively. Since $A' = \pi - a$ and $a' = \pi - A$, the conclusion follows.

§20

1a. We have

$$\cos(c) = \cos(20^\circ)\cos(35^\circ) + \sin(20^\circ)\sin(35^\circ)\cos(76^\circ) = 0.8172\ldots,$$

so $c = 35.1935\ldots^\circ$. Then $\sin(c) = 0.5763\ldots$. Using (4.13), $\sin(c)\cos(B) = \cos(35^\circ)\sin(20^\circ) - \sin(35^\circ)\cos(20^\circ)\cos(76^\circ) = 0.1497\ldots$, so $\cos(B) = 0.2598\ldots$, and $B = 74.93\ldots^\circ$. Similarly,

$$\begin{aligned}\sin(c)\cos(A) &= \cos(a)\sin(b) - \sin(a)\cos(b)\cos(C)\\ &= \cos(20^\circ)\sin(35^\circ) - \sin(20^\circ)\cos(35^\circ)\cos(76^\circ)\\ &= 0.4712\ldots,\end{aligned}$$

so $\cos(A) = 0.8176$ and $A = 35.1505\ldots^\circ$.

5a. We calculate $\sin(B) = \sin(\frac{\pi}{5}) < 1$ via the law of sines. Thus $B = \frac{\pi}{5}$ or $B = \frac{4\pi}{5}$. We eliminate the latter with part (c). If $B = \frac{\pi}{5}$, by part (d1), since a, A, b, and B are not all acute or all obtuse, there is no solution.

5b. We find $\sin(B) = 1$, so $B = 90^\circ$. Note that $a-b < 0$ and $A-B < 0$, and A and a are both acute. From (e), $\tan(\frac{1}{2}c) = \tan(-30^\circ)\sin(60^\circ)/\sin(-30^\circ) = 1$, so $c = 90^\circ$. Also, $\cot(\frac{1}{2}C) = \tan(60^\circ)\cos(60^\circ)/\cos(-30^\circ) = 1$, so $C = 90^\circ$. Of course, it is more natural to solve for c amd C via other formulas, due to the right angle at B: $\cos(c) = \cos(b)/\cos(a) = 0$ and $\cot(C) = \cot(b)/\cot(a) = 0$. See Exercise 19.

12. Let S be the south pole, A the initial position of the ship and B the final position. Then we have the measures of the sides of $\triangle^s SAB$. We have $m\,\overset{\frown}{AB} = \frac{2000}{3964}\frac{180^\circ}{\pi} \approx 28.9^\circ$. By the spherical law of cosines,

$$\cos(\sphericalangle ASB) \approx \frac{\cos(28.9^\circ) - \cos(60^\circ)\cos(45^\circ)}{\sin(60^\circ)\sin(45^\circ)} \approx 0.85,$$

so $m\sphericalangle ASB \approx 31.54^\circ$. The ship could have sailed southeast or southwest to arrive at longitude $135^\circ \pm 31.54^\circ$, i.e., $166.54^\circ W$ or $103.46^\circ W$.

16. Suppose we are given three real numbers B, C, and a between 0 and π. Then if we take the numbers $\pi - B$, $\pi - C$ and $\pi - a$, there exists a triangle with two sides having radian measure $\pi - B$, $\pi - C$ and included angle having radian measure $\pi - a$. The polar triangle has two angles with measure B and C and included side a.

§21.

5. In the northern hemisphere winter, the declination of the sun attains a minimum of -23.5°, so the full moon is within 5° of the antipode of the sun, which has declination 23.5°. Thus the moon has declination between 18.5° and 28.5°. At latitude $40^\circ N$, the altitude of the moon at midnight is between $90-40+18.5$ and $90-40+28.5$ degrees, i.e., between 68.5° and 78.5°. In the summer, the moon's declination is between -18.5° and -28.5°, so its altitude at midnight is between 21.5° and 31.5°. For a southern hemisphere location, the numbers are similar except that the altitude of the moon is from the northern horizon instead of the southern. Regardless of the northern/southern location of the observer, the moon is closer to the visible pole than the sun in the winter and further away in the summer. This means the moon is higher in altitude at midnight in locations well northern and southern, but not so

near the equator. For a northern location, the moon is furthest south when its declination is $-28.5°$, which is not visible at midnight above latitude 61.5° north. A similar value holds for a southern location.

9. Assuming the sun does not move against the background of fixed stars, it travels in a (small) circle as the earth rotates. The shadows it casts are the result of intersecting the line between the sun and the tip of the rod with a plane. The union of all these lines between the sun and the tip of the rod is roughly a circular cone, and its intersection with a flat plane then must be a conic section. If the sun does not set (i.e., if the pole distance of the sun is less than the latitude of the observer), the result is an ellipse (or circle.) If the pole distance is exactly equal to the latitude, this case produces a parabola. If the pole distance is greater than the latitude, we have a branch of a hyperbola or line. For equation derivations, see [Do1965], p. 340.

§22.

1a. $a = 40.7°$, $A = -82.04°$

1b. $a = 40.7°$, $A = 82.04°$

1c. $a = -13°$, $A = 172.1°$.

1d. $a = -13°$, $A = -172.1°$.

5. Let $X = \odot$. In $\triangle^s PZX$, $m\ \overset{\frown}{ZP} = 90° - \phi$, $m\ \overset{\frown}{PX} = 90° - \delta = 80°$ and $m\ \overset{\frown}{ZX} = 90° - a$. We have three sides of a triangle and can use the law of cosines to find the angles. We have $\cos(90° - a) = \cos(90° - \delta)\cos(90° - \phi) + \sin(90° - \delta)\sin(90° - \phi)\cos(h)$ so $\sin(a) = \sin(\delta)\sin(\phi) + \cos(\delta)\cos(\phi)\cos(h)$ and $\cos(h) = (\sin(a) - \sin(\delta)\sin(\phi))/\cos(\delta)\cos(\phi)$. Let $\phi = 40°$, $\delta = 10°$ and $a = 45°$ and solve for h. The argument is similar to find $A = m \prec PZX$. There are two possibilities for h and A, one negative and one positive. The negative values represent positions for the sun east of the observer's meridian and the positive values represent positions of the sun west of the observer's meridian.

9. Here we are given $m\ \overset{\frown}{XZ} = 90° - 25° = 65°$, $m\ \overset{\frown}{XP} = 90° - (-5°)) = 95°$ and $m \prec ZPX = 50°$. We can find A, ϕ by solving $\triangle^s XPZ$ in the case SSA. We have $\frac{\sin(65°)}{\sin(50°)} = \frac{\sin(95°)}{\sin(A)}$ so $\sin(A) = \sin(95°)\sin(50°)/\sin(65°) \approx 0.84$ so $A \approx 57.4°$ or $A \approx 122.6°$. Since $65 < 95$ we must have $50 < A$ but both candidates for A satisfy that. So there are two possible positions for the star's position in the west. We use (4.65) to find ϕ if $A \approx 57.4°$:

$$\frac{\sin\frac{1}{2}(95-65)}{\sin\frac{1}{2}(95+65)} = \frac{\tan\frac{1}{2}(57.4-50)}{\cot\frac{1}{2}(90° - \phi)}$$

so $\tan\frac{1}{2}(90° - \phi) \approx 4.1$, $90° - \phi \approx 152.5°$ and $\phi \approx -62.5°$.

If $A \approx 122.6°$:

$$\frac{\sin\frac{1}{2}(95-65)}{\sin\frac{1}{2}(95+65)} = \frac{\tan\frac{1}{2}(122.6-50)}{\cot\frac{1}{2}(90° - \phi)}$$

so $\tan\frac{1}{2}(90° - \phi) \approx 0.36$, $90° - \phi \approx 39.4°$ and $\phi \approx 50.6°$.

15. Suppose P Z, and N are the pole, zenith, and north point on the horizon and S and T are the two stars when they are both rising (so on the horizon). The time between culminations easily gives us $m \prec SPT$. Since the declinations are known, we know $m\ \overset{\frown}{SP}$ and $m\ \overset{\frown}{TP}$. Then we know the sides and angles of $\triangle^s SPT$. We need to determine the altitude of this triangle from P. To do this, we first find $\prec PTS$ by solving $\triangle^s PTS$. This gives us $\prec PTN$ in $\triangle^s PTN$. Since we have $m\ \overset{\frown}{PT}$ and $m \prec PTN$ in right triangle $\triangle^s PTN$, we can solve right $\triangle^s PTN$ to obtain $\overset{\frown}{PN}$. More information is needed to determine the sign of the latitude, such as which star culminates first.

§23.

1. For New York, Rigel: not circumpolar since $|-8.2°| < 90° - 40°$. Without refraction, $h = \pm 5.52^h$, $A = \pm 100.8°$. With refraction, $h = \pm 5.57^h$, $A = \pm 100.33°$.

§26.

1. Suppose that the polyhedron is projected to a sphere of radius r containing it (possible by convexity, in fact also for star-convex polyhedra). Suppose that for $i = 1$ to F, the ith face has p_i vertices (and edges) with angle measures θ_{ij} for $j = 1$ to p_i. The area of the ith face is

$$\Big((\sum_{j=1}^{p_i} \theta_{ij}) - (p_i - 2)\pi \Big) r^2$$

The sum of all the areas is

$$\sum_{i=1}^{F} \Big((\sum_{j=1}^{p_i} \theta_{ij}) - (p_i - 2)\pi \Big) r^2 \tag{A.9}$$

which is

$$\sum_{i=1}^{F} (\sum_{j=1}^{p_i} \theta_{ij}) r^2 - \sum_{i=1}^{F} ((p_i - 2)\pi) r^2.$$

The first sum is merely r^2 times the sum of all the angle measures in all faces; around each vertex this is just 2π, so the first sum is $2\pi r^2 V$. The second sum is

$$\sum_{i=1}^{F} ((p_i - 2)\pi) r^2 = \pi r^2 (\sum_{i=1}^{F} p_i - \sum_{i=1}^{F} (2)) = \pi r^2 (2E - 2F).$$

Thus (A.9) is $\pi r^2 (2V - 2E + 2F)$, which is the area of the sphere $(4\pi r^2)$. Thus $V - E + F = 2$.

5. Refer to Figure 6.6. We make use of two planar right triangles: $\triangle ABC$ (right angle at B), $\triangle OCB$ (right angle at C.) From the proof of Proposition 26.1, $m\angle ACB = \frac{\pi}{p}$ and $s/(2BC) = AB/BC = \tan(\angle ACB) = \tan(\frac{\pi}{p})$. But also $BC/r = BC/CO = \tan(\angle BOC) = \tan(\alpha) = \tan(\frac{\pi - I}{2}) = \cot(\frac{I}{2})$. Thus $s/2r = \tan(\frac{\pi}{p})\cot(\frac{I}{2})$. Thus $r = \frac{s}{2}\cot(\frac{\pi}{p})\tan(\frac{I}{2})$.

9. The surface area of the polyhedron is F times the area of a face. The area of its p-gon face is $ps^2\cot(\frac{\pi}{p})/4$. Since there are F faces, the total volume is $ps^2F\cot(\frac{\pi}{p})/4$. But $pF = 2E$ so this is $s^2E\cot(\frac{\pi}{p})/2$.

13. Let N be the midpoint of segment $\overline{GE}$ and P, Q the projection of M, N to the circumscribed sphere. Then the measure of arc $\overset{\frown}{PQ}$ is the angle between two adjacent faces. Let R be the midpoint of $\overset{\frown}{PQ}$. Then $\triangle^s GRP$ has a right angle at R and we seek $m\overset{\frown}{PQ}= 2m\overset{\frown}{PR}$. We have $m\prec PGR = \frac{\pi}{p}$ and $m\overset{\frown}{PG}= \frac{1}{2}m\overset{\frown}{AG}= \frac{\beta}{2}$ where $\cos(\beta) = \cot(\frac{\pi}{p})\cot(\frac{\pi}{q})$. Now

$$\sin(\frac{\pi}{p}) = \sin(\prec PGR) = \frac{\sin(\overset{\frown}{PR})}{\sin(\overset{\frown}{PG})} = \frac{\sin(\overset{\frown}{PR})}{\sin(\frac{\beta}{2})}.$$

Then $\sin(\frac{\overset{\frown}{PQ}}{2}) = \sin(\frac{\pi}{p})\sin(\frac{\beta}{2})$, so $\sin^2(\frac{\overset{\frown}{PQ}}{2}) = \sin^2(\frac{\pi}{p})\sin^2(\frac{\beta}{2})$, and

$$1-\cos(\delta) = 1-\cos(\overset{\frown}{PQ}) = \sin^2(\frac{\pi}{p})(1-\cos(\beta)) = \sin^2(\frac{\pi}{p})(1-\cot(\frac{\pi}{p})\cot(\frac{\pi}{q})).$$

Then $\cos(\delta) = \cos^2(\frac{\pi}{p})+\sin^2(\frac{\pi}{p})\cot(\frac{\pi}{p})\cot(\frac{\pi}{q}) = \cos^2(\frac{\pi}{p})+\sin(\frac{\pi}{p})\cos(\frac{\pi}{p})\cot(\frac{\pi}{q})$.

We obtain the other formula by reversing the roles of the vertices and faces in $\{p, q\}$ and using a similar argument, resulting in a formula where p and q are switched.

17. Suppose $\{p, q\} = \{3, 3\}$. Then we must have $\frac{1}{2} < \sin(\frac{\pi}{r})\frac{\sqrt{3}}{2}$, so we need $\sin(\frac{\pi}{r}) > \frac{1}{\sqrt{3}} = 0.5777\ldots$. If $r = 5$, $\sin(\frac{\pi}{r}) = \sqrt{\frac{5-\sqrt{5}}{8}} = 0.587\ldots$. Thus the inequality holds for $r = 5$, and must hold also for $r = 3, 4$ since $\sin(\frac{\pi}{r})$ is larger for $r = 3, 4$.

Suppose $\{p, q\} = \{3, 4\}$. Then we need $\sin(\frac{\pi}{r}) > \frac{\sqrt{2}}{\sqrt{3}} = 0.816\ldots$. This is false for $r = 4$ since $\sin(\frac{\pi}{4}) = 0.707\ldots$ (so also for $r = 5$) but true for $r = 3$, as $\sin(\frac{\pi}{3}) = 0.866\ldots$.

Suppose $\{p, q\} = \{4, 3\}$. Then we need $\sin(\frac{\pi}{r}) > \frac{1}{\sqrt{2}}$ which is true only for $r = 3$.

Suppose $\{p, q\} = \{5, 3\}$. Then we need $\sin(\frac{\pi}{r}) > \sqrt{\frac{2}{5-\sqrt{5}}} = 0.850\ldots$ which is false for $r = 4$ and 5 but true for $r = 3$ (as $\sin(\frac{\pi}{3}) = 0.866\ldots$.)

Suppose $\{p, q\} = \{3, 5\}$. Then we need $\sin(\frac{\pi}{r}) > \frac{\sqrt{5}+1}{2\sqrt{3}} = 0.934\ldots$. This is false for $r = 3$, 4, and 5.

§27.

1. If f is in $L(M : \mathbf{Z})$ then $(uf)(\mathbf{v}+\mathbf{w}) = u(f(\mathbf{v}+\mathbf{w})) = u(f(\mathbf{v})+f(\mathbf{w})) = uf(\mathbf{v}) + uf(\mathbf{w})$ and $(uf)(z\mathbf{v}) = u(f(z\mathbf{v})) = uzf(\mathbf{v}) = z(uf(\mathbf{v}))$, as desired.

5. We have $(\mathbf{w}-\mathbf{v})\cdot(\mathbf{w}-\mathbf{v}) = \mathbf{w}\cdot\mathbf{w}+\mathbf{v}\cdot\mathbf{v}-2\mathbf{w}\cdot\mathbf{v} = 2\lambda^2 - 2\lambda^2\cos(2\theta) = 2\lambda^2(1-\cos(2\theta)) = 2\lambda^2(1-(1-2\sin^2(\theta)))$, or $4\lambda^2\sin^2(\theta)$. Thus $|\mathbf{w}-\mathbf{v}|$ is the square root of $(\mathbf{w}-\mathbf{v})\cdot(\mathbf{w}-\mathbf{v})$, or $2\lambda\sin(\theta)$.

§28

1. If f is an isometry and $f(X) = f(Y)$ then the distance from $f(X)$ to $f(Y)$ is zero, so the distance from X to Y is zero. So $X = Y$, so f is one-one.

5. Suppose X has coordinates (ϕ, θ) and $T(X)$ has coordinates $(\phi, \theta + \tau(\bmod 2\pi))$. If $\phi \neq 0$ and $\phi \neq \pi$ then $T(X) = X$ only if $\theta = \theta + \tau(\bmod 2\pi)$, i.e., $\tau = 0(\bmod 2\pi)$, which means T is rotation by an integer multiple of 2π, so is the identity. So P and P^a are the only fixed points if T is not the identity.

9. The spherical coordinates of the point are $(\phi, \theta - \theta_2(\bmod 2\pi))$ if the $\theta = \theta_2$ meridian is made the zero meridian. Reflection in the θ_2 meridian gives the point $(\phi, -\theta + \theta_2(\bmod 2\pi))$ in coordinates where the θ_2 meridian is the zero meridian. Switching back to coordinates with the original meridian, we get the point with coordinates $(\phi, -\theta + 2\theta_2(\bmod 2\pi))$.

§29

1. Suppose the sphere has radius r. A parallel of latitude λ_1 has radius $r\cos(\lambda_1)$. The projection of this parallel is a circle of radius r via a dilation by factor of $1/\cos(\lambda_1)$. Suppose a short segment of meridional arc has length z. It forms an angle of approximately λ_1 with the vertical, so it is projected to a segment of length approximately $z\cos(\lambda_1)$. Then a small region bounded by two parallels and two meridians is approximately a rectangle whose horizontal and vertical dimensions are distorted by factors $1/\cos(\lambda_1))$ and $\cos(\lambda_1)$, so its area is distorted by a factor of approximately 1.

5. When $z = 0$ the mapping is the identity so preserves the base. The apex $(0, b, c)$ maps to $(0, 0, c)$ by using the given formula. Because the mapping is linear it maps line segments to line segments so maps the lateral surface of the first cone to lateral surface of the second cone. An invertible linear mapping also preserves planes, and in those planes conic sections are preserved because a conic section is defined by a quadratic in (x, y), which, in a linear change of variables, is again a quadratic.

§31.

1. We have $pq = a_1a_2 + a_1(b_2\mathbf{i} + c_2\mathbf{j} + d_2\mathbf{k}) + a_2(b_1\mathbf{i} + c_1\mathbf{j} + d_1\mathbf{k}) - \mathrm{Ve}(p) \cdot \mathrm{Ve}(q) + \mathrm{Ve}(p) \times \mathrm{Ve}(q)$ which gives $(a_1a_2 - b_1b_2 - c_1c_2 - d_1d_2) + (a_1b_2 + a_2b_1 + c_1d_2 - c_2d_1)\mathbf{i} + (a_1c_2 - b_1d_2 + a_2c_1 + b_2d_1)\mathbf{j} + (a_1d_2 + b_1c_2 - b_2c_1 + a_2d_1)\mathbf{k}$.

5. First we consider the case where each of q_1, q_2 and q_3 is one of $1, \mathbf{i}, \mathbf{j}, \mathbf{k}$. If any of q_1, q_2 or q_3 equal 1, the statement simply says that the product of the other two quaternions equals itself. So we assume that each of q_1, q_2 and q_3 is equal to $\mathbf{i}$, $\mathbf{j}$, and $\mathbf{k}$. It would suffice to prove the associative law in the case where $q_1 = \mathbf{i}$ because of the symmetry in the multiplicative relationship among $\mathbf{i}$, $\mathbf{j}$, and $\mathbf{k}$, where in the sequence $\mathbf{i}, \mathbf{j}, \mathbf{k}, \mathbf{i}, \mathbf{j}, \mathbf{k}, \ldots$ the product of any two in a row gives the next. Then there are nine cases to check.

For general q_1, q_2, q_3, we use the distributive property of quaternions to write $q_1(q_2q_3)$ and $(q_1q_2)q_3$ as the sum of terms each of which involves a product of a real number and three pure quaternion units ($\mathbf{i}$, $\mathbf{j}$, and $\mathbf{k}$.) Corresponding terms of these two sums are equal by the associative property for products of 1, $\mathbf{i}$, $\mathbf{j}$, and $\mathbf{k}$.

9. $\mathrm{Re}(pq) + \mathrm{Re}(p\overline{q}) = \mathrm{Re}(p)\mathrm{Re}(q) - \mathrm{Ve}(p) \cdot \mathrm{Ve}(q) + \mathrm{Re}(p)\mathrm{Re}(q) - \mathrm{Ve}(p) \cdot$

$\mathrm{Ve}(\overline{q})$. But $\mathrm{Ve}(\overline{q}) = -\mathrm{Ve}(q)$ and we get the desired result. Similarly, $\mathrm{Re}(p\overline{q}) - \mathrm{Re}(pq) = \mathrm{Re}(p)\mathrm{Re}(q) + \mathrm{Ve}(p)\cdot\mathrm{Ve}(q) - (\mathrm{Re}(p)\mathrm{Re}(q) - \mathrm{Ve}(p)\cdot\mathrm{Ve}(q)) = 2\mathrm{Ve}(p)\cdot\mathrm{Ve}(q)$.

§32.

1a. $\mathbf{e} = (-5\mathbf{i}+3\mathbf{j}-\mathbf{k})/\sqrt{35}$, $\theta \approx 71.3°$ or $\mathbf{e} = (5\mathbf{i}-3\mathbf{j}+\mathbf{k})/\sqrt{35}$, $\theta \approx -71.3°$

1b. $\mathbf{e} = (-\mathbf{i} - 2\mathbf{j} + 3\mathbf{k})/\sqrt{14}$, $\theta \approx 136.9°$ or $\mathbf{e} = (\mathbf{i} + 2\mathbf{j} - 3\mathbf{k})/\sqrt{14}$, $\theta \approx -136.9°$

5. Let $\mathbf{a}' = \mathbf{a}/|\mathbf{a}|$ and $\mathbf{b}' = \mathbf{b}/|\mathbf{b}|$. If $\mathbf{a}$ and $\mathbf{b}$ are noncollinear, so are $\mathbf{a}'$ and $\mathbf{b}'$. The angle θ between $\mathbf{a}$ and $\mathbf{b}$ is the same as the angle between $\mathbf{a}'$ and $\mathbf{b}'$. Note that $(\mathbf{a}\times\mathbf{b})/|\mathbf{a}\times\mathbf{b}| = (\mathbf{a}'\times\mathbf{b}')/|\mathbf{a}'\times\mathbf{b}'|$; we call this $\mathbf{e}$. Apply the theorem to $\mathbf{a}'$ and $\mathbf{b}'$. Then $\mathbf{b}\mathbf{a}^{-1} = \mathbf{b}'\mathbf{a}'^{-1} = \cos(\theta)+\sin(\theta)\mathbf{e}$. The degenerate case works as stated in the proof of the case $|\mathbf{a}| = |\mathbf{b}| = 1$.

9. Let $\mathbf{A},\mathbf{B},\mathbf{C},\mathbf{D}$ denote the vectors pointing to the vertices A, B, C, D of the quadrilateral. Then $(\mathbf{A}+\mathbf{B})/(|\mathbf{A}+\mathbf{B}|)$ is the midpoint of $\overset{\frown}{AB}$, $(\mathbf{C}+\mathbf{D})/(|\mathbf{C}+\mathbf{D}|)$ is the midpoint of $\overset{\frown}{CD}$, $(\mathbf{C}+\mathbf{B})/(|\mathbf{C}+\mathbf{B}|)$ is the midpoint of $\overset{\frown}{CB}$, $(\mathbf{A}+\mathbf{D})/(|\mathbf{A}+\mathbf{D}|)$ is the midpoint of $\overset{\frown}{AD}$, $(\mathbf{A}+\mathbf{C})/(|\mathbf{A}+\mathbf{C}|)$ is the midpoint of diagonal $\overset{\frown}{AC}$, and $(\mathbf{D}+\mathbf{B})/(|\mathbf{D}+\mathbf{B}|)$ is the midpoint of diagonal $\overset{\frown}{DB}$. The arc through the midpoints of $\overset{\frown}{AB}$ and $\overset{\frown}{CD}$ is represented by quaternion $Q_1 \equiv (\mathbf{A}+\mathbf{B})(\overline{\mathbf{C}}+\overline{\mathbf{D}})/(|\mathbf{A}+\mathbf{B}|)(|\mathbf{C}+\mathbf{D}|)$. Similarly, the arc through the midpoints of $\overset{\frown}{AD}$ and $\overset{\frown}{CB}$ is represented by quaternion $Q_2 \equiv (\mathbf{B}+\mathbf{C})(\overline{\mathbf{A}}+\overline{\mathbf{D}})/(|\mathbf{A}+\mathbf{D}|)(|\mathbf{C}+\mathbf{B}|)$ and the arc through the midpoints of $\overset{\frown}{AC}$ and $\overset{\frown}{BD}$ is represented by quaternion $Q_3 \equiv (\mathbf{A}+\mathbf{C})(\overline{\mathbf{B}}+\overline{\mathbf{D}})/(|\mathbf{A}+\mathbf{C}|)(|\mathbf{B}+\mathbf{D}|)$. We show that $\mathrm{Re}(\mathbf{A}+\mathbf{B}+\mathbf{C}+\mathbf{D})(Q_i) = 0$ for $i = 1, 2$ and 3. We have $(\mathbf{A}+\mathbf{B}+\mathbf{C}+\mathbf{D})(Q_1) = ((\mathbf{A}+\mathbf{B})^2+(\mathbf{C}+\mathbf{D})(\mathbf{A}+\mathbf{B}))(\overline{\mathbf{C}}+\overline{\mathbf{D}})/(|\mathbf{A}+\mathbf{B}|)(|\mathbf{C}+\mathbf{D}|)$. Now $(\mathbf{A}+\mathbf{B})^2 = -|\mathbf{A}+\mathbf{B}|^2$ so $\mathrm{Re}(\mathbf{A}+\mathbf{B})^2(\overline{\mathbf{C}}+\overline{\mathbf{D}}) = 0$, as a pure quaternion. Thus $\mathrm{Re}((\mathbf{A}+\mathbf{B}+\mathbf{C}+\mathbf{D})(Q_1)) = \mathrm{Re}(\mathbf{C}+\mathbf{D})(\mathbf{A}+\mathbf{B})(\overline{\mathbf{C}}+\overline{\mathbf{D}})/(|\mathbf{A}+\mathbf{B}|)(|\mathbf{C}+\mathbf{D}|) = \mathrm{Re}(\mathbf{A}+\mathbf{B})(\overline{\mathbf{C}}+\overline{\mathbf{D}})(\mathbf{C}+\mathbf{D})/(|\mathbf{A}+\mathbf{B}|)(|\mathbf{C}+\mathbf{D}|)$ (by (8.4)), which is $\mathrm{Re}(\mathbf{A}+\mathbf{B})|\mathbf{C}+\mathbf{D}|^2/(|\mathbf{A}+\mathbf{B}|)(|\mathbf{C}+\mathbf{D}|)$, the real part of a pure quaternion, which is zero. So $\mathrm{Re}(\mathbf{A}+\mathbf{B}+\mathbf{C}+\mathbf{D})(Q_i) = 0$ for $i = 1$; the cases $i = 2, 3$ are similar. Now $\mathrm{Re}(\mathbf{A}+\mathbf{B}+\mathbf{C}+\mathbf{D})(Q_i) = -\mathrm{Ve}(\mathbf{A}+\mathbf{B}+\mathbf{C}+\mathbf{D})\cdot\mathrm{Ve}(Q_i)$. Thus $\mathbf{A}+\mathbf{B}+\mathbf{C}+\mathbf{D}$ is perpendicular to each Q_i, which means the Q_i (the poles of the arcs connecting the midpoints) lie on the polar of $(\mathbf{A}+\mathbf{B}+\mathbf{C}+\mathbf{D})/|\mathbf{A}+\mathbf{B}+\mathbf{C}+\mathbf{D}|$, and so the great circle containing each of the arcs connecting the midpoints contains $(\mathbf{A}+\mathbf{B}+\mathbf{C}+\mathbf{D})/|\mathbf{A}+\mathbf{B}+\mathbf{C}+\mathbf{D}|$.

§33.

1. The product with $\mathbf{B}'$ produces another analogue formula. Multiplication by $\mathbf{C}'$ produces $\sin(c) = \sin(b)\cos(a)\cos(A) + \cos(b)\sin(a)\cos(B) + \sin(b)\sin(a)\sin(A)\sin(B)\sin(c)$. Multiplication by $\mathbf{A}$ and $\mathbf{B}$ both produce the formula for $\mathbf{A}\cdot\mathbf{B}\times\mathbf{C}$.

§34.

13. Multiplying both sides on the right by $\mathbf{B}$ and distributing,

$$\begin{aligned} & \cos(\frac{c}{2})\cos(E)\mathbf{B} - \sin(\frac{c}{2})\cos(E)\mathbf{C}'\mathbf{B} - \cos(\frac{c}{2})\sin(E) + \sin(E)\sin(\frac{c}{2})\mathbf{C}' \\ = & \cos(\frac{b}{2})\cos(\frac{a}{2})\mathbf{B} + \cos(\frac{b}{2})\sin(\frac{a}{2})\mathbf{A}'\mathbf{B} \\ + & \sin(\frac{b}{2})\cos(\frac{a}{2})\mathbf{B}'\mathbf{B} + \sin(\frac{b}{2})\sin(\frac{a}{2})\mathbf{B}'\mathbf{A}'\mathbf{B}. \end{aligned}$$

We take the real part of both sides. On the left side, the first and last terms are pure quaternions, and the second is $Re(\mathbf{C}'\mathbf{B}) = -\mathbf{C}' \cdot \mathbf{B} = 0$. On the right side, we may similarly show that the real parts of the first two terms are zero. Thus we obtain

$$-\cos(\frac{c}{2})\sin(E) = \sin(\frac{b}{2})\cos(\frac{a}{2})Re(\mathbf{B}'\mathbf{B}) + \sin(\frac{b}{2})\sin(\frac{a}{2})Re(\mathbf{B}'\mathbf{A}'\mathbf{B}). \quad (A.10)$$

Now

$$Re(\mathbf{B}'\mathbf{B}) = -\mathbf{C} \times \mathbf{A} \cdot \mathbf{B}/\sin(b) = -\mathbf{B} \cdot \mathbf{C} \times \mathbf{A}/\sin(b) = -\mathbf{A} \cdot \mathbf{B} \times \mathbf{C}/\sin(b)$$

by (8.21) and this equals $-\sin(a)\sin(C)$ by (8.32). Next, $(\mathbf{B}'\mathbf{A}') = -\mathbf{B}' \cdot \mathbf{A}' + \mathbf{B}' \times \mathbf{A}' = -\cos(c') - \sin(c')\mathbf{C} = -\cos(\pi - C) - \sin(\pi - C)\mathbf{C} = \cos(C) - \sin(C)\mathbf{C}$. Then $\mathbf{B}'\mathbf{A}'\mathbf{B} = \cos(C)\mathbf{B} - \mathbf{CB}\sin(C)$, which has real part

$$0 + \mathbf{C} \cdot \mathbf{B}\sin(C) = \cos(a)\sin(C).$$

Then (A.10) is

$$\begin{aligned} & -\cos(\frac{c}{2})\sin(E) \\ = & -\sin(\frac{b}{2})\cos(\frac{a}{2})\sin(a)\sin(C) + \sin(\frac{b}{2})\sin(\frac{a}{2})\cos(a)\sin(C) \\ = & \sin(\frac{b}{2})\sin(C)(-\cos(\frac{a}{2})\sin(a) + \sin(\frac{a}{2})\cos(a)) \\ = & \sin(\frac{b}{2})\sin(C)(\sin(\frac{a}{2}) - a) \\ = & \sin(\frac{b}{2})\sin(C)(-\sin(\frac{a}{2})), \end{aligned}$$

which gives us what we want.

17a. If $a + b > \pi$ then for C near π, the triangle occupies more and more of a hemisphere whose area is near $2\pi r^2$.

17b. If $a + b = \pi$ then the area of $\triangle^s ABC$ is half the area enclosed by lune CAC^aBC. The area of the lune is $2Cr^2$ so the area of $\triangle^s ABC$ is Cr^2. As C approaches π this area approaches πr^2 but never attains it because an angle cannot have measure π.

17c. Note that if $a+b<\pi$ then $\frac{a}{2}+\frac{b}{2}<\frac{\pi}{2}$ so

$$\cot(\frac{a}{2})\cot(\frac{b}{2}) > \cot(\frac{a}{2})\cot(\frac{\pi}{2}-\frac{a}{2}) = 1.$$

So $\cot(\frac{a}{2})\cot(\frac{b}{2})+\cos(C) > 1+\cos(C) > 0$ and so $\tan(E)$ is always defined and positive. The expression for $\tan(E)$ is the slope between $(\cos(C),\sin(C))$ and $(-\cot(\frac{a}{2})\cot(\frac{b}{2}),0)$. As C varies from 0 to π the point $(\cos(C),\sin(C))$ ranges over the upper half of the unit circle. The point $(-\cot(\frac{a}{2})\cot(\frac{b}{2}),0)$ is on the x-axis to the left of the unit circle. Thus the slope is maximum when the line between the two points is tangent to the unit circle. Using the fact that a tangent to a circle is perpendicular to the radius at the point of contact, this occurs for a value of $C>\frac{\pi}{2}$ such that $\cos(C) = -1/(\cot(\frac{a}{2})\cot(\frac{b}{2})) = -\tan(\frac{a}{2})\tan(\frac{b}{2})$.

17d. Let d be the length of the arc from C to the midpoint of $\widehat{AB}$. Using §16, Exercise 18a, $\cos(a)\sin(\frac{c}{2})+\cos(b)\sin(\frac{c}{2}) = \cos(d)\sin(c)$. Using the double angle formula for the sine, we find

$$\cos(a)+\cos(b) = 2\cos(d)\cos(\frac{c}{2}). \tag{A.11}$$

Now we know $\cos(C) = -\tan(\frac{a}{2})\tan(\frac{b}{2}) = -\frac{1-\cos(a)}{\sin(a)}\frac{1-\cos(b)}{\sin(b)}$ by the half-angle formula for the tangent. Substituting this value of $\cos(C)$ into the spherical law of cosines, we find

$$\begin{aligned}
\cos(c) &= \cos(a)\cos(b)+\sin(a)\sin(b)\cos(C)\\
&= \cos(a)\cos(b)-(1-\cos(a))(1-\cos(b))\\
&= \cos(a)\cos(b)-1+\cos(a)+\cos(b)-\cos(a)\cos(b)\\
&= -1+\cos(a)+\cos(b)
\end{aligned}$$

So $1+\cos(c) = \cos(a)+\cos(b)$, so $2\cos^2(\frac{c}{2}) = \cos(a)+\cos(b)$. Substituting this into (A.11) we find $2\cos^2(\frac{c}{2}) = 2\cos(d)\cos(\frac{c}{2})$ so $\cos(d)=\cos(\frac{c}{2})$. Thus $d=\frac{c}{2}$ and the triangle is diametrical.

Bibliography

[Ab1841] Richard Abbatt, *The Elements of Plane and Spherical Trigonometry; and Its Applications to Astronomy, Dialling and Trigonometrical Surveying.* Ostell and Co., London, 1841.

[AF2008] Ilka Agricola and Thomas Friedrich, *Elementary Geometry.* AMS, Providence, 2008.

[Ar2017] Michael Artin, *Algebra.* Pearson, 2017.

[AM1976] Neil W. Ashcroft and N. David Mermin, *Solid State Physics.* Saunders College, Philadelphia, 1976.

[Bl1994] F. Donald Bloss, *Crystallography and Crystal Chemistry, An Introduction.* Mineralogical Society of America, 1994.

[Br1957] Louis Brand, *Vector Analysis.* Wiley, New York, 1957.

[Br1947] Louis Brand, *Vector and Tensor Analysis.* Wiley, New York, 1947.

[BEG2012] David A. Brannan, Matthew F. Esplen and Jeremy J. Gray, *Geometry,* 2nd ed. Cambridge University Press, Cambridge, 2012.

[Br1942] Raymond W. Brink, *Spherical Trigonometry.* Appleton-Century, New York, 1942.

[Ca1916] H.S. Carslaw, *The Elements of Non-Euclidean Plane Geometry and Trigonometry.* Longmans, Green and Co., London, 1916.

[Ca1889] John Casey, *A Treatise on Spherical Trigonometry.* Hodges, Figgis & Co, Dublin, 1889; Watchmaker Publishing, 2008.

[Co1973] H.S.M. Coxeter, *Regular Polytopes.* Dover, New York, 1973.

[Do1945] J.D.H. Donnay, *Spherical Trigonometry after the Cèsaro Method.* Interscience, New York, 1945.

[Do1965] Heinrich Dörrie, *100 Great Problems of Elementary Mathematics.* Dover, New York, 1965.

[Du1981] Peter Duffett-Smith, *Practical Astronomy with Your Calculator.* Cambridge University Press, Cambridge, 1981.

[Fe1959] J.M. Feld, *An Application of Turns and Slides to Spherical Geometry.* The American Mathematical Monthly **66**, No. 8 (1959), 665–673.

[FFDFT] F. Firneis, M. Firneis, L. Dimitrov, G. Frank, R. Thaller, *On Some Applications of Quaternions in Geometry.* Proceedings of the Third International Conference on Engineering Graphics and Descriptive Geometry, Vol. 1 (Vienna, 1988), 158–164, Tech. Univ. Vienna, 1988.

[Fo2004] C.M.R Fowler, *The Solid Earth: An Introduction to Global Geophysics,* 2nd ed. Cambridge University Press, Cambridge, 2004.

[Ga1973] David Gans, *An Introduction to Non-Euclidean Geometry.* Academic Press, New York, 1973.

[Gr1985] Robin M. Green, *Spherical Astronomy.* Cambridge University Press, Cambridge, 1985.

[Ha2015] Christopher Hammond, *The Basics of Crystallography and Diffraction,* 4th ed. Oxford University Press, New York, 2015.

[Ha1942] Walter W. Hart and William L. Hart, *Plane Trigonometry, Solid Geometry and Spherical Trigonometry.* D.C. Heath and Company, Boston, 1942.

[Ha2005] Robin Hartshorne, *Geometry: Euclid and Beyond,* Springer, New York, 2005.

[H1996] David W. Henderon, *Experiencing Geometry on Plane and Sphere.* Prentice-Hall, Upper Saddle River, 1996.

[He2015] Rani Hermiz, *English Translation of the Sphaerica of Menelaus.* MS Thesis, California State University San Marcos, 2015. Available at library.csusm.edu.

[Je1994] George Jennings, *Modern Geometry with Applications.* Springer, New York, 1994.

[KKB1942] Lyman M. Kells, Willis F. Kern, James R. Bland, *Spherical Trigonometry with Naval and Military Applications.* McGraw-Hill, New York, 1942.

[KMP] L. Christine Kinsey, Teresa E. Moore and Efstratios Prassidis, *Geometry and Symmetry.* Wiley, New York, 2010.

[Ki2005] Charles Kittel, *Introduction to Solid State Physics,* 8th ed. Wiley, New York, 2005.

[Kl1945] Felix Klein, *Elementary Mathematics from an Advanced Standpoint: Arithmetic, Algebra, Analysis.* Dover, New York, 1945.

[Kr1936] M. Krause, *Die Sphärik von Menelaos aus Alexandrien in der Verbesserung von Abū Naṣr Manṣūr b. 'Alī b. 'Irāq,* Abhandlungen der Gesellschaft der Wissenschaften zu Göttingen, phil.-hist. Klasse, 3, 17, Berlin, 1936.

[LP2014] M.F.C. Ladd and R.A. Palmer, *Structure Determination by X-ray Crystallography,* 5th ed. Springer, New York, 2014.

[Ma2013] Werner J. Massa, *Crystal Structure Determination, 2nd. ed.* Springer, Berlin, 2013.

[MP1912] William J. McClelland and Thomas Preston, *A Treatise on Spherical Trigonomgetry with Applications to Spherical Geometry with Numerous Examples. Part I.* Macmillan, London, 1912.

[Me2007] Z.A. Melzak, *Companion to Concrete Mathematics.* Dover, Mineola, 2007.

[Me2008] Z.A. Melzak, *Invitation to Geometry.* Dover, Mineola, 2008.

[Mi1978] H.R. Mills, *Positional Astronomy and Astro-Navigation Made Easy.* Wiley, New York, 1989.

[MD1982] Edwin E. Moise and Floyd L. Downs, Jr., *Geometry.* Addison-Wesley, Menlo Park, 1982.

[Mo1963] Edwin E. Moise, *Elementary Geometry from an Advanced Standpoint.* Addison-Wesley, Reading, Massachusetts, 1963.

[Mu1900] Daniel A. Murray, *Plane and Spherical Trigonometry.* Longmans, Green and Co., New York, 1900.

[No1950] John A. Northcott, *Plane and Spherical Trigonometry.* Rinehart & Company, New York, 1950.

[Pa2014] Athanase Papadopoulos, *On the works of Euler and his followers on spherical geometry.* Ganita Bharati 36 (2014), no. 1, 53–108.

[Pe1852] Benjamin Peirce, *An Elementary Treatise on Plane and Spherical Trigonometry, with Their Applications to Navigation, Surveying, Heights and Distances, and Spherical Astronomy,...* Munroe and Co., Boston, 1852.

[RP2017] R. Rashed and A. Papadopoulos, *Menelaus' Spherics. Early Translation and al-Mahani/al-Harawi's Version.* De Gruyter, Berlin, 2017.

[RH1943] Paul R. Rider, Charles A. Hutchinson, *Navigational Trigonometry.* Macmillan, New York, 1943.

[Ro1988] B.A. Rosenfeld, *A History of Non-Euclidean Geometry.* Springer-Verlag, New York, 1988.

[Ro1925] David Rothrock, *Elements of Plane and Spherical Trigonometry.* Macmillan, New York, 1925.

[Ry1986] Patrick J. Ryan, *Euclidean and non-Euclidean Geometry An Analytical Approach.* Cambridge University Press, Cambridge, 1986.

[Si2008] Joel S. Silverberg, *Napier's Rules of Circular Parts,* Canadian Society for the History and Philosophy of Mathematics, 2008.

[Si1945] Harvey Alexander Simmons, *Plane and Spherical Trigonometry, 2nd ed.,* Wiley, New York, 1945.

[Sm1952] Lloyd L. Smail, *Trigonometry Plane and Spherical,* McGraw-Hill Book Company, Inc., New York, 1952.

[SG1977] W.M. Smart and R.M. Green, *Textbook on Spherical Astronomy.* Cambridge University Press, Cambridge, 1977.

[SC1949] Rolland R. Smith and John R. Clark, *Modern-School Solid Geometry, New Edition,* World Book Company, Yonkers-on-Hudson, 1949.

[Sn1993] John P. Snyder, *Flattening the Earth Two Thousand Years of Map Projections.* University of Chicago, Chicago, 1993.

[So1914] D.M.Y. Sommerville, *The Elements of Non-Euclidean Geometry.* Dover, New York, 1958.

[So1929] D.M.Y. Sommerville, *An Introduction to the Geometry of N Dimensions.* Dover, New York, 1958.

[Sp1928] Pauline Sperry, *Short Course in Spherical Trigonometry.* Johnson Publishing, Richmond, 1928.

[St1992] John Stillwell, *Geometry of Surfaces.* Springer-Verlag, New York, 1992.

[SJ1989] George H. Stout and Lyle H. Jensen, *X-Ray Structure Determination A Practical Guide, 2nd ed.* Wiley, New York, 1989.

[Ta1980] Laurence G. Taff, *Computational Spherical Astronomy.* Wiley-Interscience, New York, 1980.

[Te1952] P. Terpstra, *A Thousand and One Questions on Crystallographic Problems.* J.B. Wolters, Groningen, 1952.

[Th1942] James E. Thompson, *Elements of Spherical Trigonometry.* D. Van Nostrand Company, Inc., New York, 1942.

[To1886] I. Todhunter, *Spherical Trigonometry for the Use of Colleges and Schools, 5th ed.* Macmillan, London, 1886.

[TS2014] Donald L. Turcotte and Gerald Schubert, *Geodynamics*, 3rd ed. Cambridge University Press, Cambridge, 2014.

[VB2009] Glen Van Brummelen, *The Mathematics of the Heavens and the Earth The Early History of Trigonometry.* Princeton University Press, Princeton, 2009.

[VB2012] Glen Van Brummelen, *Heavenly Mathematics: the Forgotten Art of Spherical Trigonometry.* Princeton University Press, Princeton, 2012.

[We1894] G.A. Wentworth, *A Text-Book of Geometry, Revised Edition.* Ginn & Co., Boston, 1894.

[We1952] Hermann Weyl, *Symmetry.* Princeton University Press, Princeton, 1952.

[Wo1945] Harold E. Wolfe, *Introduction to Non-Euclidean Geometry.* Holt, Rinehart and Winston, New York, 1945.

[Wy2009] C.R.Wylie, *Foundations of Geometry.* Dover Publications, Mineola, 2009.

Index

9781032475370